WITHDRAWN

624.
193

MAI

B. Maidl, M. Herrenknecht, U. Maidl, G. Wehrmeyer
Mechanised Shield Tunnelling

2nd Edition

Mechanised Shield Tunnelling

Bernhard Maidl
Martin Herrenknecht
Ulrich Maidl
Gerhard Wehrmeyer

Prof. Dr.-Ing. Bernhard Maidl
mtc – Maidl Tunnelconsultants GmbH & Co. KG
Fuldastr. 11
47051 Duisburg

Dr.-Ing. E.h. Martin Herrenknecht
Herrenknecht AG
Schlehenweg 2
77963 Schwanau

Dr.-Ing. Ulrich Maidl
mtc – Maidl Tunnelconsultants GmbH & Co. KG
Fuldastr. 11
47051 Duisburg

Dr.-Ing. Gerhard Wehrmeyer
Herrenknecht AG
Schlehenweg 2
77963 Schwanau

Translated by David Sturge, Kirchbach, Germany

Cover: Herrenknecht-EPB-Shield S-300 Madrid M-30
By-Pass Sue Túnel Norte, Madrid, Spain
Photo: Herrenknecht AG

Library of Congress Card No.:
applied for

British Library Cataloguing-in-Publication Data
A catalogue record for this book is available from the British Library.

Bibliographic information published by
the Deutsche Nationalbibliothek
The Deutsche Nationalbibliothek lists this publication in the Deutsche Nationalbibliografie; detailed bibliographic data are available on the Internet at <http://dnb.d-nb.de>.

© 2012 Wilhelm Ernst & Sohn, Verlag für Architektur und technische Wissenschaften GmbH & Co. KG, Rotherstr. 21, 10245 Berlin, Germany

All rights reserved (including those of translation into other languages). No part of this book may be reproduced in any form – by photoprinting, microfilm, or any other means – nor transmitted or translated into a machine language without written permission from the publishers. Registered names, trademarks, etc. used in this book, even when not specifically marked as such, are not to be considered unprotected by law.

Coverdesign: Sonja Frank, Berlin, Germany
Produktion management: pp030 – Produktionsbüro Heike Praetor, Berlin
Typesetting: Reemers Publishing Services GmbH, Krefeld
Printing and Binding: Strauss GmbH, Mörlenbach

Printed in the Federal Republic of Germany.
Printed on acid-free paper.

2nd Edition
Print ISBN: 978-3-433-02995-4
ePDF ISBN: 978-3-433-60150-1
ePub ISBN: 978-3-433-60149-5
mobi ISBN: 978-3-433-60148-8
o-Book ISBN: 978-3-433-60105-1

*„A plan whatever it may be
must be made for the bad ground,
it must be calculated to meet all exigencies, all disasters
and to overcome them after they have occured."*

Marc Isambard Brunel
on suggested improvements
after the flooding of the Thames Tunnel in 1831.

The authors

1	Introduction	B. Maidl
2	Support of the cavity and settlement	U. Maidl
3	Construction and design methods	G. Wehrmeyer
4	Excavation tool and excavation process	M. Herrenknecht G. Wehrmeyer
5	Muck removal	M. Herrenknecht G. Wehrmeyer
6	The tunnel lining	B. Maidl
7	Shield tail sealing, grouting works	G. Wehrmeyer
8	Open shields	B. Maidl
9	Compressed air shields	B. Maidl
10	Slurry shields	U. Maidl
11	Earth pressure shields	U. Maidl
12	Convertible shields	B. Maidl
13	Special shields and special processes	B. Maidl
14	Guided microtunnelling processes	B. Maidl
15	Surveying and steering	M. Herrenknecht G. Wehrmeyer
16	Workplace safety	B. Maidl
17	Partnering contract models and construction	U. Maidl
18	Process controlling and management	U. Maidl
19	DAUB recommendations for the selection of tunnelling machines	U. Maidl

Foreword to the 2nd Edition

The rapid progress of mechanised tunnelling to market leadership has continued – even exceeded predictions; the general worldwide trend in construction towards mechanisation and automation clearly demanded a similar development in tunnelling. It is significant that even in Austria, the traditional home of the New Austrian Tunnelling Method (NATM), mechanised tunnelling has also established its position in the last decade. Occupational health and safety, faster advance rates, improved cost security and labour-saving opened opportunities for mechanised tunnelling on a few major projects – normally in competition with conventional construction methods.

So it is appropriate that this book should now be revised, 20 years after its first publication. The extent of innovations and practical experience led to a complete reworking. Incidentally, the book "Hardock Tunnel Boring Machines", which appeared in 2008, already offered access to the newest technology in the area of tunnel support. The team of authors has adapted the content to the latest technology and has been supplemented to provide the necessary specialist knowledge.

The original authors B. Maidl and M. Herrenknecht also worked on this edition. We have gained my son Dr.-Ing. U. Maidl and my former doctoral candidate Dr.-Ing. G. Wehrmeyer, who have particularly devoted themselves to new developments.

I am very thankful that I could still rely on the help of my former employees Herr H. Schmidt and Herr G. Kaufhold for the new revision. I would like to thank Herr Dipl-Ing. M. Griese from MTC, who helped a great deal with the detailed work and overall coordination. I would also like to thank my grandson Max Maidl for his assistance. A thank-you to all, especially the author colleagues and the publisher.

Bochum, January 2011 *Bernhard Maidl*

Table of Contents

The authors VII

Foreword to the 2nd Edition IX

1	Introduction	1
1.1	Basic principles and terms	3
1.2	Types of tunnel boring machine according to DAUB	5
	1.2.1 Categories of tunnelling machines German association for underground construction (TVM)	5
	1.2.2 Tunnel boring machines (TBM)	6
	1.2.2.1 Tunnel boring machines without shield (gripper TBM)	6
	1.2.2.2 Reamer tunnel boring machines (ETBM)	7
	1.2.2.3 Tunnel boring machines with single shield (TBM-S)	7
	1.2.3 Double shield machines (DSM)	7
	1.2.4 Shield machines (SM)	8
	1.2.4.1 Shield machines with full-face excavation (SM-V)	8
	1.2.4.2 Shield machines with partial face excavation (SM-T)	11
	1.2.5 Adaptable shield machines with combined process technology (KSM)	11
	1.2.6 Special types	12
	1.2.6.1 Blade shields	12
	1.2.6.2 Shields with multiple circular cross-sections	12
	1.2.6.3 Articulated shields	12
	1.2.7 Remarks about the individual types of tunnelling machines with diagrams	12
	1.2.7.1 Tunnel boring machines (TBM)	12
	1.2.7.2 Double shield machines (DSM)	13
	1.2.7.3 Face without support (SM-V1)	13
	1.2.7.4 Face with mechanical support (SM-V2)	14
	1.2.7.5 Face with compressed air (SM-V3)	14
	1.2.7.6 Face with slurry support (SM-V4)	14
	1.2.7.7 Face with earth pressure support (SM-V5)	14
	1.2.7.8 Face without support (SM-T1)	15
	1.2.7.9 Face with partial support (SM-T2)	15
	1.2.7.10 Face with compressed air support (SM-T3)	15
	1.2.7.11 Face with slurry support (SM-T4)	16
	1.2.7.12 Adaptable machines (KSM)	16
1.3	Origins and historical developments	16
2	**Support of the cavity and settlement**	25
2.1	Support of the face	25
	2.1.1 Natural support	25
	2.1.2 Mechanical support	26

		2.1.3	Compressed air support	26
		2.1.4	Slurry support	28
		2.1.5	Earth support	32
		2.1.6	Calculation models	32
	2.2	Support of the cavity at the shield		37
	2.3	Support of the cavity behind the shield		37
	2.4	Settlement and damage classifications		39
		2.4.1	Empirical determination of the settlement	41
		2.4.2	Numerical models for the calculation of settlement	43
	2.5	Heave and compaction		46

3 Design and calculation methods ... 47

	3.1	Constructional parts of the shield		47
	3.2	Loading on the shield		50
		3.2.1	Loading on the shield skin	51
		3.2.2	Loading on the pressure bulkhead	53
		3.2.3	Loading from the thrust cylinders	54
	3.3	Calculation of the necessary thrust force		54
		3.3.1	Resistance to advance through friction on the shield skin	55
		3.3.2	Resistance to advance at the front shield	56
		3.3.3	Resistance to advance at the face through platforms and excavation tools	57
		3.3.4	Resistance to advance with slurry support, earth support and compressed air support	58
		3.3.5	Resistance to advance from steering the shield	58
		3.3.6	Summary	59
	3.4	Empirical values for the dimensioning of the shield and the thrust cylinders		60
	3.5	Calculation and dimensioning basics		61
	3.6	Regulations and recommendations for the design of shields		62

4 Excavation tools and excavation process ... 63

	4.1	Excavation tools		64
		4.1.1	Hand-held tools	64
		4.1.2	Cutting edges	64
		4.1.3	Scrapers	65
		4.1.4	Drag picks, flat chisels, round chisels, rippers	66
		4.1.5	Disc cutters, discs	68
		4.1.6	Buckets	70
	4.2	Excavation process		71
		4.2.1	Tunnelling without cutting wheel	72
		4.2.2	Manual digging	73
		4.2.3	Partial-face mechanical excavation	73
		4.2.4	Mechanical full-face excavation	78
		4.2.5	Hydraulic excavation	91
		4.2.6	Alternative excavation processes	91

5	**Muck removal**		93
5.1	Preparation for transport		93
5.2	Removal from the face		93
	5.2.1	Open shield machines	95
	5.2.2	Shield machines with pressure chamber	95
5.3	Transport along the tunnel and up shafts		101
	5.3.1	Open transport	101
	5.3.2	Piped transport	102
5.4	Quantity determination and measuring equipment		105
5.5	Separation		106
	5.5.1	Separating process	108
	5.5.2	Separating devices	108
5.6	Suitability of the muck for landfill		115
6	**The tunnel lining**		117
6.1	General		117
6.2	Construction principles for the tunnel lining		118
	6.2.1	Single-layer and Double-layer construction	118
	6.2.2	Watertight and water draining construction	119
6.3	Segmental lining		121
	6.3.1	General	121
	6.3.2	Constructional variants	122
	6.3.2.1	Block segments with rectangular plan	122
	6.3.2.2	Hexagonal segments	126
	6.3.2.3	Rhomboidal and trapezoidal segment systems	126
	6.3.2.4	Expanding segments	127
	6.3.2.5	Yielding lining systems	128
	6.3.3	Joint details	132
	6.3.3.1	Longitudinal joints	132
	6.3.3.2	Ring joints	135
	6.3.4	Steel fibre concrete segments	139
	6.3.5	Filling of the annular gap	139
	6.3.5.1	Filling with gravel	139
	6.3.5.2	Mortar grouting	139
	6.3.6	Measures to waterproof tunnels with segment linings	141
	6.3.6.1	Gaskets	141
	6.3.6.2	Grouting	143
	6.3.7	Production	143
	6.3.8	Damage	144
	6.3.8.1	Damage during ring building	145
	6.3.8.2	Damage while advancing the machine	145
	6.3.8.3	Damage in the shield tail seal	146
	6.3.8.4	Damage after leaving the shield	146
	6.3.8.5	Repair of damage	147

6.4	In-situ concrete lining		147
	6.4.1	General	147
	6.4.2	Construction	148
	6.4.3	Concreting	148
6.5	Injected concrete, Extruded concrete		149
6.6	Shotcrete layers as the final lining		155
6.7	Structural calculations		156
7	**Shield tail sealing, grouting works**		**157**
7.1	Shield tail seals		157
	7.1.1	Plastic seals	158
	7.1.2	Steel brush seals	160
	7.1.3	Outer shield tail seals	161
	7.1.4	Elastically supported face formwork for the extrusion process	161
7.2	Grouting process		162
	7.2.1	Requirements	162
	7.2.2	Conception	163
	7.2.3	Grouting systems	164
	7.2.4	Grout	168
7.3	Grouting for ground improvement		169
	7.3.1	Machinery and equipment	169
	7.3.2	Grout	171
	7.3.3	Grouting work at the Channel Tunnel	173
8	**Open shields**		**177**
8.1	Shield construction		177
	8.1.1	Hand shields	177
	8.1.2	Part-face excavation	179
	8.1.3	Full-face excavation	181
8.2	Projects		181
	8.2.1	Example: Eurotunnel – under the English Channel, 1988 to 1991	181
	8.2.2	Arrowhead Tunnel	191
8.3	Double shields [203]		195
	8.3.1	Development	195
	8.3.2	Functional principle	195
	8.3.3	Special features	196
	8.3.3.1	Shield skin and bentonite lubrication	196
	8.3.3.2	Telescopic shield	196
	8.3.3.3	Examples	198
9	**Compressed air shields**		**201**
9.1	Functional principle		202
9.2	Compressed air facilities		203
	9.2.1	Air locks	204
	9.2.2	Compressed air supply	206
	9.2.3	Compressed air regulations	207

9.3	Air requirement		209
	9.3.1	Determination of air requirement	209
	9.3.2	Verification of safety (blowout safety)	212
	9.3.3	Special processes	213
9.4	Further developments		214
	9.4.1	Compressed air shield with unpressurised working space and full-face excavation	214
	9.4.2	Compressed air shield with unpressurised working spaces and part face excavation	214
	9.4.3	Membrane shield	216
9.5	The use of compressed air with other types of shield		216
9.6	Examples		217
	9.6.1	Old Elbe Tunnel next to the St. Pauli landing stage, 1907 to 1911	217
	9.6.2	Energy supply tunnel under the Kiel Fjord, 1989/90	219

10 Slurry shields ... 223
10.1	Development history		223
10.2	Functional principle		225
10.3	Scope of application		227
10.4	Machine types		228
	10.4.1	Full-face machines with fluid support	228
	10.4.2	Part face machines with slurry support	233
10.5	Machine and process technology		234
	10.5.1	Soil excavation	234
	10.5.2	Muck transport	235
10.6	Examples		237
	10.6.1	Westerschelde	237
	10.6.2	Lower Inn Valley railway, Münster/Wiesing Tunnel, main contract H3-4; Jenbach/Wiesing Tunnel, main contract H8, 2007 to 2009	243
	10.6.3	Fourth bore of the Elbe Tunnel	247
	10.6.4	Chongming	250

11 Earth pressure balance shields ... 255
11.1	Development history		255
11.2	Functional principle		256
	11.2.1	Support pressure measurement and control	256
	11.2.2	Soil conditioning	259
	11.2.3	Mass-volume control	259
11.3	Areas of application		262
11.4	Operating modes and muck transport		264
	11.4.1	Open mode (screw conveyor – conveyor belt)	264
	11.4.2	Semi open mode (screw conveyor – conveyor belt)	265
	11.4.3	Closed mode (hydraulic mucking circuit)	266
	11.4.4	EPB mode (screw conveyor – conveyor belt or screw conveyor – piston pump)	266
	11.4.5	Open mode (conveyor belt)	266

11.5	Components		267
	11.5.1	Cutting wheel	267
	11.5.2	Bearing and drive construction	269
	11.5.3	Excavation chamber	271
	11.5.4	Screw conveyor	271
	11.5.5	Foam conditioning	273
11.6	Examples		276
	11.6.1	Katzenberg Tunnel on the new railway line Karlsruhe – Basel, 2005 to 2007	276
	11.6.2	Madrid M-30 (Bypass Sur Tunnel Nord)	280
	11.6.3	Heathrow	284
	11.6.4	DTSS Singapore	286
12	**Convertible shields or multi mode machines**		**291**
12.1	Development strategies		293
	12.1.1	Convertible shield with integrated components for multiple operating modes	293
	12.1.2	Building block systems	295
12.2	Machine concepts		295
	12.2.1	Mixshield	296
	12.2.2	Polyshield	297
12.3	Examples		297
	12.3.1	Grauholz Tunnel, 1990 to 1993	297
	12.3.2	Zürich Thalwil contract 2.01	301
	12.3.3	Socatop	305
13	**Special shields and special processes**		**309**
13.1	Blade shields		309
	13.1.1	Face support with blade shields	311
	13.1.2	Support types with blade shields	312
13.2	Multi-face shields		315
	13.2.1	Arrangement of the cutting wheels in multi-face shields	316
	13.2.2	Tunnel support with multi-face shields	317
13.3	Enlargement of shield tunnels		319
13.4	Pipe jacking		322
	13.4.1	Pipe jacking	324
	13.4.2	Box jacking	325
13.5	New concepts in mechanised shield tunnelling		328
	13.5.1	Shield machines for flexible cross-sections	328
	13.5.2	Ultra-flexible shield	330
	13.5.3	Horizontal and vertical shield machines	330
	13.5.4	Enlargement shields	331
	13.5.5	Rotation shields	331
	13.5.6	Shield docking method	332

14	**Guided microtunnelling processes**	337
14.1	Pilot tube process	338
14.2	Auger microtunnelling	339
14.3	Shield microtunnelling	340
14.4	English Mini Tunnel system	342
14.5	New developments	344

15	**Surveying and steering**		349
15.1	Surveying		350
	15.1.1	Navigation with tunnel laser and automatic target unit	351
	15.1.2	Navigation with gyroscope system and hose water level	351
	15.1.3	Navigation with total station and automatic target unit	352
	15.1.4	Navigation with total station and prisms	353
15.2	Ring design and calculation of the ring installation sequence		354
15.3	Ring convergence measurement		354
15.4	Steering		355
15.5	Further surveying and data logging tasks		357

16	**Workplace safety**	359
16.1	General safety requirements	360
16.2	Control stations	363
16.3	Electrical cut-out and safety devices	364
16.4	Control devices and control systems	364
16.5	Towing connections	366
16.6	Laser guidance	367
16.7	Ventilation and the control of dust and gas	367
16.8	Fire protection	368
16.9	Storage of safety equipment for the personnel	369
16.10	Maintenance	369
16.11	Content of handbook	369
16.12	Evaluation of risk in mechanised tunnelling [26]	370

17	**Partnering contract models and construction**	383
17.1	Introduction	383
17.2	Requirements for the contract model	384
17.3	Contract model according to VOB	385
17.4	Time and cost drivers	386
17.5	Under-pricing as a performance killer	387
17.6	Chances and risks of partnering	388
17.7	Partnering – contractual implementation	389
17.8	Partnering – mutual process optimisation	390

18	**Process controlling and data management**	393
18.1	Introduction	393
18.2	Procedure	393
18.3	Data management	394

18.4	Target-actual comparison		395
18.5	Target process structure		397
18.6	Analysis of the actual process		399

19 DAUB recommendations for the selection of tunnelling machines 401
19.1 Preliminary notes 401
19.2 Regulatory works 402
    19.2.1 National regulations 402
    19.2.2 International standards 403
    19.2.3 Standards and other regulatory works 403
19.3 Definitions and abbreviations 404
    19.3.1 Definitions 404
    19.3.2 Abbreviations 406
19.4 Application and structure of the recommendations 406
19.5 Categorisation of tunnelling machines 408
    19.5.1 Types of tunnelling machine (TVM) 408
    19.5.2 Tunnel boring machines (TBM) 408
    19.5.2.1 Tunnel boring machines without shield (Gripper TBM) 408
    19.5.2.2 Enlargement tunnel boring machines (ETBM) 409
    19.5.2.3 Tunnel boring machine with shield (TBM-S) 410
    19.5.3 Double shield machines (DSM) 410
    19.5.4 Shield machines (SM) 410
    19.5.4.1 Shield machines for full-face excavation (SM-V) 410
    19.5.4.2 Shield machines with partial face excavation (SM-T) 413
    19.5.5 Adaptable shield machines with convertible process technology (KSM) 414
    19.5.6 Special types 414
    19.5.6.1 Blade shields 414
    19.5.6.2 Shields with multiple circular cross-sections 414
    19.5.6.3 Articulated shields 414
    19.5.7 Support and lining 415
    19.5.7.1 Tunnel boring machines (TBM) 415
    19.5.7.2 Tunnel boring machines with shield (TBM-S), Shield machines (SM, DSM, KSM) 416
    19.5.7.3 Advance support 417
    19.5.7.4 Support next to the tunnelling machine 418
19.6 Ground and system behaviour 418
    19.6.1 Preliminary remarks 418
    19.6.2 Ground stability and face support 418
    19.6.3 Excavation 419
    19.6.3.1 Sticking 419
    19.6.3.2 Wear 420
    19.6.3.3 Soil conditioning 420
    19.6.3.4 Soil separation 421
    19.6.3.5 Soil transport and tipping 421
19.7 Environmental aspects 422

19.8	Other project conditions		424
19.9	Scope of application and selection criteria		425
	19.9.1	General notes about the use of the tables	425
	19.9.1.1	Core area of application	425
	19.9.1.2	Possible areas of application	425
	19.9.1.3	Critical areas of application	426
	19.9.1.4	Classification in soft ground	426
	19.9.1.5	Classification in rock	426
	19.9.2	Notes about each type of tunnelling machine	426
	19.9.2.1	TBM (Tunnel boring machine)	426
	19.9.2.2	DSM (Double shield machines)	426
	19.9.2.3	SM-V1 (full-face excavation, face without support)	427
	19.9.2.4	SM-V2 (full-face excavation, face with mechanical support)	427
	19.9.2.5	SM-V3 (Full-face excavation, face with compressed air application)	427
	19.9.2.6	SM-V4 (full-face excavation, face with slurry support)	427
	19.9.2.7	SM-V5 (full-face excavation, face with earth pressure balance support)	428
	19.9.2.8	SM-T1 (partial excavation, face without support)	428
	19.9.2.9	SM-T2 (partial excavation, face with mechanical support)	428
	19.9.2.10	SM-T3 (partial excavation, face with compressed air application)	428
	19.9.2.11	SM-T4 (Partial excavation, face with slurry support)	428
	19.9.2.12	KSM (Convertible shield machines)	428
19.10	Appendices		429

Bibliography 449

Index 463

1 Introduction

The mined construction of underground infrastructure has made steady progress over recent years. It is now possible to construct underground works with very little impairment of buildings or traffic flow at ground level. Particularly in inner-city areas, with sensitive infrastructure and high population density, there is an enormous demand for underground structures.

The cavities created in this way have until now mostly been for underground transport routes, although there are also other possible uses such as energy extraction, storage and refuge spaces, utility tunnels and, not least, underground urban development. This has led to extensive schemes and projects, particularly in Japan due to the very restricted space availability (Figure 1-1).

Particularly in the field of shield tunnelling, the prominent role of Japan has been unmistakeable. But the development of this construction method is also at a high and internationally respected level in Germany and other parts of Europe. The shield construction process enables the production of elongated underground structures, even at shallow depths, in soil with poor load-bearing capacity or under the groundwater table, without causing any disturbance or significant settlement on the ground surface. Ground conditions with loose spherical material can be mastered, as can soft plastic or flowing soils. But the use of these machines is also practicable in temporarily stable ground, where the shield only acts as head protection. All in all, shield machines have a wide scope of application.

Figure 1-1 Japanese scheme for the exploitation of underground space in an inner-city area [155]

The shield construction process could but should not generally replace other methods of tunnelling. It can, however, offer a technically feasible and also economic alternative to other methods of tunnelling in unfavourable geological conditions, for long contract sections, high advance rate requirement or where stringent surface settlement limits apply. The essential advantages and disadvantages are summarised below.

Advantages:

- the possibility of mechanisation and high advance rate,
- precision of profile,
- minimisation of the effect on buildings on the surface,
- improved safety for the miners,
- environmentally friendly construction method,
- raising of the groundwater table,
- little noise,
- enables a high-quality and economic lining.

Disadvantages:

- long lead time for the design, production and assembly of the shield machine,
- long familiarisation time,
- elaborate and expensive site facilities (a separating plant may be required); tenders may only be competitive for longer tunnels,
- performance risk in changeable ground,
- the cross-section normally has to be round with little possibility of variation,
- high cost of altering the excavated geometry, e.g. for wider sections,
- the lining normally has to be specially designed to resist the thrust forces.

Application is therefore practicable where the advantages can be sensibly exploited and the disadvantages are taken into account as far as possible in the design and construction planning. Experience shows that a shield in the smaller diameter range can generally compete with other tunnelling methods for tunnel drives up to 2,000 m. For longer tunnels, economic applications of shield machines are possible and even cheaper than using open machines or conventional methods.

The successful use of a shield always requires meticulous design and planning of the machine, the lining and the logistics. Experience and know-how are essential for a practicable and economic scheme. According to [235], too many clients have chosen the wrong machine or construction concept for the ground conditions and have later been faced with unacceptable settlement on the surface, unexpectedly slow advance rates, spalling or failure of the lining, water ingress or other defects. For the client, only a tunnel constructed on schedule, of good quality and at reasonable cost, and with as little impact on the environment as possible is of interest. The designers of shield equipment need to take these natural concerns into consideration. Mechanical engineering issues have to be effectively linked to those of the tunnel itself. Constant exchange of experience between mechanical and civil engineers is essential, with the appropriate evaluation of experience from completed projects.

1.1 Basic principles and terms

The basic principle of a shield is that a generally cylindrical steel construction is driven along the tunnel axis while the ground is excavated. The steel construction supports the excavated cavity until temporary support or the final lining has been installed. The shield therefore has to resist the pressure of the surrounding ground and hold back any groundwater.

While the cavity along the sides of the tunnel is supported by the shield skin itself, additional support measures will be required to support the face, depending on the ground and groundwater conditions encountered. Figure 1-2 shows five different methods of stabilising the face, which are described in detail in Chapter 2. These are:

– natural support,
– mechanical support,
– compressed air support,
– slurry support,
– earth pressure balance support.

These methods of supporting the face represent the great advantage of the shield tunnelling process. In contrast to other methods of tunnelling, it is possible to provide immediate support of the ground as soon as it is disturbed.

In addition to the type of face support, the method of excavation is an important characteristic of shields. The most simple process is manual digging in hand shields, and this is still used today in exceptional cases, for example for short sections and under certain geological conditions. Mechanical excavation is, however, more usual. This can be differentiated into mechanical partial- and full-face excavation. In partial-face excavation, the face is worked in sections using machinery such as hydraulic excavators or roadheaders, which are operated and controlled either by operators or automatically. The full face can be excavated, according to the ground conditions encountered, by open-mode wheels, rim wheels (in some cases with shutters) or closed cutter heads. Further methods are hydraulic excavation using pressurised jets of fluid and extrusion excavation, where the action of the thrust cylinders on highly plastic soil forces it through closable openings in the front wall of the shield. Excavation processes are described in more detail in Chapter 4.

The removal of the excavated material requires special transport systems to move the muck from the face, through the shield and to the surface. The most suitable system depends directly on the nature of the ground encountered and the associated type of face support and excavation, since these factors have a great influence on the consistency and transport properties of the muck. Figure 1-3 gives an initial overview of the possible transport systems through the shield, which will be explained in more detail in Chapter 5. There are numerous transport methods available today, which can be categorised into the three basic groups

– dry transport,
– fluid/slurry transport,
– high-density solid pumping.

Transport along the tunnel can use pumped pipes, conveyor belts, dumpers or rail-based systems (muck trains). The transfer area to the tunnel transport system is integrated into the backup.

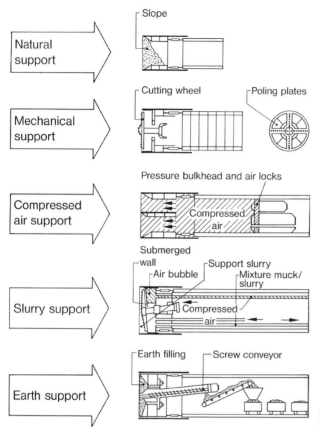

Figure 1-2 Methods of supporting the ground and holding water at the face [266]

The shield is pushed forward in the direction of the tunnel axis with the progress of excavation in order to support the resulting cavity. The required thrust forces are produced by hydraulic cylinders, normally pushing against the already installed lining. This means that the tunnel lining and boring machinery have to be finely matched. The correct function of the shield and the quality of the final tunnel lining both depend on this compatibility, which is dealt with in more detail in Chapter 6.

The cavity produced by excavation is mostly supported with precast elements called segments. There are numerous different forms, materials, possible layouts, sealing systems and installation methods, which require detailed description (Chapter 6). Other lining systems are also possible and are already in use today (Figure 1-4). The pumping of concrete under pressure into formwork (called the extrusion process) is an interesting possibility, but has not been further developed. Even shotcrete can be used in connection with shield tunnelling.

As the support is normally installed inside the protection of the shield skin, a gap remains as the shield progresses further. The gap has to be filled in order to minimise loosening and settlement. This has to be suitably backfilled or grouted and the shield must be provided with the appropriate equipment.

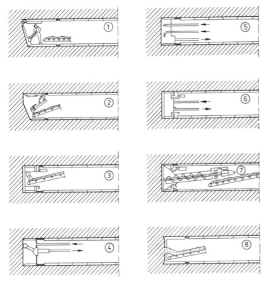

1. Manual excavation and dry removal of the spoil on a conveyor belt,
2. Mechanical partial-face excavation (in this case using a hydraulic excavator arm) and dry transport on a conveyor,
3. Mechanical full-face excavation and dry transport on a conveyor belt (the transfer to the conveyor belt can be either in the central or lower part of the cutting wheel through muck pockets),
4. Mechanical partial-face excavation and slurry transport,
5. Hydraulic excavation and slurry transport,
6. Mechanical full-face excavation and slurry transport,
7. Mechanical full-face excavation, transport with a screw conveyor and transfer to a conveyor belt,
8. Extrusion excavation and transfer of the material onto a conveyor belt.

Figure 1-3 Methods of muck removal in shield tunnelling

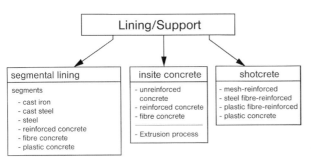

Figure 1-4 Possible types of lining in shield tunnelling

1.2 Types of tunnel boring machine according to DAUB

The recommendations of DAUB (the German Tunnelling Committee) are reproduced in their entirety in Chapter 19 [54].

1.2.1 Categories of tunnelling machines German association for underground construction (TVM)

Tunnelling machines either excavate the full face with a cutter head or cutting wheel or part of the face with suitable excavation equipment.

These can be tunnel boring machines (TBM), double shield machines (DSM), shield machines (SM) or combination machines (KSM). While the acronym "TBM" in English will be used for all types of tunnelling machines, the German DAUB reserves the abbreviation for the hard rock machines.

As the ground is excavated, the machine is pushed forward, either continuously or intermittently.

A systematic categorisation of tunnelling machines is shown in Figure 1-5 (see also Appendix 1 "Overview of tunnelling machines" in Chapter 19).

1.2.2 Tunnel boring machines (TBM)

Tunnel boring machines are used for driving tunnels through stable hard rock. Active support of the face is not required and, in any case, is technically impossible. These machines can normally only drive a circular cross-section.

Tunnel boring machines can be differentiated into those without shields (open gripper TBM), reamer or enlargement tunnel boring machines (ETBM) and shielded tunnel boring machines (TBM-S).

These machines are described in detail in [203].

1.2.2.1 Tunnel boring machines without shield (gripper TBM)

Open tunnel boring machines without shield are used in hard rock that has medium to long stand-up time. They have no complete shield skin. Economic application can be greatly influenced and limited by the high cost of wear of the excavation tools.

In order to be able to apply thrust force to the cutter head, the machine is braced radially by hydraulically driven grippers acting against the sides of the tunnel.

Excavation is carried out with little damage to the surrounding rock mass and to an exact profile by disc cutters mounted on the rotating cutter head. The machine fills a large part of the cross-section. Systematic support of the tunnel walls is normally installed behind the machine (10 to 15 m or more behind the face). In rock with a shorter stand-up time or

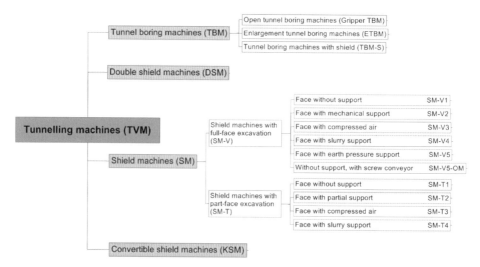

Figure 1-5 Types of tunnel boring machines

liable to rock falls, support measures such as steel arches, poling plates or rock bolts are installed at the closest possible distance behind the cutter head.

Where shotcrete lining of the tunnel is necessary, this should only be applied in the rearward part of the backup area in order to keep the mess off the machinery and control gear in the forward machine area as far as possible. In exceptional cases, however, shotcrete may have to be applied as close behind the face as possible.

If the geological forecast describes poor rock or a heterogeneous condition of the rock mass (high degree of jointing and fault zones), it is recommended to equip the machine to enable advance investigation drilling or advance rock consolidation.

The excavation of the face produces material in small pieces with the associated dust development. Machines therefore require equipment for the reduction of dust development and dedusting. This can be:

- spraying with water at the cutter head,
- a dust shield behind the cutter head,
- dust extraction with dedusting on the backup.

Material handling and disposal from the machine nowadays requires very long backup facilities.

1.2.2.2 Reamer tunnel boring machines (ETBM)

Reamer tunnel boring machines (enlargement machines) are used in hard rock to enlarge a previously bored pilot tunnel to the intended final diameter. The enlargement to the final diameter is performed in one or two working steps using an appropriately constructed cutter head.

The main elements of these machines are the cutter head, the bracing and the drive mechanism. The bracing of the specialised machines is situated in front of the cutter head with grippers in the pilot tunnel, and the cutter head of the machine is drawn towards the grippers as it bores. In faulted rock formations, measures to improve the fault zone can be carried out from the previously bored pilot tunnel in order to minimise the risks during the boring of the main tunnel.

1.2.2.3 Tunnel boring machines with single shield (TBM-S)

In hard rock with a short stand-up time or liable to rock falls, shielded tunnel boring machines are used. For this case, the installation of the lining within the shield is appropriate (segments, pipes etc.). While advancing, the machine can be supported from the lining, so bracing is not normally required. The remaining statements made about tunnel boring machines apply accordingly.

1.2.3 Double shield machines (DSM)

Double shield machines (DSM) consist of two parts arranged one behind the other. The front part is equipped with the cutter head and the main thrust cylinders, and the back part houses the auxiliary thrust cylinders and the grippers. The front part of the machine can be advanced by a complete ring length ahead of the back part using a telescopic section.

In stable hard rock, the gripper shoes resist the drive torque and the thrust forces. The secure fixing of the back part of the machine using the grippers enables the assembly of the segmental lining in the shield tail while boring is in progress. In a stable rock mass, it may also be possible to omit the installation of the lining.

In unstable ground, where the gripper shoes cannot find sufficient resistance, the thrust can be resisted from the last segment ring. The front and back parts of the machine are retracted together and the thrust forces are pushed from the segment ring by the auxiliary thrust cylinders.

It is not normally possible to actively support the face or the sides of the excavation.

Due to the rapid advance of the back part of the machine after a boring stroke has been completed and the grippers are being regripped, the rock mass has to be able to stand up independently until the annular gap has been completely filled with grout or stowed with pea gravel.

1.2.4 Shield machines (SM)

These can be shield machines with full-face excavation (with a cutter head: SM-V) and shield machines with partial-face excavation (using a roadheader boom, excavator: SM-T). Shield machines are used in loose ground above and below the groundwater table. This normally means that the ground around the cavity and at the face has to be supported. Shield machines can be further divided according to the type of face support (Figure 1-5).

1.2.4.1 Shield machines with full-face excavation (SM-V)

1) Face without support (SM-V1)

If the face will stand up, e.g. in clay soil with stiff consistency and sufficient cohesion or in solid rock, open shields can be used. The cutting wheel fitted with tools excavates the soil and the muck is removed on a conveyor belt.

In solid rock liable to rock falls, shield machines with a mostly closed cutter head fitted with disc cutters and fully protected from the unstable ground by a shield skin are normally used. The thrust forces and the cutter head drive torque are transferred through the thrust cylinders to the last ring of segments installed.

2) Face with mechanical support (SM-V2)

With tunnelling machines with mechanical support, the support of the face during excavation is provided by elastically fixed support plates arranged in the openings of the cutting wheel. In practice, however, experience shows that that no appreciable mechanical support of the face can be provided by the rotating cutting wheel. For this reason, this type of cutting wheel did not prove successful in unstable ground and is no longer in use today. The mechanical face support by the cutting wheel or the support plates should only be considered a supplementary safety measure and the supporting effect should not be taken into account in calculations to verify the stability of the face.

3) Face with compressed air support (SM-V3)

Shield machines of type SM-V3 can be used below the groundwater table even if it cannot be lowered or groundwater lowering is not allowed. In this case, the water at the face must be held back with compressed air. A precondition for the displacement of the groundwater is the formation of an air flow to the surface. Impermeable strata above the tunnelling machine can retain the applied air and prevent the effective displacement of the water (and thus the formation of an air flow). The permeability limit of the surrounding ground is therefore significant.

As no pressure difference can be built up at the face, compressed air cannot generally provide support against earth pressure, which applies particularly in permeable soil. The loss of the apparent cohesion in non-saturated soil is also possible.

For the duration of tunnelling work, either the entire tunnel is pressurised or the machine is provided with a pressure bulkhead to maintain the excavation chamber under pressure. In both cases, air locks are required. Particular attention needs to be paid to compressed air bypassing the shield tail seal and the lining. The recommendations and requirements for working under compressed air should complied with.

Any additional support of the face provided by the cutting wheel or support plates should be regarded solely as an additional security. It is not permissible to take the supporting effect into account in calculations to verify the stability of the face.

4) Face with slurry support (SM-V4)

Tunnelling machines with slurry support provide support to the face through a pressurised fluid, which is specified depending on the permeability of the surrounding ground. It must be possible to vary the density and viscosity of the fluid. Bentonite suspensions have proved particularly successful for this purpose. In order to support the face, the working chamber is closed from the tunnel by a pressure bulkhead.

The required support pressure can be regulated very precisely with an air bubble behind a submerged wall and by adjusting the output of the supply and extraction pumps. The required and the maximum support pressures over the entire length to be bored should be calculated before the start of tunnelling – referred to as the slurry support pressure calculation.

The soil is excavated from the full face by a cutting wheel fitted with tools (open-mode or rimmed wheel) and removed hydraulically. Subsequent separation of the removed suspension is essential.

If it is necessary to enter the excavation chamber, for example to change tools, carry out repair work or to remove obstructions, the support slurry has to be replaced by compressed air. The support slurry then forms a low-permeability membrane on the face, which however is of limited durability (risk of drying out). The membrane permits the support of the face by compressed air and may need to be renewed regularly. The support slurry can be completely (empty) or only partially (lowering) replaced by compressed air. The maximum partial lowering is limited particularly by the requirement for sufficient working space. This should be chosen to be large enough for safe working to be possible at all times and so that an adequately large space is available for the workers to retreat.

If an open cutting wheel is used, it should be possible to mechanically close the face with shutter segments in the cutting wheel or with plates, which can be extended from behind, in order to protect the personnel working in the excavation chamber while the machine is stopped, and this is also sensible owing to the limited duration of the membrane effect.

Stones or rock benches can be reduced to a size that can be removed by disc cutters in the cutting wheel and/or crushers in the working chamber.

In stable ground, the slurry shield can also be operated in open mode without pressurisation, with water being used for muck removal.

Any additional mechanical support of the face provided by the cutting wheel or support plates should be regarded solely as additional security, and it is not permissible to take the supporting effect into account in calculations to verify the stability of the face.

5) Face with earth pressure support (SM-V5)

Tunnelling machines with earth pressure support provide support to the face with remoulded excavated soil. The excavation chamber of the shield is closed from the tunnel by a pressure bulkhead. A cutting wheel, fitted with tools and more or less closed, excavates the soil. Mixing arms on the back of the cutting wheel (rotors/back buckets) and on the pressure bulkhead (stators) assist the remoulding of the soil to a workable consistency. The pressure is checked with pressure cells, which are distributed on the front of the pressure bulkhead. A pressure-tight screw conveyor removes the soil from the excavation chamber.

The support pressure is regulated by varying the revolution speed of the screw conveyor or through the pressure-volume-controlled injection of a suitable conditioning agent. The pressure gradient between excavation chamber and tunnel is provided by friction in the screw. Either the soil material in the screw can ensure the sealing of the discharge device or an additional mechanical device has to be installed. Complete support of the face, in particular the upper part of it, is only successful when the soil acting as a support medium can be remoulded to a soft or stiff-plastic mass. This is particularly influenced by the percentage content of fine-grained material smaller than 0.06 mm. The scope of application of earth pressure balance (EPB) shields can be extended through the use of soil conditioners such as bentonite, polymers or foam, but attention should be paid to the environmentally acceptable disposal of the material.

In stable ground, the EPB shield can also be operated in open mode without pressurisation with a partially filled excavation chamber (SM-V5-OM). In stable ground with water ingress, operation is also possible with partially filled excavation chamber and compressed air.

If the groundwater pressure is high and the ground is liable to liquefaction, the critical location of the transfer of material from the screw conveyor to the conveyor belt can be replaced by a closed system (pumped material transport).

Any additional mechanical support of the face provided by the cutting wheel or support plates should be regarded solely as additional security and it is not permissible to take the supporting effect into account in calculations to verify the stability of the face.

1.2.4.2 Shield machines with partial face excavation (SM-T)

1) Face without support (SM-T1)

This type of open mode shield can be used with a vertically or steeply sloping and stable face. The machine consists of a shield skin and the excavation tools (excavator, milling head or ripper tooth), the spoil removal equipment and the thrust cylinders. The excavated material is removed on a conveyor belt or chain scraper.

2) Face with partial mechanical support (SM-T2)

For partial support of the face, working platforms and/or poling plates can be used. In platform shields, the tunnelling machine is divided into one or more platforms at the face. Natural slopes form on these, which support the face. The ground is excavated manually or mechanically. Platform shields have a low degree of mechanisation.

A disadvantage is the danger of large settlements resulting from uncontrolled face support. In shield machines with face support, the face is supported by poling plates supported on hydraulic cylinders. In order to excavate the soil, the poling plates are partially withdrawn. A combination of poling plates and platforms is also possible. If support of the crown alone is sufficient, hinged poling plates can be fixed at the crown.

3) Face with compressed air support (SM-T3)

If there is groundwater, this can be retained by compressed air with machines of type SM-T1 and SM-T2. Either the entire tunnel is filled with compressed air or the machine is fitted with a pressure bulkhead (comparable to SM-V3). The excavated material is removed hydraulically or dry through the air lock.

4) Face with slurry support (SM-T4)

In the past, many trials have been undertaken to achieve active face support through the use of support fluid with partial face machines (e.g. Thixshield). The excavation chamber in this case has to be completely filled with support slurry. The excavation of the soil can be mechanical or using high-pressure jetting.

As the excavation of the ground cannot be sufficiently controlled, this method of tunnelling has not proved successful and is no longer used.

1.2.5 Adaptable shield machines with combined process technology (KSM)

Numerous tunnels pass through highly changeable geological conditions, which can range from rock to loosely compacted soil. This requires the excavation mode to be adapted to the geological conditions and the use of correspondingly adapted shield machines. These can be categorised into:

a) Shield machines that enable the change of excavation mode without rebuilding:
   Earth pressure shield SM-V5 ↔ Compressed air shield SM-V3,

b) Shield machines that have to be rebuilt to change the excavation mode technology. The following combinations have been used:

Slurry shield SM-V4 ↔ Shield without support SM-V1,
Slurry shield SM-V4 ↔ Earth pressure shield SM-V5,
Earth pressure shield SM-V5 ↔ Shield without support SM-V1.

The rebuilding work normally lasts a number of shifts.

1.2.6 Special types
1.2.6.1 Blade shields

In blade shields, the shield skin is split into blades, which can be advanced independently. The ground is excavated by partial face machinery, cutting wheel or excavator. An advantage of blade shields is that they do not have to be circular and can, for example, drive a horseshoe-shaped section, in which case the invert is normally open. This is described as blade tunnelling. Because of various negative experiences in the past, however, blade shields are seldom used today.

1.2.6.2 Shields with multiple circular cross-sections

These shields are characterised by overlapping and non-concentric cutting wheels. This type of machine is currently only offered by Japanese manufacturers and mostly used to drive underground station cross-sections. The machines are difficult to steer and have not yet been used in Europe.

1.2.6.3 Articulated shields

Practically all types of shields can be equipped with an articulated joint to divide their length. This is provided particularly when the length of the shield skin is longer than the shield diameter, in order to make the tunnelling machine easier to steer. The layout can also be necessary to drive very tight radius curves.

The description of the tunnelling machines is then according to one of the categories described above. A separate category of "articulated shields" is no longer usual.

1.2.7 Remarks about the individual types of tunnelling machines with diagrams
1.2.7.1 Tunnel boring machines (TBM)

The most important application of tunnel boring machines (TBM) (Figure 1-6) is competent to loose rock, in which case the ingress of groundwater and joint water can be overcome. The uniaxial compression strength σ_D should be between about 25 and 250 MN/m². Higher strengths and toughness of the rock or a high content of abrasive minerals represent limits to the scope of economic application. Problems with the bracing of the machine can also make an application questionable. In order to evaluate the rock, the splitting strength and RQD value can be measured. With a degree of fracturing of the rock mass with RQD of 50 to 100 % and a joint spacing > 0.6 m, the use of a TBM should be safe. If the fracturing is higher, the stability should be investigated. The use of a TBM is ruled out in loose ground or solid rock with properties similar to soil.

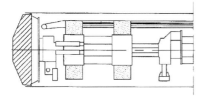

Figure 1-6 Open gripper TBM

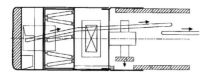

Figure 1-7 Double shield TBM

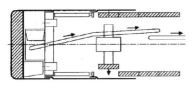

Figure 1-8 Single shield TBM, full-face excavation

1.2.7.2 Double shield machines (DSM)

Double shield machines (Figure 1-7) are mainly used on tunnel projects with shorter sections with loose to brittle rock, in addition to long sections through stable rock. In stable rock (see the requirements for the use of a TBM), the tunnel can be bored in continuous mode using the grippers. In fault zones or sections with low rock strength, in which the grippers can no longer be used, the shield is telescoped in and the tunnelling machine now supports itself off the last ring of segments.

1.2.7.3 Face without support (SM-V1)

This type of machine (Figure 1-8) can only be used in stable, mostly impermeable cohesive soil with high fines content. The stability of the face should be verified with calculations. The excavation sidewalls should be able to stand up until the final installation of the tunnel lining and this should be verified. Loosening of the ground, which could lead to a reduction of the subgrade reaction, should also be ruled out. If there are buildings sensitive to settlement on the surface, the deformation of the ground and loosening should be verified based on the usual damage classification (e.g. gradient of the settlement trough).

In hard rock, this type of machine is used in loose to brittle rock, also with groundwater and joint water. If the rock strength is high in a stable rock mass, the bonding strength can be reduced considerably. This corresponds to a joint spacing of $\approx$ 0.6 to 0.06 m and an RQD value between about 10 and 50 %. Generally, however, these machines can also be used with lower rock compression strengths below 5 MN/m^2, for example in heavily weathered rock.

The stability of the face and the excavated sides should be verified with calculations. If there is high ingress of formation water, appropriate measures should be provided.

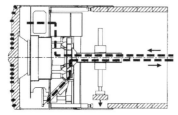

Figure 1-9 Slurry shield with air bubble support

1.2.7.4 Face with mechanical support (SM-V2)

As a result of numerous failed projects, this type of machine is no longer recommended.

1.2.7.5 Face with compressed air (SM-V3)

The application of compressed air enables the machine type SM-V1 to be used in stable ground, even in groundwater. The air permeability of the ground or the air consumption and the verification of the formation of a flow field and of blowout safety are essential criteria for the application of this machine type. The groundwater table should be above the tunnel crown with an adequate safety margin.

1.2.7.6 Face with slurry support (SM-V4)

The main areas of application for slurry shields (Figure 1-9) are in coarse and mixed-grained soil types. The groundwater table should be above the tunnel crown by an adequate margin. During excavation, a pressurised fluid, e.g. bentonite suspension, supports the face. Highly permeable soils impede the formation of a membrane. If the permeability is greater than $5 \cdot 10^{-3}$ m/s, there is a risk that the bentonite suspension can flow uncontrolled into the surrounding ground. The addition of fine material and filler or additives to improve the rheological properties can expand the scope of application. Alternatively, additional measures to reduce the permeability of the soil may be necessary (for example filling of pore cavities). Stones and blocks, which cannot be pumped, are first brought down to size by crushers. High fines content can lead to problems with separation. It should also be borne in mind that the rheological properties of the support medium are worsened by very fine material, because the separation of clay fractions and bentonite is not technologically feasible.

1.2.7.7 Face with earth pressure support (SM-V5)

Machine types with earth pressure support (Figure 1-10) are particularly suitable for soils with a fines content (< 0.06 mm) of over 30 %. In coarse and mixed-grained soils and rock, thrust force and cutting wheel torque increase over-proportionately with increasing

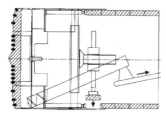

Figure 1-10 Earth pressure balance shield

support pressure. The hydraulic behaviour of the excavated soil can be improved with the addition of a suitable conditioner, e.g. bentonite, polymer or foam. For active control of the support medium and to ensure low-settlement tunnelling, soil conditioning with foam is recommended outside the predefined area of application.

Earth pressure balance shields have the advantage that operation in open mode (SM-V5-OM) with partially filled and non-pressurised excavation chamber, i.e. without active face support, is also possible without alteration of the process technology. It should, however be noted that in this mode, due to the design of the cutting wheel and screw, the excavated soil/rock is grind much more than is the case with removal by conveyor through the centre (SM-V1). If the ground tends to stick, then obstructions and increased wear have to be expected. To improve the material flow and to reduce the tendency to sticking, conditioning measures should be planned. Particularly unfavourable for earth pressure shields, in soil as well as in rock, is a combination of high support pressure, high permeability, high abrasiveness and difficult breaking of the grain structure.

1.2.7.8 Face without support (SM-T1)

This type of machine can be used above the groundwater table, if the face is continuously stable – see also SM-V1.

Partial face machines always offer easy access to the face, so the process can offer great advantages if a lot of obstructions are expected.

1.2.7.9 Face with partial support (SM-T2)

This type of machine (Figure 1-11) can be used when the support of the material spreading on the platforms at its natural angle of repose is adequate for the driving of the tunnel with limited control of deformation. Poling plates can be used to provide additional support at the crown and on the platforms. The main areas of application are in weakly- to non-cohesive gravel-sand soils above the groundwater table with the relevant friction angle.

1.2.7.10 Face with compressed air support (SM-T3)

The use of this type of machine (Figure 1-12) is appropriate when types SM-T1 and SM-T2 are used in groundwater. The entire working area, including the already excavated tunnel, or just the excavation chamber, is pressurised.

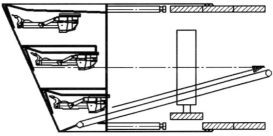

Figure 1-11 Partial-face excavation with mechanical face support

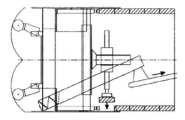

Figure 1-12 Partial-face excavation with compressed air face support

1.2.7.11 Face with slurry support (SM-T4)

Partial-face excavation machines with slurry support are no longer used.

1.2.7.12 Adaptable machines (KSM)

Adaptable machines (Figure 1-13) combine the areas of application of the relevant machine types in changeable ground conditions. Their application spectrum is therefore extended to both criteria.

The number of rebuilding changes from one tunnelling mode to another should be kept as low as possible, because rebuilding is time- and cost-intensive.

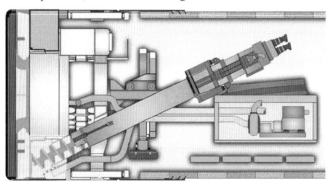

Figure 1-13 Adaptable shield machine

1.3 Origins and historical developments

The human race has been building tunnels for about 5,000 years. Tunnels have been excavated for the protection of goods and people, for secret passages to forbidden locations and for the exploitation of resources or to accelerate transport.

Tunnellers soon learnt to support rock liable to falling and in loose ground with timbering, followed by a brickwork lining. This method could also succeed in ground with percolating or joint water, and this continued into the 19th century, but was not practical below the groundwater table, in loose ground or particularly under open water. The situation altered in 1806, when the ingenious engineer Sir Marc Isambard Brunel invented and later patented the principle of shield tunnelling in London. The purpose was the construction of a crossing of the River Neva in St Petersburg, which could remain open in winter, needed because the piers of the bridge were badly damaged each year by pack ice from Lake Ladoga. Brunel developed a tunnel solution for this project, although a suspension bridge had originally been proposed.

1.3 Origins and historical developments   17

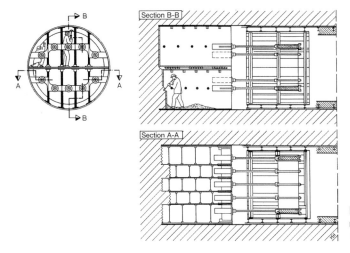

Figure 1-14 Box shield used by M. I. Brunel, 1806 [268]

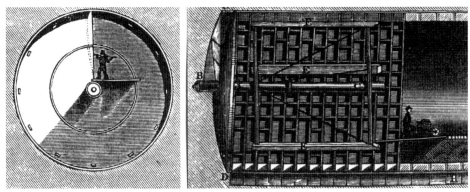

Figure 1-15 Screw shield used by M. I. Brunel, 1818 [251]

The original shield construction of M. I. Brunel featured a division into cells, each with a worker working independently and safely (Figure 1-14). In one method, the cells were fixed in the shield structure and the entire shield was driven forward by hydraulic cylinders after the excavation of a section; in another method, the single cells could be advanced independently. All closed full shields in use today are based on the former method; the latter was never put into practice, unless the blade shield could be considered as a further development of it.

A completely different process with closed shield skin and full-area, screw-shaped excavation with immediate lining is shown in Figure 1-15. This shield can be regarded as the predecessor of the earth pressure shields.

The Thames Tunnel project in London finally gave M. I. Brunel the opportunity to put his ideas into practice (Figure 1-16). The shield was rectangular and consisted of 12 adjacent frames, each divided into three chambers. One miner worked in each of these chambers – 36 miners altogether. The system worked like this: first the top poling boards were driven into the ground with screw jacks. The timbering to the face was then removed starting from the top, and the soil was excavated for about 150 mm, the face was timbered again and supported with screw jacks. The entire frame was supported from the brickwork lining, which had been filled in behind. Work on the Thames Tunnel started in 1825 amid great difficulties, and the tunnel was only completed in 1843 after more than five severe water inflows. It is interesting that after the first serious flood had stopped work (Figure 1-17), Callodan suggested the use of compressed air in 1828, which Brunel however turned down [268].

In 1869, the engineer James Henry Greathead used a circular shield for a further tunnel under the Thames, which also marked the first use of circular cast iron segments for the lining [19]. The construction of the 402 m long tunnel with an external diameter of 2.18 m was completed without great difficulties, as the tunnel passed through impermeable clay along its entire length, and there were no water problems. The circular Greathead shield was the pattern for most subsequent designs. Figure 1-18 shows one of the two Greathead shields used for the construction of the Rotherhithe Tunnel (1904–1908) with 9.35 m diameter. The tunnel under the Thames linked Rotherhithe to Ratcliffe.

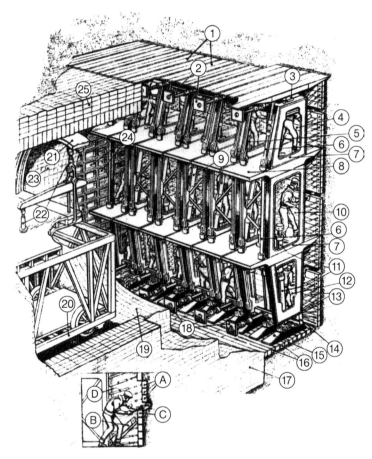

Figure 1-16 Shield used by Sir M. I. Brunel for the Thames Tunnel, 1825/43 [213]

1 top poling boards
2 screw jacks
3 abutment
4 upper chamber
5 jack
6 strengthening
7 side framing
8 floor (upper chamber)
9 bracing
10 middle chamber
11 support
12 lower chamber
13 timbering
14 jack
15 jack base plate
16 floor timbering
17 brickwork
18 thrust jacks
19 invert
20 wheeled working platform
21 vault falsework
22 jack
23 western side wall
24 shield
25 brickwork in the crown

A forward timbering
B timber board
C initial timbering
D jack

Figure 1-17 Flooding of the tunnel under the River Thames on 12 January 1828 [213]

The problem of controlling water in the construction of underwater tunnels in loose ground was first solved by Admiral Sir Thomas Cochrane through the use of compressed air, following the suggestion from Callodan to Brunel from 1828. In 1830, he invented the compressed air lock, which enabled access to a working space under increased pressure [8], [268]. The first uses of compressed air in a tunnel occurred almost simultaneously in 1879 in Antwerp and 1880 in New York, although these were worked without shields.

In 1886, Greathead achieved the application of a shield in combination with compressed air for the first time for the construction of the London Underground [128]. The use of compressed air considerably simplified tunnelling in water-bearing strata. This was the birth of the compressed air shield, and a critical gap was closed in tunnelling capability, leading to a considerable increase in the number of shield tunnels all over the world. At the start of the 20th century, most tunnels were being excavated using Greathead-type shields.

Brunel's invention naturally soon led to the idea of replacing the manual digging of the ground with mechanical excavation. The first patent for such a mechanised shield was applied for by the Englishmen John Dickinson Brunton and George Brunton (Figure 1-19) and was granted in 1876 [268]. The shield had a hemispherical rotating cutting head built up of single plates. The cut material was intended to fall into mucking buckets mounted radially on the cutting head. The buckets threw the excavated material onto a conveyor belt, which transported it backwards out of the shield. The cutting head itself was turned by six hydraulic cylinders, which worked against a ratchet ring fixed to the cutting head. This idea was later repeated for the construction of the underground railway in Kiev.

A better design was the Price shield, named after its inventor, J. Price, and patented in 1896 (Figures 1-20, 1-21). This machine was used with great success in London clay from

1897. It was the first machine to make use of the performance of a rotating cutting head based on simple principles inside a Greathead shield. The cutting head consisted of four arms arranged like spokes, on which the cutting and raking tools were fitted. The cutting head was also fitted with basin-shaped buckets, which picked up the excavated soil and deposited it down a slide onto waiting trucks for carting way. The cutting head was electrically driven through a long shaft [268].

In soil with higher permeability, it is difficult to support the face with compressed air. Greathead therefore developed a shield in 1874 with fluid-supported face to avoid the disadvantages of compressed air. The soil was intended to be removed hydraulically by a fluid and transported hydraulically as slurry (Figure 1-22).

Figure 1-18 Greathead shield, 9.35 m external diameter, Rotherhithe Tunnel, 1904–1908 (Markham)

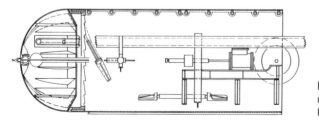

Figure 1-19 Mechanised shield used by J. D. and G. Brunton, U.K. Patent, 1876 [268]

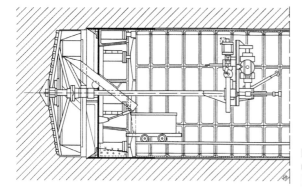

Figure 1-20 Mechanised shield used by J. Price, U. K. Patent, 1896 [268]

Figure 1-21 Mechanised shield used by Price, 1902 (Markham)

In 1896, Haag applied for a patent in Berlin for the first shield with fluid-supported face in Germany, with the excavation chamber full of fluid being hermetically sealed as a pressure chamber (Figure 1-23).

But the idea of a slurry-supported face was not successfully tested until 1959 with the design of Elmer C. Gardner for a sewage tunnel of 3.35 m diameter. In 1960, Schneidereit introduced the idea of active face support with bentonite suspension. H. Lorenz patented the stabilising effect of bentonite under pressure for face support. The first use of a slurry shield, with the excavation of the face by a cutting wheel and hydraulic removal of the muck, was a 3.1 m diameter machine in 1967 in Japan. In Germany, Wayss & Freytag AG developed and used the first shield with bentonite-supported face in 1974.

The development of earth pressure balance shields started much later, although the screw shield of Brunel (Figure 1-15) can be considered the predecessor of the basic idea. The first design was developed in 1963 by the Japanese Sato Kogyo Company Ltd. (Figure 1-24), who were looking for a method of tunnelling through soft and flowing soil beneath the groundwater table. The development of the EPB shield was surprising, because compressed air and slurry shields were already in use in Japan at the time. The reason for the development lay in the stringent environmental regulations and laws, which already applied in many of the larger cities in Japan. These related to groundwater and air pollution and the tipping of the spoil, but also required precautionary measures to avoid illness and accidents to workers under compressed air.

The journey from the historical origins of shields to the highly mechanised machines of today, as shown for example by multi-face shields, was a long and sometimes arduous or even dangerous path. To describe single developments in still more detail would exceed the space available in this book. Interested readers are recommended to read the handbook by Barbara Stack [268], which gives a detailed description of the individual principles in shield tunnelling and the associated patents.

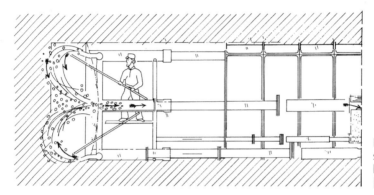

Figure 1-22 Slurry shield used by Greathead, patented 1874 [268]

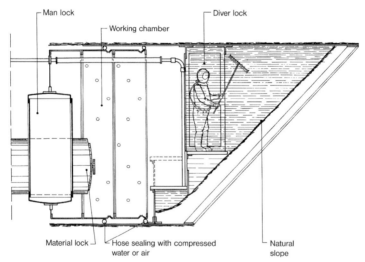

Figure 1-23 Fluid shield used by Haag, patented 1896

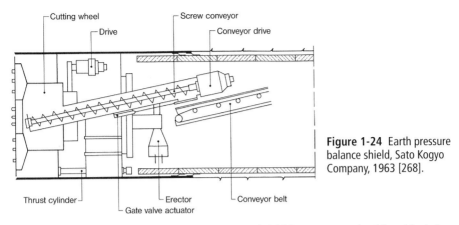

Figure 1-24 Earth pressure balance shield, Sato Kogyo Company, 1963 [268].

Some classic shield tunnels from the years 1826–1914 are summarised in table 1-1.

Table 1-1 Classic shield tunnels 1826–1914, excerpt from [128]

Year	Project	Length [m]	Diameter [m]	Daily advance [m/d]	Remark	Lining
1826–1842	Thames Tunnel (London)	460	11.40 × 7.10	1.50	Brunel shield (rectangular)	brickwork
1869–1870	Broadway (New York)	90	2.85		Beach shield (abandoned)	brickwork
1869–	Tower Subway (London)	403	2.20	2.60	Greathead shield	cast iron segmental lining
1886–1890	City South Subway (London)	10,200 various	3.10–3.45	4.00	first use of compressed air	cast iron segmental lining
1890–1893	Glasgow harbour road tunnel	580	5.20	1.00	compressed air	cast iron segmental lining
1892–1894	Sewage tunnel (Clichy)	465	2.50	2.00–3.00	compressed air to 2.9 bar	cast iron segmental lining
1896–1899	Spree road tunnel (Berlin)	375	4.00	1.40	compressed air	rolled steel profiles with concrete
1898	Orleans railway (Paris)	1,230	9.75		segment shield	Brickwork
1899–1904	Sewage tunnel (Hamburg)	2,150	3.05	1.30	compressed air (0.6–1.5 bar)	
1907–1911	Elbe Tunnel I (Hamburg)	920	5.95	1.70	compressed air (2.0–2.7 bar)	steel profiles with concrete
1911–	Sewer tunnel (Wanne-Eickel)		2.85			Brickwork
1911–	Sewer tunnel (Gelsenkirchen)	670	3.90	5.20	compressed air	Concrete

2 Support of the cavity and settlement

The purpose of a shield is to act as a mobile support unit for the excavated cavity and to enclose it reliably until the installation of support or of the final lining. It has to resist pressure from the surrounding ground and also hold back groundwater if necessary. In built-up areas, it also has to provide adequate support to avoid dangerous settlement at ground level. Settlement is the result of stress redistribution and alteration of the pore water conditions in the ground. Active support of the cavity at the face, next to the shield and behind the shield can limit surface settlement to permissible dimensions (Figure 2-1).

2.1 Support of the face

There are various ways of supporting the face depending on the type of machine chosen and the ground conditions. The selection of the optimal process will have to take into account the level of support pressure, the ground properties, the requirements for surface settlement, and process-related and economic aspects. The recommended first step is to analyse the stability of the face with calculations and determine the level of support pressure required.

2.1.1 Natural support

The term "natural support" means that the face is in equilibrium with no active support measures. The precondition is that the soil has sufficiently high shear strength parameters. The resistance pressure of a slope within a shield may be assumed, as long as this can be implemented with the process used. In larger shields, one or more intermediate platforms can be installed to reduce the failure body (Figure 2-2). Natural support cannot be assumed in water-bearing soil.

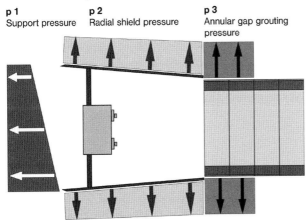

Figure 2-1 Active ground support in shield tunnelling

Figure 2-2 Model of the shield with intermediate platforms, Elbhang tunnel Hamburg, 1968/74 (works photo Philip Holzmann AG)

2.1.2 Mechanical support

For partial-face excavation, a timber breasting of planks can be used to provide mechanical support inside the shield. In this case, the face is excavated manually from top to bottom with the breasting being installed at each new level. This type of support is, however, cumbersome and hinders the use of mechanical excavation. Due to the low advance rate and high wage costs, the method is only used in exceptional cases. Much more flexible and practical is the use of poling plates, which are pressed against the face hydraulically (Figure 2-3). The part of the face to be excavated has to be exposed (see Chapter 8).

Closed cutting wheels provide safety against falling ground for the crew in the excavation chamber. It is not, however, correct to assume for purposes of calculation any mechanical support effect from the rotating cutting wheel during excavation. Fan-shaped shutters or spring support plates (Figure 2-4) have not proved successful and are also prone to defects. Neither can the contact forces of the excavation tools be assumed to support the face for the purposes of calculation. It should, however, be considered that much higher effective stresses are transferred to the ground by the excavation tools under high contact forces, which can cause compaction of the soil at the face (see Section 2.5).

If the tunnel is below the groundwater table and the soil is permeable, the mechanical support has to be supplemented by additional measures to hold back water (compressed air support) or be replaced by different and more suitable systems. Alternatives are slurry support or earth pressure support.

2.1.3 Compressed air support

An effective method of preventing the ingress of water into underground structures is the use of compressed air (Chapter 9). Compressed air support should not, however, be assumed to provide resistance against ground pressure in coarse-grained soils. The soil must therefore possess sufficient stability. For partial-face machines, the mechanical support of poling plates may be taken into account.

This process has been used since the start of the 19th century. The principle is shown in Figure 2-5. The water pressure increases linearly with the distance below the (ground)water level. At a free surface, the water is under zero pressure, but in the case of artesian groundwater (under positive pressure), the value increases from a value above zero. In order that water ingress into the tunnel is prevented, the air pressure must be set equal to or greater than the highest water pressure at the face. This will be at the deepest point of the tunnel, the invert.

The air pressure inside the tunnel acts equally on all parts of the face. As a result, it exceeds the water pressure in the upper part of the tunnel, which leads to air escaping. If the overburden is shallow, there is a danger that a blowout could occur due to this air flow. This is explained in detail in Chapter 9 and in [128]. Variations with partial groundwater lowering or partial exclusion of water from the shield have proved successful, given sufficient expert knowledge and experience.

Figure 2-3 Shield with active mechanical support of the face with hydraulically powered poling plates, in this case in combination with mechanical partial-face excavation (Robbins)

Figure 2-4 Full-face machine with mechanical face support

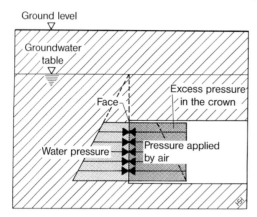

Figure 2-5 Principle of compressed air support

Increasing values of water permeability, greater than $k_w = 10^{-4}$ m/s, make the use of this process more difficult, as the air can more easily displace the water from the increasingly large voids and then escape in increasing quantities. A minimum overburden above the crown is still necessary in order to be able to ensure safety against blowouts. When compressed air is used for maintenance purposes, the face can be sprayed with bentonite suspension in order to avoid blowouts. The face is then practically sealed to form a membrane, and higher pressures can be applied than the prevalent water pressure. This makes it possible to resist ground pressure with compressed air.

2.1.4 Slurry support

Slurry support is the support of the face by a fluid under pressure (Figure 2-6). Ground pressure is resisted as well as the water pressure. Pure water is only suitable in impermeable, fine-grained soils. In coarse-grained, permeable soils, water is not suitable, because it would escape into the soil. Because of its thixotropic properties, bentonite suspension is particularly suitable as a support medium.

Either the membrane model or the penetration model is applicable depending on the permeability of the soil (Figure 2-7). If the permeability is relatively low and the bentonite content is sufficient, the suspension penetrates into the soil under the differential pressure and seals the face with the suspended solid material (plugging grain) with a mostly impermeable membrane (filter cake), through which the support pressure can be applied. This process takes place in a short time of about 1–2 seconds [151], [169].

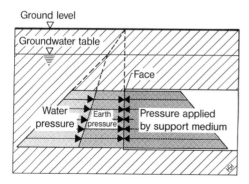

Figure 2-6 Principle of slurry support

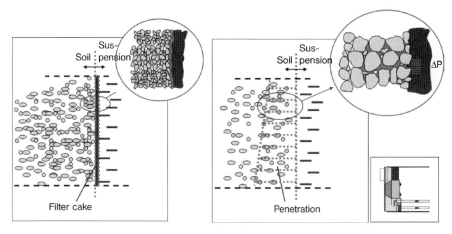

Figure 2-7 a) Membrane model   b) Penetration model

In coarse-grained, more permeable soils, a filter cake cannot always be formed, even with a high bentonite content. The bentonite suspension penetrates into the face and, due to its thixotropic properties, transfers shear forces into the grain skeleton. The penetration depths results according to Figure 2-8 from the initial shear strength of the suspension τ_F, the differential pressure acting Δp and the permeability d_{10} and can be determined according to the following equation:

$$s = \frac{\Delta p \cdot d_{10}}{2 \cdot \tau_F}$$

Soil element permeated by support medium

Figure 2-8 Penetration depth according to DIN 4126

where:

Δp	the pressure difference between supporting fluid and groundwater
d_{10}	the decisive grain size at 10 % by mass of the grading curve
τ_F	the liquid limit of the clay suspension

The type of bentonite used, the grading distribution and compaction of the soil and the pressure gradient have a decisive effect on the formation of a filter cake and the penetration depth. It is recommended to determine the penetration depth for each project through laboratory tests. The following illustrations show the test rig for the determination of the penetration and a graph of penetration behaviour over time as an example with a bentonite content of 5 % (Figure 2-9).

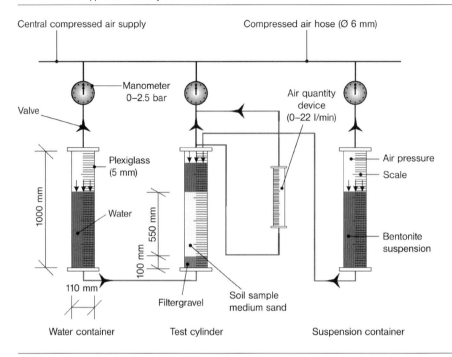

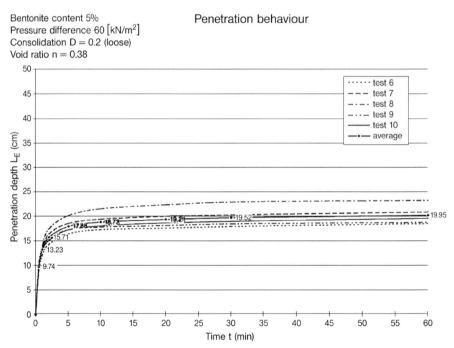

Figure 2-9 Test rig to determine the penetration depth and graph of the penetration behaviour over time [140]

In order to prove the internal stability of the face, it must be verified that no failure lens moves vertically downwards at the face. The minimum liquid limit τ_F of the suspension can be calculated according to DIN 4126, Chapter 6.3 "Safety against the sliding of single grain sizes or fractions" as follows:

$$\tau_F \geq \frac{\gamma_k'' \cdot \gamma_G \cdot d_{10} \cdot \gamma_\varphi}{2 \cdot \eta_F \cdot \tan \varphi_k'}$$

with:

γ_k'' characteristic value of the density of the soil under the buoyancy of the supporting fluid
η_F adaptation factor for the liquid limit (here: 0.6)
φ_k' characteristic value of the friction angle in the soil layer under consideration
γ_G partial safety factor for permanent actions (here: 1.00; GZ1C, LF2; DIN 1054)
γ_φ partial safety factor for the friction angle $\tan \varphi_k'$ of the drained soil (here: 1.15)
d_{10} decisive grain size at 10 % by mass of the grading curve

The explanations in DIN 4126 should also be observed.

As a simplification, the minimum liquid limit τ can be determined as follows:

$$\tau = \frac{d_{10} \cdot \gamma'}{\tan \varphi}$$

with:

τ minimum liquid limit of the suspension
d_{10} decisive grain size in the tunnel section
γ' submerged density
φ friction angle

In water-impermeable fine-grained soils (clays and silts), the use of bentonite can be omitted. Water, which is loaded with fine grains in the slurry circuit, is used as a support and transport medium.

Polymer suspensions belong to the fluids described as pseudo-plastic (shear thinning) and normally possess, in contrast to thixotropic fluids, no initial shear strength. As a result of this, the penetration process never completely stops. The penetration depth is dependent on the penetration time, in addition to the viscosity, the permeability and the differential pressure. With increasing penetration depth, however, the flow gradient reduces and thus the penetration speed. In order to demonstrate the internal stability, it is necessary to verify that, taking the advance rate into account, the support medium is not located within the sliding body (failure lens) in front of the face.

The support of the face with a fluid represents an increased level of technical elaboration compared to mechanical or compressed air support, particularly considering the separation plant. It does, however, open up additional and otherwise problematic applications for shields and represents a considerable improvement in working conditions compared to the usual compressed air shields. The limits to the scope of application of fluid support are explained in Chapter 10.

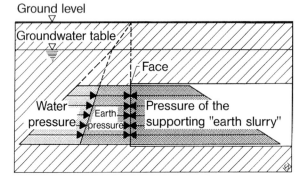

Figure 2-10 Principle of earth pressure support

2.1.5 Earth support

In soils with clay content, it is possible to use the soil itself as a support medium. The excavated material, if necessary with the addition of a conditioning agent (water, bentonite, chemical additives as foam), is remoulded to an "earth plug" and provides support to the face (Figure 2-10). The remoulded earth has to be under sufficient pressure to reach equilibrium with the forces working on the face which gave the name of this shield type (EPB-Shield = Earth Pressure Balance-Shield). The support principle is thus the same as with slurry support, but there are great differences in support pressure control and mucking equipment.

The earth plug can either be removed from the excavation chamber with a screw conveyor or, as with blind shields, be squeezed through closable extrusion openings into the shield by the thrust forces. Active support pressure control using foam conditioning equipment is also possible and is described in Chapter 11.

High density and viscosity of the support medium are favourable for earth support. In contrast to fluid or compressed air support, the completely filled excavation chamber rules out any large-scale sudden collapse.

2.1.6 Calculation models

In order to calculate the required support pressure p1 (Figure 2-1) at the face, there are a number of procedures.

Failure body model

The so-called kinematic methods investigate the equilibrium state of two- or three-dimensional failure bodies. As early as 1961, Horn [136] developed a failure body model (Figure 2-11) consisting of a wedge immediately in front of the face, which reaches up to the tunnel crown, and a prism supported on the wedge. The failure surfaces are assumed to be flat. In the kinematic method, based on the earth pressure theory, the equilibrium state of the wedge-shaped body ABCDEF is investigated, which is liable to move into the tunnel under loading from the block of soil (CDEFKLMN). The circular tunnel face is approximated by a square with sides corresponding to the tunnel diameter. The weight of the block is considered without reduction. The slope angle of the failure body is iterated according to the active earth pressure theory until the maximum support pressure resultant is reached.

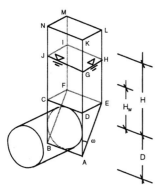

Figure 2-11 Failure body model according to Horn [136]

The model of Kovari and Anagnostou [2] (Figure 2-12) is a further development of the Horn failure body model, except that a vertical surcharge load V' or σ_V', determined according to the silo theory of Janssen/Terzaghi, acts on the sliding wedge (ABCDEF). According to the silo theory, the vertical surcharge can be reduced by the shear stresses that can be transferred to the sliding surfaces at the side of the prism (CDEFKLMN). The reduced vertical stress σ_V' at depth H can be determined using the formula:

$$\sigma_V' = \underbrace{\frac{\frac{F}{U} \cdot \gamma' - c'}{K \cdot \tan\varphi} \cdot \left(1 - e^{-K \cdot \tan\varphi \cdot \frac{\Delta H}{F/U}}\right)}_{\text{(Term n)}} + \underbrace{\sigma_{V,n-1}' \cdot e^{-K \cdot \tan\varphi \cdot \frac{\Delta H}{F/U}}}_{\text{(Term n-1)}}$$

with:

F   base area of the prism
U   circumference of the prism
φ   friction angle
γ'   submerged density
c'   drained cohesion
K   = 1
ΔH   layer thickness under consideration

Figure 2-12 shows the forces pushing (actions) and holding (reactions) the sliding body.

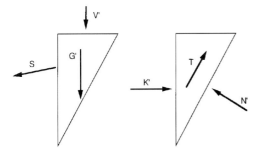

Figure 2-12 Force model according to Anagnostou and Kovari [2]

V' vertical force on prismatic failure body from silo theory = $(\sigma_V' \cdot D^2)/\tan \omega$
G' self weight under buoyancy $(D^3 \cdot \gamma_{wedge})/(2 \cdot \tan \omega)$
S   resultant of the hydraulic mass forces
K' resultant of the effective support pressures
T   resultant forces from shear stresses in the side surfaces and the inclined surface (N') of the sliding wedge activated through movements:
    $T_G + 2 \cdot T_S = N \cdot \tan \delta + (c'_{wedge} \cdot D^2)/\sin \omega + 2 \cdot (H_y \cdot \tan \delta + c' \cdot A_S)$
    ignoring the friction forces at the sides:
    $T = N \cdot \tan \delta + (c'_{wedge} \cdot D^2)/\sin \omega + 2 \cdot c' \cdot A_S$
N' normal force in the sliding joint = $(K \cdot \sin \omega) + (V'+G) \cdot \cos \omega$

The resulting holding force K' at the wedge on the face is then:

$$K'(\omega) = (V' + G') \cdot \tan(\omega - \delta) - \frac{\left(\frac{\gamma'_{wedge} \cdot D^2}{\sin \omega} + \frac{D^2}{\tan \omega} \cdot \left(K_0 \cdot \tan \delta \cdot \left(\sigma'_V(z=H) + \frac{D \cdot \gamma'_{wedge}}{3} \right) + \gamma'_{wedge} \right) \right)}{\cos \omega \cdot (\tan \omega \cdot \tan \delta + 1)}$$

with:

A_S side area of the failure wedge
ω angle of the sliding surface
δ wall friction angle = φ'
K_0 lateral pressure ratio
γ_{wedge} average soil density of wedge
H depth H
D diameter of face

Procedure according to DIN 4085

The procedure according to DIN 4085 is based on the 3D failure body model of Piaskowski/Kowalewski (Figure 2-13).

The model was originally developed for the investigation of the structural stability of diaphragm wall panels. The failure body is divided into horizontal slices, for each of which the active earth pressure is calculated. The calculation is based on the ratio z/b and the shape coefficient μ_{agh} (Table 2-1). Intermediate values may be interpolated linearly.

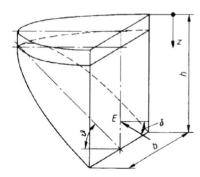

Figure 2-13 Failure body model according to Piaskowski/Kowalewski [75]

2.1 Support of the face

Tabelle 2-1 Shape coefficients according to DIN 4085

z/b	0	1	2	3	4	6	8	10
μ_{agh}	1	0.82	0.70	0.59	0.50	0.37	0.30	0.25

with:

z   depth
b   excavation diameter

The horizontal earth pressure per spacial area spa_{ca} [kN/m²] at depth z is determined according to DIN 4085 with the following relationship:

$$spa_{ca} = g' \cdot z \cdot K_{ag} \cdot \mu_{agh} + p \cdot K_{ap} \cdot \mu_{aph} + c' \cdot K_{ac} \cdot \mu_{ach}$$

= soil pressure component + imposed load component + cohesion component

with:

$\mu_{agh} = \mu_{aph}$
$\mu_{ach} = 1$
$K_{ag} = K_{ap} = (1 - \sin\varphi)/(1 + \sin\varphi)$
$K_{ac} = -2 \cdot \sqrt{K_{ag}}$

The calculation of the horizontal earth pressure coordinates starts at the tunnel crown (z = 0). Similarly to the failure body model from Horn, the weight of the overburden ground is applied fully as surcharge in the determination of the horizontal earth pressure and considered in the loading term σ_V'. Surface loads σ_o without limits at the sides are also applied fully in the loading term σ_V'.

$$\sigma_V' = \sigma_o + \gamma' \cdot H$$

with:

σ_V' vertical stress
σ_o surface load
γ' submerged density
H   overburden above the tunnel crown

The integral of the earth pressure per unit area over the area of the face gives the required support pressure resultant from earth pressure E with a safety factor of 1.0. For the determination of the necessary support pressure, the component from water pressure and the safety supplements still have to be added. According to the ZTV-ING [41], the following verifications should be carried out:

against sliding at a surface: support force $S \geq \eta_W \cdot W + \eta_E \cdot E$ [kN]
against ground deformation: $\sigma_{v,\,crown} + w \geq s$ [kN/m²]
against water ingress into the invert: $s \geq w \cdot 1.05$ [kN/m²] (with compressed air support)

The following additional verifications are required as part of the detailed design:

support pressure crown: $S_{crown} \geq (\sigma_{h,crown} \cdot \eta_E + w_{crown} \cdot \eta_W)$ [kN/m²]
against water ingress into the invert: $s \geq w + 10$ [kN/m²]

with:

S	support force [kN]
W	sum of water pressure force [kN]
E	sum of earth pressure force [kN]
η_W	safety factor for water according to ZTV-ING [-]
η_E	safety factor for earth pressure according to ZTV-ING [-]
s	support pressure [kN/m^2]
w	water pressure [kN/m^2]
σ_v	vertical stress [kN/m^2]
σ_h	horizontal stress [kN/m^2]

Safety factors

The specification of safety factors is not always consistent in the current standards and is dealt with differently for a specific project – mostly based on a global safety concept. A safety factor of 1.0 is the limit between the intended load-bearing effect and failure or insufficient load-bearing capacity. Safety factors greater than 1.0 denote a safety reserve.

Because of the mostly non-linear relationships in the determination of actions/loadings and resistances, it is not recommended to apply partial safety factors to the characteristic shear parameters. It is more sensible to determine the magnitude of the forces on the resistance and also the action side with the characteristic soil parameters according to DIN 4020 and then apply partial safety factors. For the calculation of earth/rock mass pressure, the characteristic values of the soil parameters should be specified in the geotechnical report. Imposed loads should be taken from the DIN specialist report "Actions on bridges".

It should also be taken into account that the calculation of the stability of the face is an investigation of a construction state for shield tunnelling and does not have to have any effect on a structure.

For the investigation of the structural stability of a structure in the area influenced by the tunnelling machine, the calculation of face stability based on kinematic methods is not applicable because no statements are possible about the deformations to activate the earth pressure.

The ZTV-ING [41] recommends that for the verification of horizontal equilibrium against the formation of a sliding surface, a partial safety factor $\eta_W = 1.05$ should be applied for actions resulting from water pressure and $\eta_E = 1.50$ for earth pressure, corresponding to the formula given above.

According to the new terminology in Eurocode 7 (or DIN EN 1997), the support force in the excavation chamber is to be considered a characteristic resistance. In the ZTV-ING, no division by a partial safety factor is required for the support force, which consequently leads to an additional increase of support pressure. It is, however, recommended to additionally consider a standard tolerance of 10 kN/m^2 or 30 kN/m^2 (slurry support or earth pressure support respectively) for the support pressure in the excavation chamber. The verification procedure proposed in the ZTV-ING thus corresponds more to a global safety concept. The characteristic loading from earth pressure E is increased by a factor of 1.5 for the consideration of equilibrium, independent of loading case, origin, duration of the actions and partial safety factors for soil parameters.

Therefore the verification procedure proposed in the ZTV-ING does not yet correspond to the new rules of Eurocode 7. In the currently valid version of Eurocode 7, however, tunnelling is not explicitly mentioned in the scope of application [131]. But in the literature, tunnelling is often associated with the retaining structures of Eurocode 7. It is recommended in the future to observe the rules in DIN 1054 "subsoil verification of the safety of earthworks and foundations" and also Eurocode 7 (EN 1997-1) "Geotechnical design – Part 1: General rules." The partial safety factors, characteristic soil parameters and loading cases should be agreed in collaboration with the soil mechanics specialist.

2.2 Support of the cavity at the shield

With open shields, the shield skin provides support to the sides of the excavated cavity. The gap between the ground and the segments can be calculated from the overcut and the taper of the shield skin. Ground deformation in the annular gap can lead to settlement at the surface, depending on the ground parameters and the overburden.

With EPB and slurry shields, active support of the sides of the excavation with the associated high support pressure means that reduced settlement can be assumed. Analyses of the support pressure distribution along the shield skin have confirmed the existence of pressure communication within the three pressure areas (pressure model, Figure 2-1).

The quality of pressure communication depends mainly on the rheological properties of the support medium in the excavation chamber and the material in the annular gap. With a slurry shield with air bubble support, the support and grouting pressures communicate, if the tunnel is driven with sufficiently high overcut and support pressure. With an earth pressure shield, the support pressure at the skin (p2 area in Figure 2-1) can be controlled actively through an independent grouting and pressure monitoring system. Appropriate connections are provided along the shield. The grout should not be too thin and liquid, which would result in flow into the excavation chamber. On the other hand, it must remain sufficiently fluid to completely fill the constantly altering shape of the gap as the tunnel advances. Grouting quantities are normally small. Piston and screw pumps are suitable, avoiding excessively stringent demands on the pressure control. Grout pressure can, however, be controlled pneumatically through a pressure bubble.

Vibration from the machinery and equipment operating inside the shield (e.g. cutting wheel, crushers, pumps) and transferred through the shield skin has the effect of consolidating the loosely consolidated, low-cohesion strata. The associated settlement leads to a wide settlement trough at the surface. Through the active support pressure control at the shield, the settlements around the shield resulting from dynamic actions can be reduced or even compensated.

2.3 Support of the cavity behind the shield

When precast segments are used, the continued advance of the shield leaves behind an annular gap bounded externally by the surrounding ground, internally by the segments and at the front by the shield tail seal (Figure 2-14). The width of the annular gap is essentially determined by the construction width of the shield tail seal, in addition to its support construction and the thickness of the shield tail itself. The theoretical gap width for the common types of shield tail seal today is 70–120 mm, with a working range of the seal of ± 20 to ± 40 mm. It should be noted that this measurement is independent of shield diameter, i.e. more stringent requirements are placed on the constructional tolerance of larger tunnel linings than for smaller linings. Larger gap widths of up to 250 mm have also been implemented with the so-called "extrusion process" (Chapter 6.5).

The gap can also be enlarged by the factors already mentioned – the taper of the shield skin, ground displacement in curves and overcut – unless the displaced ground immediately closes the resulting cavity. Changes of gap width around the perimeter result from the unavoidable eccentricity of the lining to the shield tail and through the deformation of the support and shield, which is to be expected.

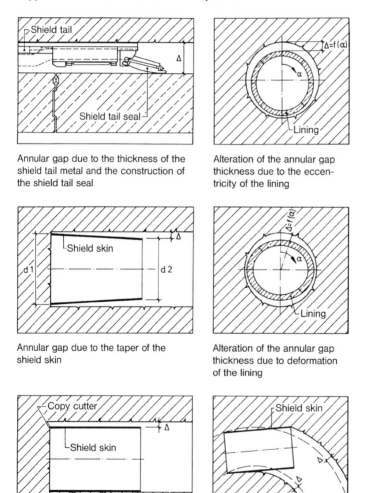

Annular gap due to the thickness of the shield tail metal and the construction of the shield tail seal

Alteration of the annular gap thickness due to the eccentricity of the lining

Annular gap due to the taper of the shield skin

Alteration of the annular gap thickness due to deformation of the lining

Annular gap due to eccentric overcut

Annular gap due to ground displaced by driving a curve

Figure 2-14 Causes of the annular gap behind the shield

Stratified ground can cause very variable gap widths around the shield perimeter. In addition, creeping of a polygonal lining (Chapter 6) within the working range of the seal cannot be avoided, particularly when driving tight curves, i.e. the annular gap width alters with the advance of the tunnel.

To avoid or reduce settlement behind the shield, the annular gap should be grouted as soon as it is created during the advance of the shield. The natural stress state of the soil can thus be largely maintained. The lower the stress transfers in the soil are thus maintained, the smaller are the resulting soil movements, which appear above the tunnel as surface settlement. Particularly with shields with slurry or earth pressure support, the component of settlement due to the reduction of stress at the face has been reduced so greatly that complete, pressure controlled grouting of the annular gap is of greater significance. Annular gap grouting with bicomponent grout is being increasingly used.

The processes and equipment for grouting the annular gap are extensively described in Chapter 7.

2.4 Settlement and damage classifications

Even if active support is used according to Figure 2-1, changes to the primary stress state can lead to an alteration of the consolidation and pore water conditions, which then appear as surface settlement. The magnitude of ground movements at the surface can be limited by the correct selection of process technology and meticulous construction.

In the course of the construction of a tunnel by shield tunnelling, the principal factors of influence for settlements are as illustrated in Figure 2-15:

- Groundwater lowering in soil, which alters its volume with water content,
- stress build-up by transfers due to the wedging effect in front of the shield (ground heave),
- alteration of the pore water pressure conditions,
- excavation of the face with unintentional removal of soil,
- insufficient support to the face,
- alteration of the grain structure due to curve driving, and vibration caused by the shield due to the consolidation of the grain structure around the shield; insufficient support of the sides of the cavity around the shield skin,
- backfilling of the shield track when the annular gap being insufficiently filled, or compensation of assumed settlement values with increased grouting pressure,
- deflation of the compressed air in the tunnel and thus reduction of the tunnel diameter through the increase of the axial force in the support.

Damage limits can be set approximately by means of the gradient of the settlement trough, i.e. differential settlement of 1:500 for superstructures capable of deforming and 1:1,000 for stiff, buildings susceptible to damage. The type of foundation, the dimensions, deformation rates and many other factors should be taken into account in the calculation of the damage to be expected. Further information can be found in [184].

In order to estimate any potential damage, the process of Boscardin & Cording [34] is often used as a basis.

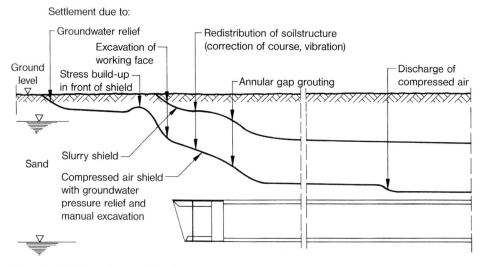

Figure 2-15 Schematic curve of settlement for a compressed air shield with groundwater lowering and manual excavation and for a slurry shield

The angular rotation tan α represents the difference of the vertical displacement of two adjacent foundations, related to the horizontal spacing "*l*" (Figure 2-17). As can be seen in Figure 2-16, the next higher damage level can also be reached more quickly due to additional horizontal strain.

Investigations of settlement troughs (Figure 2-19, settlement trough with failure function curve) also enable the determination of the maximum gradient caused by volume loss. In contrast to local investigation, as with the angular rotation between two adjacent foundations, this is a global consideration with classification into a damage class.

The classification of possible damage ranges from "insignificant" over "light" to "very severe". The classification is according to the angular rotation tan α [°] and the horizontal strain [mm]. The diagram can be used for buildings with a length of 30 to 40 m according to the recommendations of the National Coal Board. Shorter or longer lengths of the building under investigation lead to an over- or underestimation of the damage to be expected [34].

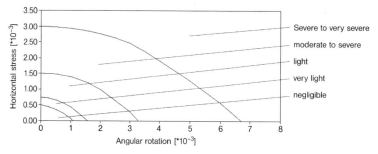

Figure 2-16 Damage analysis according to Boscardin & Cording [34]

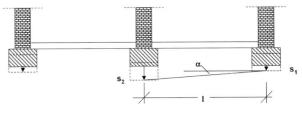

Angular rotation: $\tan \alpha = \dfrac{\Delta s}{l} = \dfrac{s_2 - s_1}{l}$

Figure 2-17 Angular rotation

2.4.1 Empirical determination of the settlement

Empirical processes for the estimation of the settlement occurring due to tunnelling are based on values from experience and the evaluation of the measurement data from completed tunnel projects [212]. In addition to the depth of the tunnel and the type of ground, the volume loss is a decisive factor for the calculated settlement. Typical empirical values for the volume loss lie between 0.5 % and 1.5 %. These values are, however, strongly dependent on the type and quality of the tunnel drive.

Peck in 1969 [233] evaluated statistics from many tunnels completed with various types of shields and produced soil-dependent guideline values for the width of the settlement trough.

The empirically determined curves are based on the ratio of the overburden "c" to the diameter of the cavity "d", depending on the soil encountered, the dimension "i" at which the point of inflexion "WP" lies on the curve representing the settlement trough at right angles to the tunnel axis. With this dimension, "i", and the Equation 1 shown in Figure 2-18, the settlement can be determined at any point x at right angles to the tunnel axis. The area

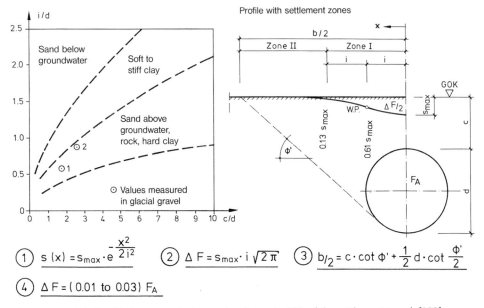

① $s(x) = s_{max} \cdot e^{-\dfrac{x^2}{2i^2}}$ ② $\Delta F = s_{max} \cdot i \sqrt{2\pi}$ ③ $b/2 = c \cdot \cot \phi' + \dfrac{1}{2} d \cdot \cot \dfrac{\phi'}{2}$

④ $\Delta F = (0.01 \text{ to } 0.03) F_A$

Figure 2-18 Relationship between relative overburden and width of the settlement trough [233]

ΔF enclosed by the ground surface and the settlement curve can be determined according to Equation 4. The combination with Equation 2 delivers the maximum settlement s_{max}.

The range of variation in the ratio of the settlement trough area ΔF to the cross-sectional area of the cavity F_A of 1–3 % given in [233] for the evaluated tunnel projects reflects the diligence of the tunnel drive, with predominantly manual excavation or as partial face excavation. Fully mechanical excavation restricts the range of variation of the ratio. Slurry and earth pressure support reduce the ratio of areas considerably to the smaller value.

Equation 3 according to Figure 2-18 then gives the area influenced by settlement at right angles to the tunnel axis. The graph at the left-hand side shows the relationships c/d to i/d. This makes it clear that the influence of settlement to be taken into account extends further outwards in the soft-plastic range, but also in non-cohesive sand under the groundwater.

O'Reilly and New (1982) [229] further developed the work of Peck into an extended empirical method of forecasting settlement, which in addition to the volume loss also takes into account stress redistribution in the ground. The volume loss factor includes the settlement components due to face support, tunnelling machine passage and annular gap grouting.

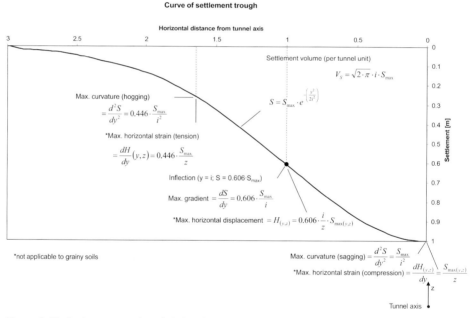

Figure 2-19 Settlement trough with failure function curve [229]
with:
i   settlement trough factor
z_0 ground cover over tunnel axis
S_{max} maximum settlement
y   horizontal distance from the tunnel axis
V_S = $(2 \cdot \pi)^{0.5} \cdot i \cdot S_{max}$
K   = i/z

Using data from completed tunnel drives in clay soil and loose ground and taking stress redistribution into account, the measured settlement curves were investigated and evaluated.

This gave settlement trough factors according to O'Reilly and New for
a) fine-grained soil: $i = 0.43 \cdot z + 1.1$
b) coarse-grained soil: $i = 0.28 \cdot z - 0.1$

The settlement trough factor and the equations shown in Figure 2-19 allow us to determine, for any location "y" in the area of the settlement trough:
– maximum settlement
– horizontal and vertical ground strains

$$\varepsilon_v = \frac{V_s}{\sqrt{2\pi} \cdot K \cdot z^2} \cdot e^{\frac{-y^2}{2 \cdot (K \cdot z)^2}} \cdot \left[\frac{y^2}{(K \cdot z)^2} - 1 \right]$$

$$E_h = \frac{V_s}{\sqrt{2\pi} \cdot K \cdot z^2} \cdot e^{\frac{-y^2}{2 \cdot (K \cdot z)^2}} \cdot \left[1 - \frac{y^2}{(K \cdot z)^2} \right]$$

– horizontal and vertical ground displacements

$$S_{(y,z)} = S_{(max,y,z)} \cdot e^{\frac{-y^2}{2 i_z^2}} = \frac{V_s}{\sqrt{2\pi} \cdot K \cdot z} \cdot e^{\frac{-y}{2(K \cdot z)^2}}$$

$$H_{(y,z)} = \frac{y}{z} \cdot S_{(max,y,z)} \cdot e^{\frac{-y^2}{2 i_z^2}} = \frac{V_s \cdot y}{\sqrt{2\pi} \cdot K \cdot z^2} \cdot e^{\frac{-y}{2(K \cdot z)^2}}$$

– inflexion point
– maximum curvatures

Another method for calculating the area affected by settlement at right angles to the tunnel axis makes use of a model that essentially consists of an area bounded by two borderlines angled at $\Phi' = 45°$, which meet the cross-section of the cavity at axis height [168].

Because significant geotechnical and process-related factors remain totally ignored by empirical methods, the results produced and the suitability of these forecast methods have to be questioned.

2.4.2 Numerical models for the calculation of settlement

Investigations using the finite element method (FEM) have largely superseded empirical methods. The use of a numerical model makes it possible, in contrast to empirical methods, to simulate the progress of construction in detail and also to include consideration of the behaviour of the ground through appropriate constitutive laws. Usually, the linear-elastic ideal-plastic constitutive law of Mohr-Coulomb is used. If appropriate and more comprehensive soil mechanical parameters are available, then the description of the soil behaviour can be improved with the use of better quality constitutive laws, such as, for example, the hardening soil model [39]. In contrast to empirical methods, ground improvement measures such as, for example, compensation grouting or jet grouting (Figure 2-22) for the protection of building can be considered in the model.

Currently, 2D FE models are mostly used, but the use of three-dimensional models is increasing. In 2D calculations, two different processes are mostly used. In the first, the tunnel is first given the expected volume loss resulting from the passage of the shield as a contraction, and then the tunnel lining is installed. In the second process, the tunnel lining is installed after the ground has relaxed by a given proportion (β method) [211]. The advantage of 2D calculations is that they require little computing time in comparison to 3D calculations. On the other hand, the spatial vault effect (load-bearing system) is neglected and the construction process is also represented by a considerable simplification.

3D calculations offer considerable advantages in comparison to 2D calculations. The stress state in the ground is considered in three dimensions, and the advance of the tunnel drive can also be represented in detail. The geometrical dimensions of the tunnel boring machine, the support pressure at the face and the grouting pressure and the grouting of the annular gap between ground and segments are included in the calculation model. The taper of the shield skin is mostly simulated by a contraction [39]. Kasper and Meschke [160] show a very detailed model of a shield, in which the friction between shield skin and ground and also the individual thrust cylinders are simulated (Figure 2-20). The limits of sensible application are less the model formation but rather the realistic representation of the constitutive law for the ground.

Modern process technology makes it possible to actively control the support pressure at the face and at the sides of the excavation. In the pressure model, the analysis of deformation does not make use of contraction (volume loss), but applies a support pressure along the face and the sides of the excavation. Using the measured pressure distribution over the whole area of the shield machine, Maidl/Ruse [199] have developed a new calculation concept for the three-dimensional simulation of a tunnel drive.

The deformations and convergences during the passage of the shield are given by this model independent of the actual pressure distribution according to Figure. 2-1. Accord-

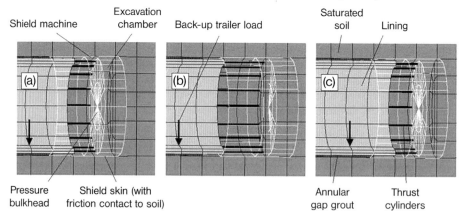

Figure 2-20 Representation of thrust cylinders in the FE model [160]
a) End of the previous excavation step
b) Advance of the TBM
c) Excavation of the soil and activation of the elements for annular gap grouting and segmental lining

ingly, if pressures are low, then large deformations occur. These result in corresponding settlement or heave at the surface.

Finite element calculation methods thus enable a sound analysis of the system behaviour and the interaction between ground and tunnelling machine. Using powerful calculation programs, ground, tunnelling machine, terrain profile and also buildings on the surface can be represented in an integrated 3D model and the advance of the tunnel drive can be simulated. As an end result, FE methods deliver a complete picture of stress and deformation in the area affected by the tunnel. The efficiency of additional measures (support walls, anchoring, piled foundations) or ground improvement measures (grouting, soil exchange) to protect buildings can be planned qualitatively and quantitatively by numerical methods (Chapter 18).

An FEM simulation is carried out in the following phases:

1. geometrical and process-related modelling with realistic discretisation of all mechanical elements (shield tail grouting, shield gap grouting, face support etc.)
2. material modelling with the selection of a suitable constitutive law (consideration of excess pore water pressures, creep effects, non-linear elastic behaviour etc.)
3. step-by-step analysis to consider the effects of different phases of construction on the stress conditions in the ground
4. verification of the results and plausibility check

Figure 2-21 shows a 3D model of a shield tunnelling machine in the ground. The areas of the excavation chamber, shield, annular gap grouting and segment lining can be recognised. As part of the calibration of the model, the forecast results are compared with actual measured results (for example surface settlements, earth pressure measurements, inclinometer and extensometer measurements). The forecasts should be checked with comparative calculations during the construction phase in accordance with the observation method [74].

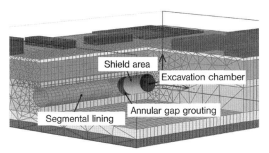

Figure 2-21 3D modelling of a shield machine in the ground

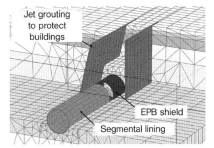

Figure 2-22 3D modelling of ground improvement measures in shield tunnelling

A disadvantage of FEM simulation is the very laborious but still uncertain calibration of the model. The uncertainty concerning the actual stratification of the ground, the soil parameters and the constitutive laws used, but also the numerous aspects that cannot be modelled (time-dependent, process-dependent, person-dependent) leads to the fact that the simulation is indeed very suited to the qualitative improvement of understanding, but it has to be regarded with reservations concerning quantitatively correct results.

2.5 Heave and compaction

If the sum of the support forces at the face exceeds the undisturbed earth pressure, then compaction and heaving of the ground in front of the face can occur, the compaction of the ground being caused by the excavation process. In this case, the following process-related factors can together or independently lead to an increase of the density of the undisturbed soil (densification = compaction).

a) introduction of effective stresses by the cutting tools
b) introduction of effective stresses by the support pressure and/or filter cake
c) introduction of effective stresses by the cutting wheel (steel poling plates, rim, spokes, tool boxes)
d) introduction of fine-grained material in the pore cavity by the suspension or the "earth slurry"
e) ground/disintegrated corn fractions in the excavation zone

The mechanical loading from the excavation equipment and the pressure transferred through the support medium and penetration into the ground can therefore alter the properties of the ground and the characteristic parameters of density, grading distribution, content of broken components, fines content and the shear parameters of the undisturbed ground at the face.

Figure 2-23 Compaction at the face in shield tunnelling with slurry-supported face

3 Design and calculation methods

The shield is a "mobile", temporary support providing protection for the working space for the miners and the technical equipment and machinery for tunnel excavation. All works from the excavation of the ground to the installation of the lining are carried out inside the protection of the shield. The construction of the shield must therefore be designed according to the pertaining geological and hydrological conditions in order to avoid unintended surface settlement or disturbance to the groundwater and ensure the safety of the workers.

3.1 Constructional parts of the shield

The shield skin is normally a circular tube; deviations from a circular cross-section are not normally possible for full-face excavation with cutting wheels. With partial-face excavation, on the other hand, individual layouts of the tunnel profile (vault with flat invert, box shape) or the shield cross-section are possible.

The diameter of the shield is determined by the later use, i.e. from the required structure gauge, the necessary wall thickness of the lining and the required minimum annular gap thickness. The latter is determined by the intended theoretical ring installation clearance, the thickness of the deflector bar and the shield tail sheet metal and the taper of the shield skin.

The arrangement of the most important constructional elements of a shield is shown in Figure 3-1 through the example of a Mixshield. A more detailed description of different support principles and excavation tools can be found in Chapters 2 and 4.

In principle, a shield machine is divided into three sections: front shield, middle shield and shield tail or tailskin. The front shield as the foremost part with its perimeter shield blade forms the external edge of the excavation chamber, in which the excavated ground material is fed to the transport system. A further part of the front shield is the pressure bulkhead, which separates the pressurised excavation chamber from rest of the tunnel under atmospheric pressure. In the centre of the front shield are the drive motors for the chosen excavation device. Ideal is a separation of the drive components from the shield construction in order that no distortion is caused by the transfer of loads from the shield construction into the drive system. Regarding the decoupling or overload protection for the cutting wheel drive, there are various types of support (Table 3-1), in this example open-centre, peripheral drives (Figure 3-2 a to e).

Each of these constructional elaborations of the bearing increases complication and cost along with the improved degree of overload protection for the cutter wheel drive.

3 Design and calculation methods

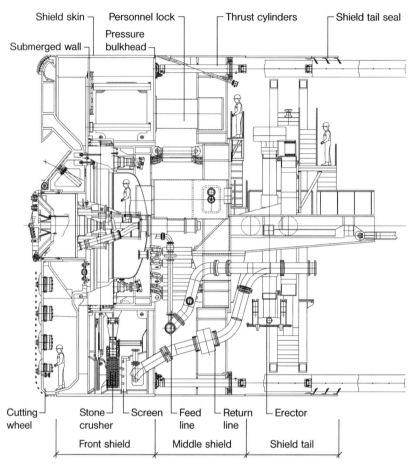

Figure 3-1 Constructional elements of a shield, in this example a Mixshield, fourth tube of the Elbe tunnel 1997/99 (Herrenknecht)

Table 3-1 Decoupling or overload protection of the cutting wheel drive with peripheral drive [123]

Bearing type	Decoupling or overload protection
direct installation (Figure 3-2 a)	no protection
longitudinally adjustable installation in fixed sliding guide (Figure. 3-2 b)	axial load protected
longitudinally adjustable installation in elastic sliding guide (Figure 3-2 c)	axial load protected; breakdown torque and radial force partially protected (damping by elastic sliding guide)
longitudinally adjustable installation in articulated bearing (Figure 3-2 d)	axial force and breakdown torque fully protected
fully hydraulic mounting (Figure 3-2 e)	axial force, radial force and breakdown torque fully protected

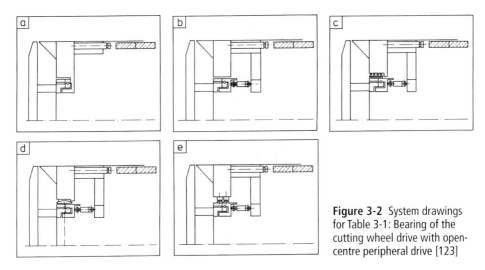

Figure 3-2 System drawings for Table 3-1: Bearing of the cutting wheel drive with open-centre peripheral drive [123]

The thrust cylinders are supported axially by the segments and transfer their load to the shield. The erector with its travelling beam is connected as a cantilever construction to the erector support crossbeam in the middle shield of the shield skin. Except with blade shields (Chapter 13), the internal shield tail area has to be kept free without supports for the thrust cylinder guides to extend and retract the cylinders and for the installation of the lining.

The thickness of the shield (Figure. 3-3) is greatest (50 to 100 mm) in the front shield, because this is where the mechanical loading on the skin is highest. The frequently used wall thickness of 80 mm results from the requirements of various specifications. If the shield blade has the function of cutting the in-situ soil, as is the case with internal cutting wheels, point loads can also occur when obstructions are encountered. Radial stiffening of the shield blade is possible in some cases when the cutting wheel is at the front. The front shield is stiffened by the pressure bulkhead and any further separation walls like the submerged wall of a Mixshield. If possible and necessary, ribs and gusset plates can stiffen the construction. In the middle shield, the axial forces are transferred from the cutting wheel drive through the supporting crossbeam and the ring box into the skin and from there into the perimeter reaction ring, on which the thrust cylinders are mounted with a hinged connection. The shield tail, which provides protection for the installation of the tunnel lining, is connected to the middle shield.

The metal thicknesses in the middle shield range from 40 mm for smaller shields with 6 to 8 m shield diameter and increase to a wall thickness of 60 mm in larger shields of more than 10 m diameter. Shield tail metal thicknesses start from 40 mm for shields of 6 m diameter. Shield tail diameters larger than 10 m are constructed with a single skin 60 mm thick. The transition to the tunnel lining is closed by the shield tail seal (Chapter 7).

For tight radius curves and shields, which are very long in relation to their diameter, the following measures are possible to avoid distortion in the shield tail:

– tapered shield with decreasing external diameter from the shield blade to the shield tail,
– division of the stiff shield skin with a articulation joint of the shield tail to the middle shield (in this version, the shield tail follows the track of the shield with free movement, with a floating connection through a passive joint),

- further division of the forward part of the shield into a front or steering shield controlled by articulated cylinders and a following middle shield
- creation of an overcut, either through extending copy cutters or by tilting the cutting wheel with an articulated bearing. In the latter version, the overcut can be produced as required on one side by using the tools in the gauge area of the cutting wheel, and can also be kept to an absolute minimum to minimise settlement.

3.2 Loading on the shield

The loading on the shield can be divided into external and operational loads, and these can act at right angles or also along the shield. The design of the shield is determined by the least favourable combination of simultaneously occurring loading groups. Secondary effects like the abrasiveness of the ground, uncertainties in load assumptions and significant deviation from symmetrical loading state make the application of safety factors unavoidable. Empirical values based on experience of previous projects are also important.

The following loads should be considered in the design of a shield machine:

- self weight,
- radial external loading: earth pressure, water pressure, reaction forces from steering motion,
- axial external loading: support pressure in the excavation and possibly also in the working chamber, tool pressure on the face, thrust cylinder forces, towing forces from the backup,
- further loading: skin friction, reaction forces at the shield blade, reaction forces of the erector when installing rings, reaction forces from the cutting wheel torque, thrust forces from the auxiliary hydraulic cylinders, reaction forces from the brush seal and grease chamber pressure, grouting pressures from ground improvement measures carried out from the shield.

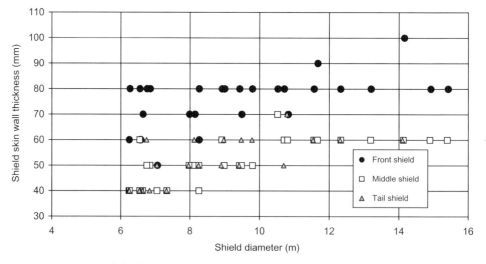

Figure 3-3 Diagram of shield skin wall thicknesses related to shield diameter, divided into front shield, middle shield and shield tail in single-skin construction [123]

The shield construction together with all its stiffening elements represents a highly structurally indeterminate and asymmetric shell structure. The moments and forces acting on sections are determined today with elaborate FEM calculations, which will not be described in detail here. As a simplification, the shield skin is assumed to be a circular cylinder with elastic bedding. Stiffening elements (pressure bulkhead, ring beams) can be considered additionally.

The shield tail is flexible in bending compared to the cutting wheel and middle shields, because no further support of the shield skin is possible in the thrust cylinder area and in the lining installation area, which has to be kept free. The shield tail is calculated as a three-dimensional pipe, which is either attached to the middle shield with a joint or in short shields is welded to the middle shield to provide a stiff connection.

3.2.1 Loading on the shield skin

The shield is loaded by external earth and water pressure at right angles to the surface of the skin. The load assumptions for earth and water pressure can be assumed according to Figure 3-4 as a simplification, unless a more precise determination is demanded. The usual calculation methods for tunnel linings [184] apply, although it should be noted that these assumptions are theoretical for a shield skin. The shield machine is actually only temporarily subjected to loading from the ground. For reasons of geometry or as a result of the process (overcut, shield taper, filling of the gap between ground and shield skin with plastic earth material of support slurry, and also the asymmetrical location of the shield in the excavated cross-section), the ground only comes into contact with the shield skin after considerable deformation, normally resulting in three-dimensional arch-type stress redistribution with reduced loading from the ground. Unfavourable effects, like for example reduced or increased elastic subgrade reaction, are compensated through load transfers to a radially symmetrical condition (K = 1.0), so the stated load assumptions normally suffice for the structural analysis of the shield construction and lie on the safe side.

The lateral earth pressure coefficient K can be taken as K = 0.45 to 0.50 in accordance with the DGEG recommendations for tunnels in loose ground [82], unless more precise data is available.

Detailed methods for the structural design of shield tunnelling machines, particularly regarding questions of vertical earth pressure or its reduction, actions resulting from steering movements and special cases like closely spaced tunnel tubes, adjacent excavations, ground susceptible to swelling or karstified ground, can be found in [57].

In contrast to reinforced concrete tunnel tubes, for which one design cross-section is normally decisive and is subjected to relatively high bending loading and comparatively low axial forces, the decisive design cross-sections for shield machines constructed of steel are subject to high bending and also high axial force. The decisive comparative parameters for the determination of the decisive design cross-section are the absolute magnitude of the loading from earth and water pressure, the difference between vertical and horizontal loading and the subgrade reaction applicable to the relevant section.

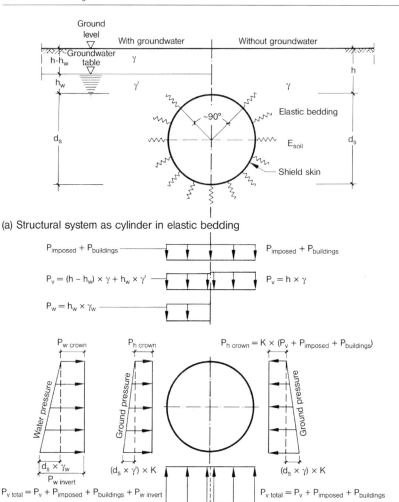

(a) Structural system as cylinder in elastic bedding

(b) Loading assumptions for ground and water pressures perpendicular to shield

Figure 3-4 Structural design basis for shields

vertical earth pressure:	p_v
total vertical pressure:	p_{vges}
building load:	p_{Beb}
imposed load:	p_{Verk}
water pressure:	p_w
horizontal pressure:	p_h
overburden depth:	h [m]
or, if appropriate, reduced overburden depth according to Terzaghi:	h
shield diameter:	d_s
bulk weight density of the soil:	γ or γ' under buoyancy
bulk weight density of water:	γ_w
groundwater table above crown:	h
lateral earth pressure coefficient:	K = 0.45 to 0.5 (approximation according to [82])

3.2.2 Loading on the pressure bulkhead

The wall separating the pressurised excavation chamber in a shield machine with face support from the part of the shield under atmospheric pressure is called the pressure bulkhead. The pressure bulkhead transfers the proportion of the thrust force required to support the face into the support medium (air, support slurry, earth plug).

For the design of the pressure bulkhead, the design cross-section with the highest support pressure is decisive. The pressure can lie above the limit for working under pressure of 3.6 bar and thus require additional measures regarding entrance to the chamber (e.g. grouting bodies, ground treatment).

The determination of the necessary support pressure p in the excavation chamber is normally undertaken by investigating the structural stability of possible sliding wedges. The support pressure must be in equilibrium with or exceed the prevalent earth and water pressure at each level of the face in order to rule out loss of stability or water ingress. If the support medium has a lower density than the prevalent groundwater, then the support pressure in the invert is decisive. The support pressure should if possible correspond to the actual pressure including a safety supplement and should not reach the passive earth pressure in order to rule out heaving of the ground surface. This requirement often leads to problems under low overburden pressure, such as danger of blowouts. If the support pressure falls below the required value, there is a danger of a loss of stability or the draining of the soil in front of the face.

Figure 3-5 shows the loading diagrams for slurry shield, earth pressure balance shield and compressed air shield. The suspension pressure in a slurry shield is controlled either through the slurry feed pipe (Figure 3-5 a) or through a compressed air bubble in the crown between the submerged wall and the pressure bulkhead (Figure 3-5 b). Opening a door in the top of the submerged wall also makes it possible to work with a lowered suspension level at the face and compressed air support to the upper part, but without hindering the hydraulic transport of material (Figure 3-5 c).

When the excavation chamber is partitioned by a submerged wall, the loading pattern on the pressure bulkhead can vary from the load on the face. This load resultant must be transferred by the submerged wall into the shield (Figure 3-5 b).

For earth pressure and compressed air support (Figure 3-5 d and 3-5 e respectively), the load distribution on the face corresponds to that on the pressure bulkhead. Because the density of air is negligible, the pressure in the excavation chamber of a compressed air shield is constant (Chapter 2).

In addition to this area loading from the support medium, the pressure bulkhead is also loaded by the reaction forces from the excavation equipment.

In case of a stoppage without support pressure in the excavation chamber, the earth pressure acting on the cutting wheel, and on partition wall or breasting plates if present, has to be transferred into the pressure bulkhead. In order to determine the earth pressure, the active earth pressure coefficient may be used.

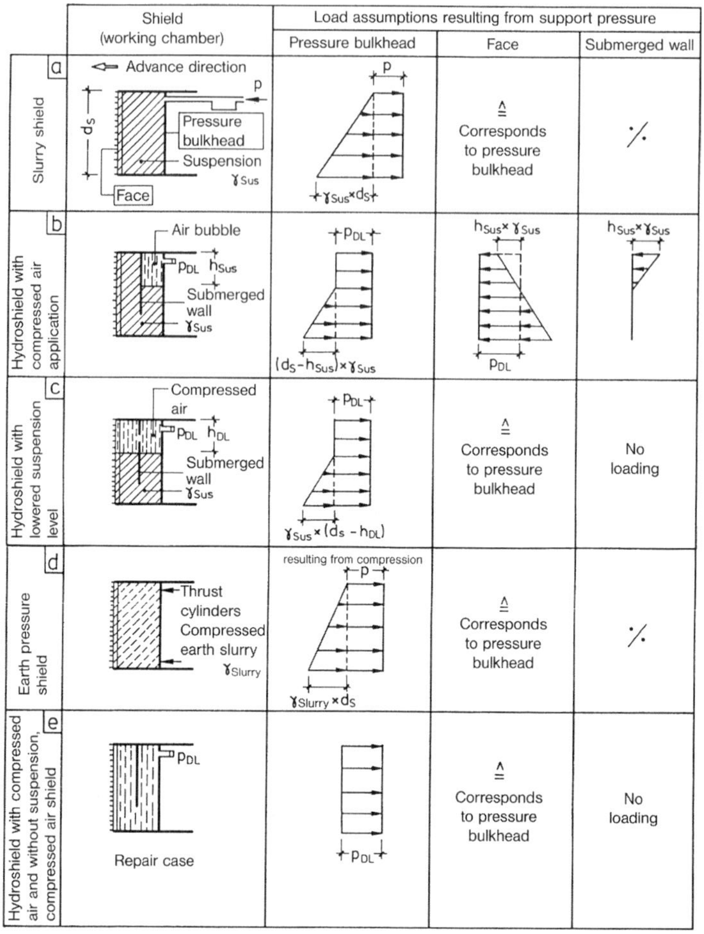

Figure 3-5 Examples of load assumptions for the pressure bulkhead of a shield

3.2.3 Loading from the thrust cylinders

The thrust cylinders push the shield forward in the direction of drive against the friction forces in the ground and against the necessary support pressure (with face support). The shoes of the thrust cylinders are placed against the last ring to be installed, which thus serves as abutment. At the forward end of the cylinders, forces are transferred through the pressure bulkhead into the support medium or into the shield skin (Figure 3-6).

3.3 Calculation of the necessary thrust force

The calculation of the necessary thrust force is of particular importance in the mechanical design of a tunnelling shield. Under-dimensioning or the occurrence of unexpected resistance forces can in the worst-case lead to expensive technical modification work underground and should be avoided by careful advance calculation, including the evaluation and analysis of completed projects.

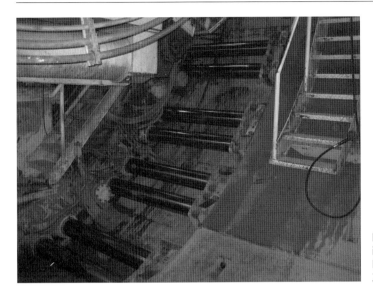

Figure 3-6 Thrust cylinders and thrust pads, Mixshield Finne Tunnel, 2009 (Herrenknecht)

3.3.1 Resistance to advance through friction on the shield skin

The radial, or horizontal and vertical, loading from overburden, buildings on the surface and imposed loads as well as the self-weight of the shield produce friction forces around the shield, which have to be overcome by the forces in the thrust cylinders. These friction forces can be reduced by tapering the shield, overcutting by the front shield or by lubrication (e.g. with bentonite) (Table 3-2).

Table 3-2 Friction coefficient μ between shield skin (steel) and type of soil [125]

Soil type	Friction coefficient μ [-]
gravel	0.55
sand	0.45
loam, marl	0.35
silt	0.30
clay	0.20

The composition and type of soil of the ground to be passed through [125] determine the friction coefficients μ [-], resulting in the approximate friction force at the shield skin for calculation purposes W_M:

$$W_M = \mu \times [2\pi \times r \times l\,(p_v + p_h) \times 0.5 + G_S]$$

friction force on the shield skin: W_M [kN]
friction coefficient $\mu = \tan \delta$ as a function of the wall friction angle δ [-] (Table 3-2)
perimeter of the shield skin: $2\pi \cdot r$ [m]

soil surcharge: p_v [kN/m²] (Figure 3-4)
length of the shield skin: l [m]
vertical load: $p_{v\,ges} = p_v + p_{Beb} + p_{Verk}$ [kN/m²]
horizontal load: $p_h = K_0 \cdot p_{vges}$ [kN/m²]
coefficient of undisturbed earth pressure: K_0 [-]
self-weight of the shield: G_S [kN]

In sandy and gravelly ground, the shield skin can be lubricated with a bentonite or clay suspension, which can lower the friction coefficient µ to 0.1 to 0.2. For the determination of the vertical surcharge, the reduced overburden height h' as a result of arch and silo effects may be used.

3.3.2 Resistance to advance at the front shield

The shield skin is driven through the ground with a shield blade. The peak resistance p_{Sch} [kN/m²] according to soil type is given in Table 3-3 [125].

The peak resistance is independent of the actual overburden and other loading assumptions. The coefficient of lateral earth pressure K is mostly greater than the passive earth pressure coefficient Kp:

$p_{Sch} > K_p \cdot p_{v\,ges}$ [kN/m²]

According to [125], this gives for the unreduced blade resistance W_{Sch} at the perimeter of the shield blade:

$W_{Sch} = 2\pi \cdot r \cdot p_{Sch} \cdot t$ [kN]

Blade resistance: W_{Sch} [kN]
Peak resistance according to soil type (Table 3-3): p_{Sch} [kN/m²]
Circumference of the shield blade: $2\pi \cdot r$ [m]
Blade wall thickness: t [m]

Figure 3-7 Extendable gauge cutter of the Mixshield, RandstadRail Rotterdam (Herrenknecht)

Table 3-3 Peak resistance p_{Sch} of the ground depending on soil type [125]

Soil type	Peak resistance p_{Sch} [kN/m²]
rock-type soil	12.000
gravel	7.000
sand, densely consolidated	6.000
sand, medium consolidated	4.000
sand, loosely consolidated	2.000
marl	3.000
Tertiary clay	1.000
silt, Quaternary clay	400

On reaching a critical value p_{Sch}, a local ground failure occurs at the shield blade, so that the shield can penetrate the ground. The blade resistance can be reduced by intentional overcut, either by extending over cutters or by the cutting wheel being located at the front with a larger boring diameter (Figure 3-7).

3.3.3 Resistance to advance at the face through platforms and excavation tools

The presence of working platforms, e.g. in manual excavation shields, has a similar effect to blades and leads to resistance to the advance of the shield. The load assumptions are to be made similarly to blades (Section 3.3.2).

The contact pressures of the excavation tools to loosen the soil depend on the type of soil encountered. For full-face machines with closed cutting wheels, part of the contact pressure may be considered to support the face.

The load assumption for the resistance W_{BA} for the contact pressure of the excavation tools while excavating can be approximately determined as follows for loose ground:

$$W_{BA} = A_{BA} \times K \times p_{v\,ges} \; [kN]$$

Resistances of the excavation tools: W_{BA} [kN]
Contact area of the excavation tools: A_{BA} [m²]
lateral earth pressure coefficient K with: $K_a < K < K_p$ [-]

For the contact pressures required for disc cutters in rock, different assumptions have to be made (Section 4.1.5).

For a cutting wheel with horizontally adjustable mounting, the forces in the cutting wheel displacement cylinders correspond to the contact force required for the excavation tools to excavate the ground.

3.3.4 Resistance to advance with slurry support, earth support and compressed air support

The resistances to advance from earth and water pressure and from the support pressure resultant have to be applied by the thrust cylinders (Figure 3-6).

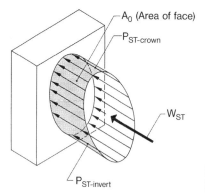

Figure 3-8 Support force W_{ST} from the integral of the support pressure on the face

As shown in Figure 3-8, the resultant support force W_{ST} is the integral of the support pressure over the area of the face.

$$W_{ST} > W_E + W_W$$

Resultant earth pressure resistance: W_E from the investigation of failure body
Resultant water pressure resistance: $W_W = A_0 \times (p_{W\,crown} + p_{W\,invert}) \times 0.5$
Resistance for face support: $W_{ST} = A_0 \times (p_{ST\,crown} + p_{ST\,invert}) \times 0.5$
Water pressure at the crown: $p_{W\,crown}$ (Figure 3-4)
Water pressure at the invert: $p_{W\,invert}$ (Figure 3-4)
Support pressure at the crown: $p_{ST\,crown}$
Support pressure at the invert: $p_{ST\,invert}$

Face area: $A_0 = \dfrac{\pi \cdot d_s^{\,2}}{4}$

When compressed air is used, the support pressure is to be assumed constant over the entire area of the face.

3.3.5 Resistance to advance from steering the shield

Shields are driven round curves by differential activation or extension of the thrust cylinders. With small-diameter shields, the thrust cylinders can be controlled individually; with larger shields, control is through thrust groups.

When tight curves are driven, secondary bending moments occur, which are larger, the longer the shield is in relation to its diameter (Figure 2-15). These effects can be reduced by intentional overcut, tapered shape of the shield and by lubrication with bentonite through injection nozzles in the shield skin.

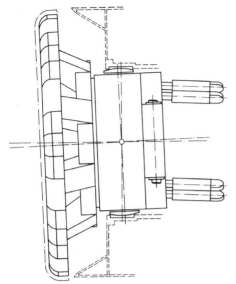

Figure 3-9 Schematic diagram of overcut functionality with tilting cutting wheel (articulation bearing)

With full-face machines with tilting mechanisms for the cutting wheel and extending gauge cutters, the desired overcut can be positioned exactly (Figure 3-9). Progammable logic control (PLC) are used.

The provision of a articulation joint between middle shield and shield tail can also reduce secondary moments resulting from driving curves. Modern large-diameter shield machines are normally equipped with both systems.

The remaining resistances to advance from the steering of the shield should be considered empirically.

3.3.6 Summary

The dimensioning of the thrust cylinders is based on the sum of the individual resistances including the consideration of a safety margin:

$P_V = \Sigma W +$ safety margin

The least favourable combination of individual resistances is decisive. The safety margin is an empirical factor and includes all the forces listed below, which cannot be calculated exactly.

– Towing force for the backup,
– friction force between shield tail seal and tunnel lining,
– increased blade resistance on encountering obstacles,
– increased skin friction and increased blade resistance in grouted zones,
– increased skin friction force through ground swelling pressure,
– increased skin friction force through curve driving and steering.

The calculated individual resistances can be subject to considerable variations due to the geological situation. The reduction or complete lack of one variable (e.g. skin friction, support pressure leads), with the thrust force remaining constant, must lead to the increase of another variable (e.g. contact pressure of the cutting wheel). In order to avoid overloading individual assemblies, constant control of the thrust force is therefore necessary.

Horizontally adjustable cutting wheels make it possible to calculate the contact force of the cutting wheel from the pressure of the displacement cylinder (Figure 3-2) and thus optimise the contact pressure of the disc cutters to suit the geological conditions encountered. This can avoid overloading of the main bearing.

The shield skin friction, which is always an unknown variable, can also be determined with articulated shield tail systems. If the support pressure is measured at the same time and a sufficiently precise division of the force components is possible, then the interpretation of unusual events is made easier.

3.4 Empirical values for the dimensioning of the shield and the thrust cylinders

Shield manufacturers collect data from each project for the dimensioning of shields and their technical equipment. This practical experience has resulted in the minimum thicknesses of the shield skin and special construction principles. Subsequent alterations and the fitting of additional equipment during a shield tunnel project are expensive and waste time [3]. Most structural verifications demanded by clients are recalculations or are based on the transfer of dimensioning from other projects, where they have proved successful over time (e.g. the metal thickness of the front shield skin 60 to 80 mm, for the middle shield 40 to 60 mm and for the shield tail 40 to 60 mm).

This also applies for the determination of the necessary thrust cylinder forces. Krause [171], Herzog [125], Anheuser [7], Kuhnhenn [174] and Krabbe [167] have collected em-

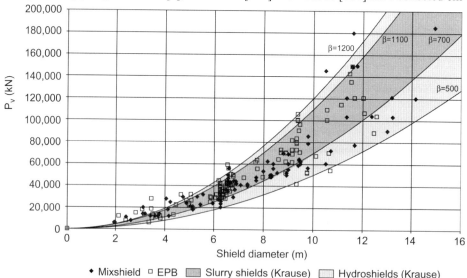

Figure 3-10 Relationship between the maximum thrust force P_V and the shield diameter d_S for 397 slurry and 12 Hydroshields (investigation of Krause [171]), with additional values for 250 EPB and Mixshields [123]

pirical figures for thrust cylinder forces. Average values for the thrust cylinder forces have been structurally analysed without special consideration of the constraints applicable for a particular project, like for example the effect of the prevailing geological and hydrogeological conditions, any lubrication of the shield skin surfaces etc. Krause [171] investigated 397 slurry shields in Japan and 12 Hydroshields in Germany and describes the relationship of thrust cylinder force to shield diameter with the following formula (Figure 3-10):

$$P_V = \beta \times d_S^2 \ [kN]$$

empirical factor: β $[kN/m^2]$
shield diameter: d_S $[m]$

Figure 3-10 shows the installed thrust cylinder forces for the 397 suspension shields investigated in Japan with the empirical factor β = 700 to 1,200 shown shaded dark grey:

$$P_V = (700 \text{ to } 1{,}200) \cdot d_S^2 \ [kN]$$

For the 12 Hydroshields investigated in Germany (Figure 3-10, shaded light grey), values were found for the maximum thrust forces P_V with β = 500 to 1,200:

$$P_V = (500 \text{ to } 1{,}200) \cdot d_S^2 \ [kN]$$

The ranges determined by Krause are confirmed by the listed maximum thrust cylinder forces of the EPB and Mixshields. As with the cutting wheel drive, the thrust cylinders of the EPB shields with the same diameter also have more powerful thrust forces, which are necessary for the excavation performance of the EPB cutting wheels, which are almost completely fitted with disc cutters.

Szechy [277], working some time ago, proposed general values for the total resistances for use in the dimensioning of thrust cylinder forces of approx. 60 Mp/m² = 600 kN/m², related to the area of the face. Anheuser [7] gives, according to soil strength and shield friction, thrust loads of 1.0 to 1.5 MN per metre of circumference and increases the thrust force for fluid support of the face or compressed air support with separate compressed air compartment additional to the support pressure acting on the area of the pressure bulkhead.

3.5   Calculation and dimensioning basics

Regulations and requirements to be observed

Design is based on the following regulations and requirements, which have to be observed:

– DIN 18 800-1: Steel structures, design and construction,
– DIN 18 800-2: Stability cases, buckling of struts and framed structures,
– DIN 4114: Stability in steel construction,
– DIN 4100: Welded steel structures,
– Loading data for earth and water pressure from site investigation reports,
– Recommendations for the design of tunnels in loose ground (1980) [82],

- Recommendations for the structural design of shield tunnelling machines (2005) [57],
- Loading data from shield operations of the shield manufacturer,
- Constructional drawings of the shield manufacturer.

Materials and permissible stresses

The construction material is, unless otherwise stated, steel (S355J2 +N according to EN 10 025; formerly St 52-3). External loads (earth pressure, water pressure, support pressure) are increased by partial safety factors $\gamma_F = 1.35$ or 1.5. The calculated stresses should not exceed the yield limit of steel reduced by the partial safety factor $\gamma_M = 1.1$.

3.6 Regulations and recommendations for the design of shields

Loading assumptions, like maximum operational positive pressure, coefficient of lateral earth pressure $K \leq 0.5$ or $K \geq 0.5$ and possible loading combinations due to the operational processes of shield advance and shield stoppage, should be agreed at an early stage with the shield manufacturer, tunnel consultant, supervisory engineer, soil mechanics expert, contractor and client. The few regulations for shield construction and the recommendations for shield tunnelling should be, as with the specification requirements (Chapter 17), should be complied with in the design of the shield:

- Regulations for the rail tunnels of German Railways, supplementary regulations for DS 853, shield tunnelling EzVTU 16 [63],
- Recommendations for the design of tunnels in loose ground [82],
- Recommendations for the design of shield-driven tunnels [81],
- Recommendations for the structural design of shield tunnelling machines [57],
- guidelines for tunnel structures according to the Regulations for the construction and operation of tramlines [246],
- accident prevention regulations of the technical accident insurers,
- CEN/TC151/N8 "Tunnelling Machines" from the European Committee for Standardisation [46].

The recommendations, however, include few details about the actual construction of a shield. Detailed methods for the structural design of shield machines, particularly concerning vertical earth pressure and its possible reduction, actions due to steering movements and special situations like closely spaced tunnels, adjacent excavations ground liable to swelling and karstified ground, can be found in [57].

According to the supplementary regulations to the DS 853 EzVTU 16 [63], adequate safety reserves must be included in the design of a shield because the actual forces acting cannot be determined with sufficient precision, as is stated by the explanatory notes to the DS 853. For example, the middle shield of the shield should be able to bear all active loads.

The BOStrab (German tram tunnel guidelines) [246] points out the necessity of the structural verification of the face support during advance and stoppages.

4 Excavation tools and excavation process

Excavation denotes the loosening of the ground from its natural consolidation.

Excavation, face support and removal of the loosened spoil should be considered together, particularly for the unstable soils that require the use of a shield machine. Excavation should also proceed with as little disturbance as possible to the natural bedding structure in temporarily or partially stable ground, in order to keep the destruction of the structural fabric and resulting settlement to a minimum.

In shield tunnelling, only the face remains accessible for attack by excavation tools, and no further alteration of the excavation cross-section is possible behind the shield blade. Any overcut has to be excavated in front of the shield blade.

Particularly at membrane-supported excavation fronts, where the support pressure converts to support force in a shallow filter cake, undercutting or the removal of large areas should be avoided. Localised damage to the filter cake is not dangerous and recovers quickly.

The working face is defined as the excavation front spanned by the tips of the excavation tools (Figure 4-1). It can be, according to excavation tools and process, geometrically unchangeable (full-face with cutting wheel or cutter head), cyclically changeable (program-controlled partial face), freely changeable (with manually controlled partial face) or even changeable depending on material (e.g. with hydraulic excavation).

 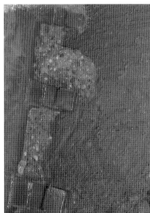 

Figure 4-1
a) Face cut by disc cutters in rock
b) Slurry-supported face in coarse gravel, excavation with drag picks and scrapers
c) Slurry-supported face in sand, excavation with scrapers and disc cutters

4.1 Excavation tools

Considering the numerous developments by manufacturers of shield machines, which have sometimes reflected very specific tunnelling experience, an overview of mechanical excavation tools can hardly hope to be comprehensive. There is no reliable data available for the wear rates of the various types of tool, which can almost solely be determined through laboratory tests [254] (for cost reasons, only a small range of factors affecting wear are investigated), although this would certainly be interesting. Wear rates in practice belong to the well-guarded store of experience of tunnelling contractors.

There now follows a description of the usual tool types and their main applications.

4.1.1 Hand-held tools

The intended use of hand-held tools is, for economic reasons, only sensible under very special conditions today (e.g. very short length of tunnel, stable but easily excavated face and low wage level).

The classic hand tools of miners are chisel, spade and drill, generally as compressed air tools. The pneumatic hammer or drill still has a certain usefulness when solid obstructions cannot be dealt with by the mechanical excavation tools and have to be removed from the face or broken to smaller size in the excavation chamber.

4.1.2 Cutting edges

Cutting edges are the continuous, slightly projecting edges of steel plates, as used for example in some cutting wheels. The basic idea is the mechanical support of the face through the provision of adjustable support plates together with fixed cutting plates. Ideally, the soil should have homogeneously plastic properties or consist of rounded, loosely consolidated sand or gravel without larger stones. As such ideal conditions are naturally not very common and stones can jam the support plates in practice, thus ruling out the simultaneous support function, this support principle has not been used since the introduction of the membrane-supported face (Figure 4-2).

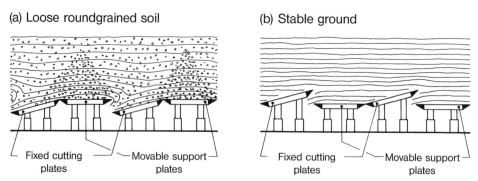

Figure 4-2 Diagram of support plates with cutting edges (Bade)

The basic principle of a continuous cutting edge corresponds to the shield blade at the perimeter of a shield skin.

The cutting bars sometimes found in hard rock cutter heads have a similar appearance to cutting edges. They have the function of assisting material transport into peripheral mucking channels by preventing the material already in the channel falling back in front of the cutter head. Since they only serve to raise the sides of the mucking channel, the description as cutting bars is rather misleading.

4.1.3 Scrapers

If the cutting edges are discontinuous, they are termed scrapers. All scrapers work together to cut the entire working face. Scrapers should ideally have a cutting action, with chips peeling off over the front edge of the blade.

There are numerous types for use in various soil types, generally with hard metal blades and additional wear protection (Figure 4-3).

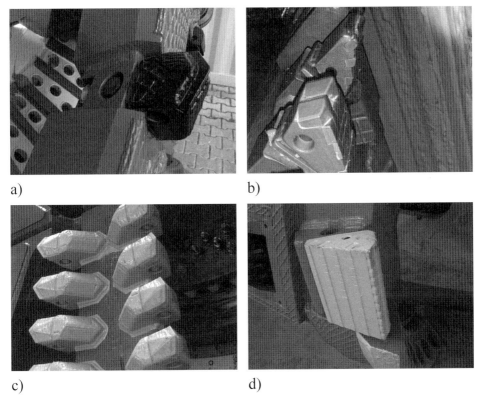

Figure 4-3 Types of scraper for various soils and applications
a) Narrow scraper with inserted holder in cutting wheel (Mikrotunnelling, 2009)
b) Narrow scraper with variable setting angle and back protection in a surface-mounted holder (S-381 Mixshield H8 Tunnel, Jenbach, 2008)
c) Narrow scraper with surface-mounted holder (S-340 EPB shield, Malmo, 2006)
d) Wide scraper with hardened strips as back protection in surface-mounted holder
(S-440 Mixshield U4 Hamburg, 2008)

Figure 4-4 Square-section drag picks (80 mm x 80 mm), S-100 Mixshield Bewag Tunnel, Berlin, 1995

According to the type of soil to be excavated, scrapers can be the sole excavation tool or can be mounted slightly behind disc or chisel cutters to work the intermediate parts of the face. Originally designed for use in clay, scrapers have now developed into universal tools for use in a wide range of loose ground.

4.1.4 Drag picks, flat chisels, round chisels, rippers

Drag picks are blunt tools of round or square section set at right angles to the face. Figure 4-4 shows square-section drag picks on the central ridge and the spokes of a cutting wheel before service.

Drag picks work by destroying the structure and tearing in silt, sand and gravel soils, although they only knead cohesive soil. In order to increase the lifetime, drag picks are often built up by welding or equipped with hard metal inserts. The orientation at right angles to the face makes drag picks independent of the direction of rotation, i.e. they can excavate in each direction.

Chisels are narrow tools suitable for breaking hard soils, generally with an exchangeable tip. They are used as breaking tools in percussive hammers and a special type with hard metal core and round shaft is used in an angled arrangement in the milling cutting heads of road headers or open shields (Figure 4-5 a). The additional transverse loading resulting from this angled arrangement with a round section free to rotate in its mounting leads to a longer life time. Details of the cutting performance and wear of various types of chisel can be found in [106].

a)   b)

Figure 4-5 Round-shafted chisels with tungster carbite insert bits
a) arranged on the transverse cutting head of a cutter arm
b) arranged as a repair solution on a cutting wheel

The described angled arrangement of round-shafted chisels can also be used in the calibre or gauge area of a cutting wheel. In the flat face area of a cutting wheel, narrow round-shafted chisels are subjected to single-sided loading, which leads to a short lifetime and justifies their use only as a repair measure (Figure 4-5 b).

Figure 4-6 Flat chisels as the advance tools of a slurry shield [130]

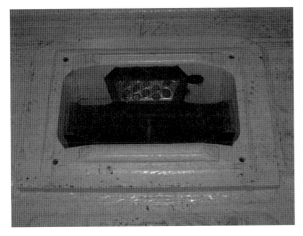

Figure 4-7 Heavy-duty ripper mounted in a disc cutter housing of a disc cutter (Herrenknecht)

Narrow flat chisels are often mounted at right angles to the face on Japanese cutting wheels for the localised loosening of a consolidated soil structure before the arrival of scrapers (Figure 4-6). Square-section drag picks in the early Hydroshields and Mixshields were often mounted on tool holders on the front of the cutting wheel with a transverse securing bolt (Figure 4-4). The limited accessibility for tool changing and the disturbance to the material flow of the excavated soil into the muck openings have reduced the use of square-section drag picks in favour of scrapers.

Chisels are increasingly installed in current cutting wheel designs as heavy ripper heads on a support construction, set into the housings of previously removed disc cutters (Figure 4-7).

The forces transferred by the individual teeth, chisels or drag picks cannot be determined in advance. An upper loading limit is determined by the failure strength of the contact zone between tool and obstacle, unless the tool has already broken.

"Activated" teeth oscillating longitudinally penetrate better into hard material. The increased expense of the mechanism, however, limits the possible number of tools with this mechanism.

4.1.5 Disc cutters, discs

Disc cutters (also known as roller cutters or discs) are rotating tools of hardened metal (Figure 4-8). As the machine bores, the discs are pressed against the face and roll with the rotation of the cutter head. The disc cutters are arranged so that they run on concentric tracks, mostly with a constant spacing.

On account of the extremely high contact forces, the rock under the cutter ring is ground and is also subject to splitting failure. The resulting cracks from adjacent tracks join up and rock chips fall out of the face.

The high contact forces, which can be transferred, the continuous improvement of material quality and the increasing size of the disc cutters have made discs an economic alternative to drilling and blasting [249].

Disc cutters are more susceptible to changing properties of the rock mass than the previously described tools. Insufficient tangential force at the perimeter of the disc or bearing damage due to overloading leads to the roller jamming and thus to rapid, one-sided wear.

The spacing of the disc tracks is approx. 70 to 100 mm, depending on rock behaviour. Double or triple discs on one roller make installation easier in restricted space, but also reduce the specific excavation performance. Special forms of the tool surface (toothed discs, discs with hard metal inserts, button discs) have been developed for special purposes.

Rock hardness, spacing and contact force are the essential factors determining the penetration depth into the rock. The relationships shown in Figure 4-9 between rock mass (rock strength), contact force and penetration show clearly that an initial critical contact force has to be exceeded in order to achieve penetration in an economic working range. Below this critical tool tip loading, disc cutters only work as rock grinders.

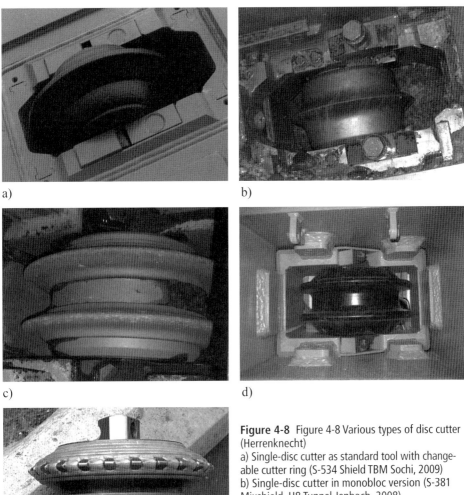

Figure 4-8 Figure 4-8 Various types of disc cutter (Herrenknecht)
a) Single-disc cutter as standard tool with changeable cutter ring (S-534 Shield TBM Sochi, 2009)
b) Single-disc cutter in monobloc version (S-381 Mixshield, H8 Tunnel Jenbach, 2008)
c) Two-disc cutter with changeable cutter rings (S-290 Mixshield, Silberwald service tunnel, 2005)
d) Two-disc cutter in monobloc version (S-440 Mixshield, U4 Hamburg, 2008)
e) Two-disc cutter with tungsten carbite insert bits (HTS)

The increase of nominal contact force from 90 kN to about 312 kN for disc cutters increased in diameter from 280 mm (11") to 483 mm (19") enables not only a considerable increase of the average contact force, but also leads to a significant improvement in the lifetime of the disc cutters. With this increase of diameter, the permissible rolling speed also increased up to about 190 m/min, which due to the outermost gauge cutter is decisive for the maximum cutting wheel revolution speed.

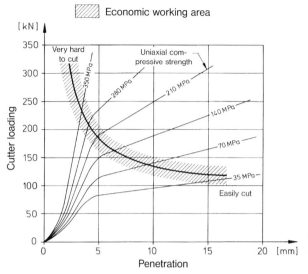

Figure 4-9 Diagram of the relationships between penetration and disc cutter contact force depending on rock strength [203]

In loose ground, discs are the most suitable tool to deal with the occurrence of stones and blocks. As long as stones are firmly fixed in the matrix of the face, they can be broken up by disc cutters. Depending on the support principle, the provision of disc cutters may under some circumstances be the only equipment capable of overcoming stones and blocks. Disc cutters are therefore arranged slightly ahead of other types of tool, generally scrapers, in the layout of the cutting wheel with regard to penetration depth.

Disc cutters for use in closed mode originally had the same basic construction as those for use in hard rock. Later, improved sealing systems with reduced friction were developed, which are used successfully today up to a chamber pressure of about 4 bar. For higher pressures, pressure-compensated disc cutters should be provided.

Good results have been achieved with monobloc disc cutters with extended durability, which dispense with changeable cutter rings (Figure 4-8). These are indeed more expensive, but much less susceptible to secondary wear or damage. When the ground conditions permit, two-disc cutters can also be used, with a lower risk of jamming in softer soils.

At the moment, the strength of the disc material, the space required for installation and the weight of a cutter to be exchanged place a practical limit on the further increase of disc diameter. [117].

4.1.6 Buckets

Muck buckets are massive tools arranged in the outer part of the diameter and are intended to transfer the excavated muck with their bucket edge behind the steel front of the cutting wheel of a full-face machine. The soil cut from the face by the excavation tools falls under gravity down in front of the tool carrier (cutting wheel or cutter head) and collects in the invert. As a result of the rotation of the tool carrier, the buckets collect the material and scoop it backwards against the advance direction (Figure 4-10).

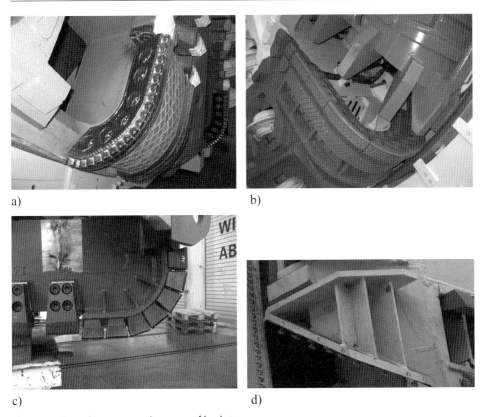

Figure 4-10 Various construction types of buckets
a) Bucket with hard metal blade; single-row fixing (S-510 EPB shield, Sabadell, 2009)
b) Bucket with applied hardening as bucket lip, two-row fixing (S-502 combined shield, Lake Mead, 2009)
c) Compartmented buckets (S-397 EPB shield, Thessaloniki, 2008)
d) Bolted bucket lip, inclined backward (double shield TBM, Guadarrama, 2003) [173]

In stable rock, buckets should not have an excavation function, but should only remove excavated material in order to prolong their lifetime. When a mixture of tools is used with discs and scrapers, the buckets in the gauge area undertake a limited contribution to excavation between the disc tracks, which are concentrated with closer spacing.

4.2 Excavation process

In the development of shield tunnelling, machinery displaced the miner from his dangerous workplace at the open face at an early date and this also enabled much higher excavation performance. But non-mechanical excavation is still practised sometimes, depending on ground conditions, tunnel length, required advance rate and availability of technical equipment. Health and safety aspects are dealt with in Chapter 16.

4.2.1 Tunnelling without cutting wheel

Tunnelling without a cutting wheel requires very uniform, homogeneous soil with no solid obstructions for the successful advance of a shield.

Homogeneous loose ground with low shear strength and pronounced plastic properties can be simply squeezed into the inside of the tunnel through one or more openings in the front wall of the shield (extrusion excavation). Under the effect of the thrust forces, the soil emerges from the closable extrusion openings as from a sausage machine and falls directly onto the conveyor. Division into lengths suitable for transport may be required as it falls. These types of shield are described as blind shields (Figure 4-11) or extrusion shields (see Chapter 11).

In order to avoid settlement, the support pressure exerted by the shield advance must exceed the undisturbed pressure of the plastic soil encountered.

Apart from pure extrusion excavation, grained soil capable of pouring can also enter the shield without particular excavation equipment. When support is provided by horizontal platforms, the soil is loosened from its natural consolidation by controlled ground failure. The material then either flows independently or is cleared from the back edge of the triangular support slope onto the conveyor.

The junctions of the working platforms with the shield skin often cause stress concentrations, resulting in considerably increased thrust force being required. Figure 4-12 shows an open shield with excavation platform.

The coordination of shield advance and removal of the excavated material requires experience and careful attention in order to avoid large settlements.

Figure 4-11 Blind shield of Japanese manufacture

Figure 4-12 Compressed air shield with excavation platform, escape tunnel, fourth tube of the Elbe Tunnel, 2000 (Herrenknecht)

4.2.2 Manual digging

Manual digging has almost disappeared from shield tunnelling. It can only be economical for very short distances and small but still accessible shield diameters. It is still practised above all in low-wage countries with low requirements for the level of equipment.

As the diameter of the remote-controlled and -operated small shields, which can be used without extensive starting provisions, has now clearly exceeded the 2 m mark, it can be expected that manual digging will become still more unusual.

The classic excavation tool of the miner is the compressed air spade, which can be used for digging in stable ground. In ground, which does not stand up, blockages at the corners of the support platforms can be dug out to reduce the thrust force required. Skill in the handling of tools and a feeling for the behaviour of the ground are essential qualities for manual digging. This can avoid excessive removal of soil with the associated settlement.

Even if the use of machinery is making the planned use of manual digging ever less common, the versatility and flexibility of manual digging can still be useful, even in fully mechanised tunnel drives, to remove obstructions, which cannot be dealt with by the machinery in use. Access to the face should therefore always be provided in the design of a shield for fully mechanical excavation.

4.2.3 Partial-face mechanical excavation

There are many methods of mechanical excavation depending on ground conditions, tunnel length and advance rate. The excavation equipment described here digs at a point under direct control, enabling experienced machine operators to remove hard localised bands and rock banks in ground with changeable strata, or also to remove obstructions from the face without breaking them up. When the excavation area is large or the view is restricted, e.g. with the cutter excavator in the slurry-filled excavation chamber of a Thixshield (Section 10.4.1.1), programmed passes along the excavation front are recommended.

The advance rates with partial-face excavation are slower than with full-face excavation due to the localised attack. The use of partial-face excavation is therefore concentrated on easily excavated but stable ground, short tunnel lengths and non-circular cross-sections, like crosscuts and enlargements.

The changing of tools is normally simple and the low unit weight of the teeth, chisels and rippers used makes access and removal easy. The tools are mostly wedged in place and installed or removed with a hammer. Because the excavator arm can be retracted to a safe position away from the face, tools can be changed quickly. The versatility of such machinery is improved by the possibility of changing individual tools or the entire excavation head (e.g. hammer or bucket).

The cutting contour is not restricted to circular profiles. Egg-shaped, horseshoe and square sections have also been excavated successfully.

Figure 4-13 Open shield with working platforms and breasting plates, S-194 Dublin Port Tunnel, 2002

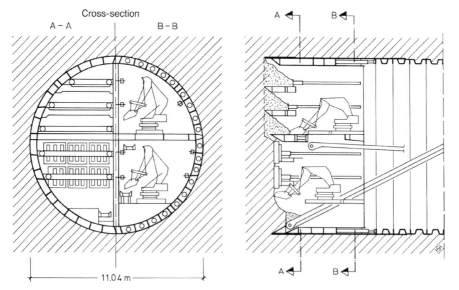

Figure 4-14 Open shield with four fixed mini-excavators, Elbe Bank Tunnel, Hamburg, 1968-1974 (Howaldswerke, Deutsche Werft)

Figure 4-15 Excavation device with integrated conveyor belt for clamping in the front pipe in pipe jacking (Herrenknecht)

Excavators with standard or special equipment

The adaptation of conventional excavation machinery for the special conditions in a shield has resulted in a variety of types of machine, which normally only attack one small part of the face at any time. Horizontal and vertical movement of the tool carrier makes it possible to excavate the face locally between two fixed support devices. Support elements however considerably reduce the performance of the machine.

Large tunnel sections may permit the use of standard excavators, as long as sufficient space is available on the invert or on intermediate platforms. As hardly any travelling is required, the carriage can often be replaced by skids or a fixed mounting to save space (Figures. 4-13 and 4-14). A number of smaller machines is also possible.

Depending on the ground conditions, universal excavators can be fitted with ripper tooth, hydraulic hammer or loading bucket. The boom supporting the tools and any support construction have to be chosen carefully to keep any obstruction as slight as possible. As the excavated spoil is normally cleared from the invert, the machines working in the upper parts of the shield often only have a breaking function (ripping tooth, hammer).

Special excavator arms are made for small diameter shields. Cutting and loading tools are combined in one unit, which can be supported by clamping in the shield, or in the first concrete pipe equipped with cutting apparatus in case of pipe jacking (Figure 4-15).

The various available sizes of these machines cover the range of sizes from just accessible sections up to about 4 m diameter. A machine, which can be rotated about the axis, eases the precise cutting of consolidated soil around the cutting contour.

Road headers

While loading buckets and ripping teeth are primarily used in cohesive soil, rotating excavation devices equipped with chisels have proved successful in solid, cemented ground. Also in this case, the palette ranges from excavation arms mounted on universal excavators to specialised machines. The independently mobile versions of these machines have become known as road headers (Figure 4-16) [183].

The standard tool for the cutting head rotating about the longitudinal or transverse axis is the round-shafted chisel with hard metal point, which is available as standard in many sizes.

Performance data and achievable excavation quantities depending on rock and grain strength can be found in the manufacturers' descriptions. Road headers equipped with

a)

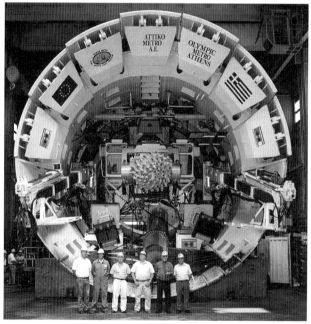

b)

Figure 4-16 Road headers installed in shields
a) with transverse cutting head (Bessac)
b) with cutting head rotating about the longitudinal axis (S-148 open shield, Olympic Metro Athens, 1998)

chisels can economically excavate rock with strengths of up to about 80 MPa (some manufacturers state up to 100 MPa). The presence of joints and cavities increases the performance.

It should be noted that excavation by tearing with chisels normally produces a considerable amount of dust.

The excavation tools can be electrically or hydraulically driven. Direct electric drive of rotating machinery has the advantages of a better degree of efficiency and lower environmental risk in comparison to hydraulic drives, which can leak oil, which still applies if hydraulic drives are mostly filled with biologically degradable types of oil. Hydraulic motors, on the other hand, offer a very high power density, i.e. small motors on the machine and problem-free overload behaviour. The pump motor to feed the linear motors (cylinders), which could hardly be implemented otherwise, can also be used to power the drive and the rotating machinery.

Cutter excavator

One special form of excavator is the cutter excavator, whose rotating cutting head with edges fitted with chisels loosens and also grades the material. As the cutter is only used in combination with hydraulic material transport, which rules out being able to view the process, the face is worked in a program-controlled pattern. The openings in the cutting head only pass the grain sizes suitable for hydraulic transport to the suction connections of the slurry pipe in the middle of the excavator arm. As a parallel development to the Hydroshield, the Thixshield disappeared from the market after a few applications after the Hydroshield patent had lapsed.

Special machines for rock

The increasing demand for protection and safety in changeable rock formations has also led to the use of shield machine constructions and similar for dedicated rock machines.

A tunnelling machine developed in France with individual drilling arms equipped with discs and with a protection roof found some applications. The discs excavate the face in spiral tracks. Increased tool wear due the flexibility of the tool carrier led to the abandonment of this machine concept.

Figure 4-17 Cutter excavator in "System Holzmann" Thixshield , East Collector, Hamburg 1978 (Ph. Holzmann/Westfalia Lünen)

Figure 4-18 Mobile Miner (Robbins)

A similarly constructed machine, which excavated the face by undercutting, was built in 1993 by the Wirth company, Erkelenz. After a trial application in a Canadian copper/nickel mine, the undercutting technology was also used as excavation system on the enlargement cutter head of the Uetliberg machine.

The Robbins Company offered a partial-face machine (Mobile Miner, Figure 4-18), whose grippers also acted as protective shield. The excavation wheel fitted with discs around the perimeter is slewed backwards and forwards across the working face with a horizontal axis. This produces vaulted excavation cross-sections with a flat invert.

4.2.4 Mechanical full-face excavation

A circular shape for a tunnel was recognised early as structurally favourable and is also suitable for full-face excavation by a machine with a tool carrier working the entire face with each revolution. This type of excavation has the following advantages:

– the tunnel profile is cut out exactly, without additional excavation,
– the contour of the face remains unaltered during the tunnel drive (except for the thickness removed with each revolution),
– the shape of the face can be optimised for stability, i.e. a forward-facing arch is also possible,
– comparatively high advance rates are possible.

This has to be balanced against the disadvantages of an expensive specialised machine unless a machine is available for reuse, which is very unusual due to the specific requirements of each project. While the excavation performance of a partial-face machine or road header in inhomogeneous rock depends greatly on the skill of the operator, this is insignificant with full-face machines. The progress of boring is determined rather by the suitability of the tools and the excavation energy to the prevailing ground. Performance variations in changeable conditions are more pronounced and any rebuilding of the excavation machinery is expensive.

This consideration of increased performance requirements despite changeable ground conditions along the length of the tunnel has led to a trend in shield machine design towards universal machines with ever higher installed power.

The most important characteristic of the full-face tunnelling machine is the integration of excavation, support and muck removal.

The excavation devices of full-face machines are differentiated into cutting wheels and cutter heads. The term cutting wheel denotes devices, in which the excavated material is transported backwards to a transport organ positioned in the invert. Material transport only takes place against the advance direction. Cutterheads, on the other hand, have buckets to pick up the excavated material from the invert and transport it down chutes into the centre, where it is transferred by a hopper (often described as a muck ring) onto the transport device, typically a conveyor belt.

Construction types

The tool carrier in full-face machines is the cutting wheel. The tools, selected to suit the prevailing soil conditions, are mounted on the front. In order that tools on adjacent tracks following behind all contribute to the excavation process, they are fixed to the cutting wheel in a spiral pattern.

Spoked cutting wheels

The spoked cutting wheel was used in the first slurry-supported machines with compressed air bubble in the late 1970s and early 1980s (Figure 4-19). The basic idea of this cutting wheel design was to enable the best possible exchange and contact of the support fluid to the tunnel face. They were used in permeable sands and gravels below the groundwater table. The tools fitted were simple square-section drag picks at the centre of the spokes or scrapers at the sides of the arms (spokes). For use in mixed soils, additional disc cutters were later installed on the arms.

Figure 4-19 Cutting wheel with four free-standing spokes equipped with drag picks; Mixshield S-12, HERA Hamburg 1985/87 (Herrenknecht)

As a result of the construction principle, the provision of back-loading disc cutters or accessibility to the central area are scarcely practical – particularly for small to medium diameters.

For larger diameters, the spoked construction could be built with accessible arm cross-sections, which permitted the complete use of back-loading disc cutters.

Rim cutting wheel

An external perimeter rim, which can be provided as part of a cylindrical or tapered shape, distributes locally occurring loads to more than one spoke. The resulting increase of friction surface at the perimeter considerably increases the required torque in comparison with the open star of a spoked cutting wheel.

Rim cutting wheels (Figure 4-20) generally run in advance of the shield blade. The perimeter rim provides mechanical support to the sides of the excavation until the shield skin arrives. The desire to be able to close the remaining openings in the cutting wheel if required is fulfilled by fixed and movable inserts plates between the spokes and the rim. This has various objectives. For example, some earth pressure shield designs have tried to make use of the material flow into the cutting wheel for the support of the face through the installation of adjustable shutters or flood doors. But the requirements for face support were not fulfilled until the introduction of foam conditioning, as the unconditioned soil in purely mechanical solutions was not able to provide the resilience necessary for face support.

In shields with slurry face support, movable inserts were installed with the objective of providing mechanical protection for workers entering the chamber, particularly for shield diameters larger than 6 to 7 m. This was implemented by the development of breasting plates, which could be extended during stoppages.

The limitations concerning access to the tools and the central area, which are applicable to spoked cutting wheels, also apply for rim cutting wheels, as the basic construction of these is identical.

Full-surface cutting wheels as closed panels with opening slots

The cutting wheel most often used today is the more or less closed full-surface cutting wheel with various degrees of openings. Different designs are used according to size. It is interesting that the use of these cutting wheels in closed mode is now almost independent of the support principle. A full-surface cutting wheel can be used successfully for slurry or in earth pressure operation. The former differentiation into earth pressure and Mixshield cutting wheels is therefore invalid or irrelevant.

With the increasing filling of the front surface, the support and muck removal functions of the cutting wheel are becoming more significant. Full-surface cutting wheels with narrow passage slots have a grading effect and hold back the larger grain sizes at the face until broken, which under some circumstances can cause considerable disturbance of the structure of the undisturbed soil. The fitting of disc cutters, which is normal today in soil containing stones, may counter this problem as the stones are already broken during the excavation process.

a) b)

Figure 4-20 a) Spoked cutting wheel with closable sectors (Lovat)
b) Spoked cutting wheel at rest with extending breasting plates S-110/118
main line tunnel Berlin, 1997 (Herrenknecht)

For hard rock machines equipped with discs, the completely closed cutter head has become the accepted solution. Buckets installed at the perimeter of the cutter head deal with the removal of the excavated material (Figure 4-21).

The assessment and differentiation criteria used today for the evaluation of a cutting wheel/cutter head design are the opening ratio, material flow, the handling of stones and boulders and tool changing.

Opening and opening ratio

The opening ratio is the area of material intake openings as a percentage of the entire cutting wheel area. This parameter is of greater significance for EPB shields than for slurry shields and normally lies between 25 and 35 % (Figure 4-22). The size and location of the material inlets in a cutting wheel is the result of a compromise between a number of requirements:

– The openings must be sufficiently large to permit the entry of the excavated and under some circumstances also already preconditioned soil in the tool gap in front of the cutting wheel into the rear excavation chamber. The support pressure in shields for closed operation is normally measured with pressure cells installed in the pressure bulkhead. Larger pressure differences of earth pressure in front of and behind the cutting wheel are not acceptable for earth pressure shields, if precise control of the support pressure is required. The use of additional earth pressure cells in the front plate of the cutting wheel has already been tried out a few times, but the precision and reliability of the measured values is only adequate for limited use as the primary parameter for the control of support pressure.

– The cross-sectional areas of the intake openings should limit the grain size of boulders, which can enter the excavation chamber. In general, only such grain sizes should enter the excavation chamber, which can be processed by the subsequent material transport equipment (open shields: conveyor belt; slurry shields: stone crusher; EPB shields: screw conveyor). This requirement usually leads to conflicts for smaller to medium-sized earth pressure balance shields, which only have room for a certain size of screw conveyor for geometrical reasons. The use of grain size limiters is usual in these cases, but often leads to sticking and stoppage of the small remaining muck inlets.

Considerably more important than the global proportion of the area of openings through a cutting wheel is their location, or the distribution of the opening ratio along the radius. Particularly the central area of a cutting wheel is susceptible to sticking. For earth pressure machines, there are scarcely any mixing dynamics in the centre. For slurry machines, this can be partially compensated by central flushing. In each case, the openings at the centre merit considerably more attention than those in the outer area.

Figure 4-21 Almost completely closed rock cutter head with discs and outer buckets, Gotthard Base Tunnel 2001 (Herrenknecht)

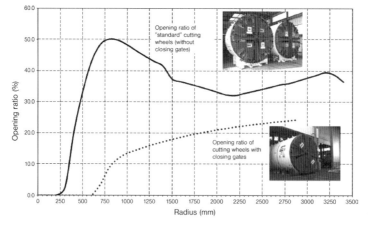

Figure 4-22 Curves of the opening ratio of EPB cutting wheel designs with comparable specifications – full installation of scrapers as well as discs in the outer area

Material flow

The material flow in the excavation chamber is an important aspect, both for slurry-supported and for EPB machines. Sticking of the cutting wheel is one of the significant risks to the performance of shield tunnelling in the affected soils. Wherever possible, the shape of the cutting wheel should avoid "dead corners" or pockets, which encourage caking or bridge formation. The shape of the cutting wheel should be as favourable to flow as possible, both at the front and at the back. In particular, the central areas are at risk, as the cutting speed here and the mixing dynamical are low.

For these reasons, active centre cutters have been developed for Mixshields, sometimes with their own slurry circuit. This measure achieves an increase of advance rate of up to 30% in cohesive soil, while simultaneously reducing the required torque by about 25% (Figure 4-20 b).

The idea of an independent inner cutting wheel was implemented for the first time in a 15.2 m diameter earth pressure shield for the M30 project in Madrid (Figure 4-23).

The independent central cutting wheel (active centre cutter) has a diameter of 7.0 m and permits a higher speed for the tools and mixing dynamics in the centre of this largest ever earth pressure shield. An additional effect is the partial compensation of the torque, because the inner and outer cutting wheels turn in opposite directions.

Cutting wheel designs with the same direction of rotation are better for the optimisation of material flow, but lose the opportunity of countering rolling of the shield through contra-rotation. Some cutting wheels with rotation in one direction are in successful use, but it has to be accepted with such designs that a certain part of the torque reaction has to be transferred to the segment rings through inclined thrust cylinders. As this places additional requirements on the tunnel lining, the most usual solution is still contra-rotating cutting wheels.

a)

b)

Figure 4-23 Earth pressure balance shield S-300 M30 with independent inner and outer cutting wheels
a) during assembly at the workshop
b) breakthrough

Boulders and stones

For both earth pressure and slurry-supported machines, encountering and boring through stones and boulders is a challenging but quite normal occurrence. The use of disc cutters in combination with scrapers is common practice under such conditions. The decision to install disc cutters in the centre is based on the estimation of risk and the expected frequency of stones. The use of disc cutters in the centre normally only leaves little room for central openings to ensure optimal material flow.

Experience from a number of various tunnel drives shows that boulders can be dealt with by disc cutters as long as they are solidly bonded into the surrounding matrix. Depending on the size of the boulder, the breaking process may be by normal chip formation, as is usual in hard rock tunnelling, or by splitting the boulder into pieces. For mechanised tunnelling under such conditions, it is important to understand that the permissible penetration will be determined by this stone-breaking process, even if it is only a partial process. Even if a hard boulder only occupies a few per cent of the face area, a penetration of only 15 to 25 mm can well destroy the affected discs due to the extreme shock loading. This occurs frequently – sometimes causing severe consequent damage to the cutting wheel – through "grinding down" of boulders.

Tool changing

A common requirement today is to be able to change tools from behind or from the side. Cutting wheel designs with closed full-surface wheels fulfil this requirement, even for the centre of the cutting wheel, with unhindered access from behind.

Tool changing with shields in closed mode (slurry or earth pressure support), however, still requires the workers to enter through a man lock and work under compressed air.

The tool changing situation is made more difficult by the high water pressures of more than 4 bar, which often have to be overcome nowadays. With increasing water pressure, reliable support of the face becomes more difficult and tool wear worse, particularly with earth pressure shields. Entries into the excavation chamber of the shields become much

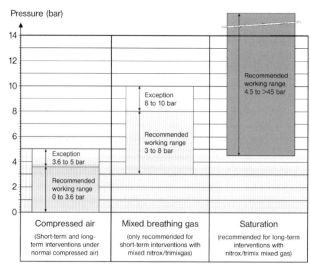

Figure 4-24 Recommended pressure ranges for the use of normal compressed air, mixed breathing gas and saturated breathing gas for TBM tunnelling [132]

more elaborate because additional compressed air installations and special breathing gas mixtures are required. In addition, the permissible working time under compressed air reduces significantly with increasing air pressure.

There are recommended pressure ranges for working under compressed air, mixed breathing gas and saturated breathing gas (Figure 4-24). Work under normal compressed air is permitted up to 3.6 bar under the German compressed air regulations [25], whereas in the UK and large parts of the USA, the upper limit for the use of compressed air is 3.0 bar. Exemptions have been applied for and issued for single projects like the fourth tube of the Elbe Tunnel and the Weser Tunnel, allowing the use of compressed air at up to 5.0 bar in selected sections and under special conditions [132].

The use of mixed breathing gases (mixtures of helium and oxygen (Heliox) or helium, nitrogen and oxygen (Trimix)) can achieve somewhat longer working times than for normal compressed air. Mixed gases can therefore be used for short interventions, e.g. for inspection, at pressures of up to 8 bar, in exceptional cases up to 10 bar.

For longer repair work or for multiple tool changing, saturated mixed gases should be used at pressures of more than 4.5 bar. If saturation interventions last a long time, then a transfer shuttle and an accommodation chamber (Habitat) are also necessary [132].

The available working time reduces considerably with increasing pressure. A maximum decompression time of 2 hours is normally planned, and this should not be exceeded in order to limit the required time in the comparatively small man lock and the wearing of oxygen masks during decompression.

Depending on the level of groundwater pressure, the abrasiveness, the stand-up time of the ground and the length of the relevant tunnel section, shields for closed mode working (slurry or earth pressure balance support) should therefore have the necessary equipment installed for interventions under normal compressed air and mixed or saturated breathing air.

a)

b)
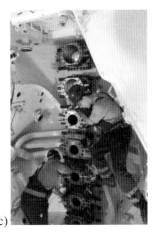
c)

Figure 4-25 Accessible cutting wheel of the Mixshield S-317-318, Chongming Shanghai, 2005
a) construction　　　　　　　　b) front view　　　　　　　　c) from inside

One alternative offered by the cutting wheels of very large diameter tunnelling machines is pressure-less tool changing, in which the enclosed inside of the arms can be maintained at atmospheric pressure for tool changing. The implementation of accessible arm cross-sections and the complete use of back-loading disc cutters like scrapers in combination with single locks for the tools to be changed was implemented for the first time on the "Fourth Tube of the Elbe Tunnel" project and further developed with the tunnelling machines for the Chongming project (Figure 4-25).

Undercutting technology

The principle of milling tools for a full-face machine was reintroduced in 1958 by Wohlmeyer [183]. A number of milling discs fitted with fixed chisels were attached to a cutting wheel. These have their own drives and rotate in the opposite direction to the cutting wheel and therefore describe cycloids. The thrust forces required for the millers are relatively low but the load on the tools is high, depending on rock conditions. The process is based on the multi-plate miller introduced in 1866 by Brunton, which was used for a trial under the English Channel.

A shield machine was equipped with a Wohlmeyer cutting wheel for the construction of the Seikan Tunnel in Japan (Figure 4-26).

Figure 4-26 Cutterhead according to the Wohlmeyer principle

Figure 4-27 Undercutting cutter head of the reamer machine for the Uetliberg Tunnel [33]

The latest application of the undercutting technology featured disc cutters arranged on sliders for the equipment of an enlargement cutter head to bore the Uetliberg Tunnel (Figure 4-27). The cutter head of the enlargement machine (reamer) consists of the basic structure of a cutter head and six cutting arms. The cutter head turns about an inner kelly braced in the section of the pilot tunnel. The disc cutters are displaced axially and radially to the tunnel axis on radially extendible sliders on the cutting arms. Each disc cutter describes a spiral track about the tunnel axis as the cutter head turns and the sliders are moved.

The course of the outer cutters in each case results in a stepped face, on which each disc cutter shears off the rock against a free area (undercutting principle). At the start of boring a so-called "round" (axial excavation length per radial stroke of the sliders), for example, the inner discs start in the pilot bore and the next discs outward in the step just bored by the inner cutter. The length of the round is limited by the maximum axial travel of the disc cutters on a slider, although shorter rounds may be bored depending on rock strength. When the six-armed cutter head with six disc cutters per arm is turned, 36 disc cutters are simultaneously rolled on six spiral tracks each displaced by 60° from one inner to one outer diameter.

After the design diameter has been reached, the sliders are retracted radially back to the smallest radius. Then the cutter head is pushed forward by one round (e.g. maximum 200 mm) and the next step begins until the advance permitted by the bracing is reached and the machine has to be regripped.

Because of the radially directed introduction of the disc thrust forces, the force components of the thrust are cancelled by the diametrically opposed arrangement of the boring arms. The low number of disc cutters (six per boring arm) with a contact pressure of about 100 to 120 kN/disc considerably reduces the torque required at the cutter head. Optimisation potential currently remains regarding the reduction of the mechanical, hydraulic and electrical components inside the cutter head. The advantages of the undercutting principle (lower cutting forces and thus lower energy consumption) would also be much more apparent if this machine concept was integrated into a shielded TBM with segmental lining, as the limitation of the advance rate through the installation of support measures from conventional tunnelling would thus be eliminated.

Bearings in the shield

The cutting wheel forms a constructional unit with the cutting wheel drive containing seals, main bearing and drive motors. The loads resulting from excavation (forces and torques) are resisted by the shield construction. The connection between the highly precise excavation machinery and the shield manufactured to relatively coarse tolerances should be made simple. Any longitudinal or tilting movement of the cutting wheel to achieve a one-sided overcut requires an elaborate connection through an articulation bearing.

The conventional bearing type is central shaft construction with bearings at the perimeter. An intermediate solution is offered by open-centre compact bearings (Figure 4-28).

The former version offers the advantage of the smaller diameter of roller bearings and seals, but requires a certain installation depth for guidance. Axial movement of the entire drive unit (e.g. to withdraw the cutting wheel from the face for tool changing) is simple to implement. The central shaft drive is typical for the spoked cutting wheels of the early Hydroshields.

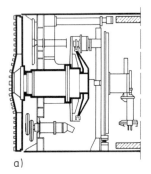 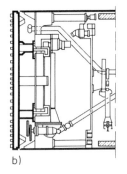 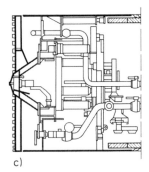

a)   b)   c)

Figure 4-28 Construction types of cutting wheel bearing, in this example a slurry shield (Mitsubishi)
a) Central shaft bearing
b) Perimeter bearing
c) Open-centre compact bearing

Bearings at the perimeter make axial movement almost impossible, but this construction type enables the installation of much greater cutting wheel torque. Perimeter bearings are used in the rim or drum cutting wheel.

In an intermediate solution compared to those listed above, the main bearing diameter in the centre leaves sufficient space for the integration of special centre flushing or muck removal equipment but is small enough to provide for axial movement of the excavation machinery.

The three bearing solutions are shown in Figure 4-28 and in Figure 11-15 for the specific requirements of an earth pressure shield.

One special case is the bearing of the so-called Mixshield type (Chapters 10, 12). The excavation machinery block is located in an articulated bearing/axial displacement bearing construction (Figure 4-29 ⑥). Three groups of supporting cylinders ⑤ bear the axial forces and permit axial displacement and controlled tilting movement through control of the stroke. This achieves one-sided overcutting in any direction in a very elegant way. Torque support cylinders arranged at the side of the cutting wheel drive resist the torque resulting from the rotation and excavation process.

In the bearing housing, the main bearings ③ and the drive pinion run in an oil bath. The drive motors ④ with flanged gears seal the housing shut at the back. The cutting wheel connection ① projecting at the front has to be sealed reliably against the entry of soil particles or the exit of bearing oil by a multi-stage seal ②.

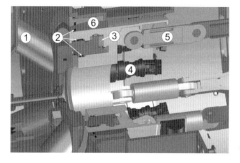

Figure 4-29 Centre-free bearing of the cutting wheel in the Mixshield (Herrenknecht)

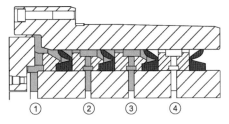

Figure 4-30 Basic construction of a four-row sealing system (Herrenknecht)
1) HBW sealing medium 2) grease
3) oil 4) leakage chamber

The constructional complexity of a sealing system increases with the requirements from the excavation side (dry, wet, under pressure). In closed mode shields (slurry and earth pressure shields), there is often a mechanical labyrinth and behind that a multiple arrangement of sealing profiles with intermediate chambers for sealing media (HBW, environmentally acceptable grease and/or oil), which are kept under constant pressure (Figure 4-30). Higher relative speeds and greater deformations of the edges of the seals are the main disadvantages of the large bearing type compared to the central shaft. Open access to the excavation chamber in the centre of the bearing also demands a second, internal seal. Particularly in soils with highly abrasive content, the sealing system is a component susceptible to problems for construction types b and c in Figure 4-28.

Even with the most meticulous design and construction of this component of the machine, constant checking in operation and the possibility of changing a seal in the tunnel are essential.

Cutting wheel drive

In machines for loose ground, the electric drive is taking over from the hydraulic drive prevalent until now. The characteristics of modern electric drives are higher efficiency due to the omission of the intermediate step of converting the electrical energy into an oil flow, the lack of waste heat, the protection of the drive motors with hydraulic clutches (Safe-Sets) or slipping clutches and the adaptation of the torque-revolution speed curve through the use of frequency converters. The hydraulic drive is the most compact type and remains in use for the installation of high installed power in the minimum of space. Other advantages of hydraulic drive are controllability and not being susceptible to overloading.

The operational revolution speed of the cutting wheel / cutter head is designed to be as slow as possible for the achievable penetration, considering that tool wear increases over-proportionately with speed. Different ranges of revolution speed are chosen according to the type of face support and the machine diameter. Pure slurry shield machines, for example, rarely exceed three cutting wheel revolutions per minute. Because of the lack of flywheel weight compared to tunnel boring machines, changes of resistance, and thus also of torque, are transferred to the drive with almost no damping. Earth pressure shields are operated at a rather higher range of revolutions, while the operator of a shielded tunnel boring machine in hard rock will try to achieve the maximum cutter head revolution speed (limited by the permissible maximum speed of the outermost gauge cutter of 190 m/min) in order to maximise the advance rate (penetration x revolutions). The flywheel moment of the heavy cutter head evens out most load variation.

Even if only one direction of rotation is recommended to ensure the optimal function of cutting tools and material flow, rotation in both directions should also be possible to allow for unforeseen situations.

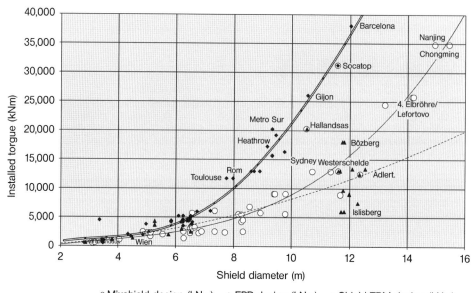

Figure 4-31 Relationship between shield diameter and installed torque (Herrenknecht)

The magnitude of the torque to be installed at the cutter head is essentially determined only by the excavation process for wheels with a pure tool carrier function. For rim and closed wheels, the share of friction increases very strongly, and one-sided spoil clearance and the ground-bearing friction are further factors that are not to be neglected [171].

In general, only a third to a half of the maximum torque of a power-regulated hydraulic drive is available at the maximum revolution speed. Increasing resistance at the cutting wheel in this case reduces the revolution speed automatically until the wheel jams.

Figure 4-31 gives a general relationship between shield diameter and installed torque for Mixshields, EPB shields and TBMs. The higher torque required by earth pressure shields compared to other types is shown clearly.

Cutting profile and overcut

If the shield turns on a central bearing, then the cutting wheel will produce a circular cavity centred on the shield. The diameter of the cavity is determined by the gauge tools at the perimeter of the cutting wheel.

Extended gauge tools will produce an intentional overcut (outside the shield diameter). In soil with solid obstructions, this overcut must be provided to protect the shield blade. Moving the drive to one side or tilting of the cutting wheel will produce a one-sided overcut eccentric to the shield.

The same effect can be achieved in some cases by a fixed excavation mechanism installed in the shield without the possibility of adjustment and extending copy cutters controlled according to rotation angle. Overcut reduces the thrust forces of the shield and if one-sided, aids steering movement in that direction. The unfavourable effect on soil support around the shield and the associated soil movements up to ground level should, however, not be forgotten (Chapter 2).

4.2.5 Hydraulic excavation

Excavation by liquid jets, which is a proven technology in opencast mining, was already included by Greathead in a shield excavation patent of 1874 (Figure 1-22). Attempts have been made ever since to excavate the face in loose ground with forward-facing, fixed fluid jets, but no durably successful application has been possible due to the uncontrolled effect, even in the latest application in the Hydrojet shield.

Flushing jet as an aid to excavation

This method has proved useful to a certain extent as an aid to excavation in shields with platform support or for the sinking of caissons, in which the sliding of the foot of the slope is supported by hand-guided water jets.

Excavation by high-pressure water

The destructive effect of high-pressure water jets (up to approx. 600 bar) on internal structure can be used to remove local obstructions (tunnelling through consolidated sections, starting and reception in shafts, cutting through piles) [50]. In order to support the considerable recoil forces, the jet must be mechanically guided in order to avoid accidents.

The attempt to improve the cutting function of chisels and discs with the assistance of high-pressure water jets has only been extended to a few applications [183]. The feeding of the high-pressure fluid to the rotating cutting wheel has proved very problematic. As an example, Figure 4-32 shows a milling head of a road header equipped with high-pressure water jetting.

Medium-pressure fluid jets are, however, often used in soils susceptible to sticking to cut the centre free and clean tools of adhesive material.

4.2.6 Alternative excavation processes

The excavation methods described here have their justification locally for shield tunnelling under special geological conditions, although these remain unusual.

Figure 4-32 Milling head of a road header equipped with round-shafted chisels and high-pressure water jets

Drilling, blasting, sawing

Drilling and blasting are used in shield tunnelling to overcome localised sections of hard rock, which the tunnelling machinery has not been designed to cope with (for cost reasons).

Open shields without pressure chamber generally enable the use of drills from inside the shield. In the restricted space of excavation chambers, only hand-held drills can normally be used.

Blasting is often used to loosen rock in order to be able continue excavation and muck clearance with the available machinery.

If sufficient time is available (e.g. night shift without tunnelling), a non-explosive expansion agent can be used to break out the drilled rock by expansion while the machine is stopped.

In order to effectively damp the propagation of blasting vibration in the ground, advance trimming blasting can be undertaken with weak charges (weak blasting). The propagation of vibration is damped still more effectively by a profile saw cut as deep as the round. (Figure 4-33).

If the saw cut (in temporarily stable ground) is then concreted, the subsequent blasting can be undertaken under the protection of the resulting vault (Pre-Voute process). Individual obstructions can be removed by drilling and mechanical splitting with a rock splitter.

Shuttle excavation

Shuttle operation of circular cutting wheels is sometimes practised. The frequent alternations of the rotation direction of the wheel do, however, lead to much greater disturbance to the structure of the ground than a rotating drive.

A shuttle excavation machine in a rectangular shield of Japanese production has become well known. An excavation comb swinging backwards and forwards about a vertical axis excavates the soil with teeth fitted at the front.

Application has so far been restricted to small shield dimensions and uncritical ground conditions.

Figure 4-33 Profile saw on its guide frame (Sipremec)

5 Muck removal

The muck removal system in shield tunnelling machines undertakes the transport of the material excavated from the face to above ground. According to the type of shield used, the muck can have very different consistencies and thus require quite different transport equipment:

- dry removal at natural or low water content for open shields and compressed air shields, also for earth pressure shields operating in open or compressed air mode,
- slurry removal of muck suspended in the support medium for slurry shields,
- high density slurry removal of muck remoulded plastically with the addition of water or conditioning agent for earth pressure shields.

There can be no explicit assignment of shield type to muck removal system or vice versa. With the addition of water or conditioning agent, the muck from the excavation chamber of an open or compressed air shield can be made plastically workable or also removed as a suspension, when the specific advantages of this transport method predominate.

Occupational safety aspects of muck removal are dealt with in Chapter 16.

5.1 Preparation for transport

The equipment used along the transport route places various requirements on the material being transported. The essential parameters are, in addition to the grading distribution, the deformability (consistency) and the uniformity (homogeneity) of the excavated muck.

If the excavation tools have not already produced the required properties, special equipment will have to be provided for processing. Some examples are:

- hydraulic hammers to break up obstructions in the face,
- injection nozzles for fluids and flow agents into the excavation chamber,
- mixing and kneading arms on the cutting wheel and in the excavation chamber,
- roller crusher/stone crusher before the intake to the transport system (Figure 5-1).

The design and dimensioning of devices working under such extreme conditions requires great experience. The continuity of transport flow during tunnelling must be maintained at all times.

5.2 Removal from the face

Picking up the material coming from the face and feeding it to a continuous transport system is often the cause of hindrance to tunnelling in changeable ground conditions.

5 Muck removal

a)

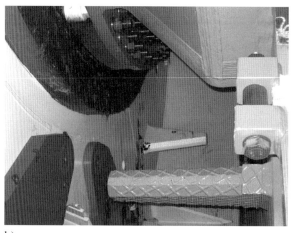

b)

c)

Figure 5-1 Equipment for the preparation of the excavated soil
a) Hydraulic demolition hammer
b) Mixing and kneading arms in the excavation chamber
c) Roller crusher (Herrenknecht)

5.2.1 Open shield machines

The open access to the excavation front in open shield machines makes the design of muck removal equipment easier. If an excavator is used, this can also deliver the material to the next stage of transport, in this case normally a conveyor belt (Figure 5-2).

When the invert is sufficiently wide, loading equipment ("lobster claw" gathering arms, scraper conveyors) collects the material falling to the side to the central extraction conveyor. With full-face excavation, the cutter head of the shielded TBM lifts the muck with buckets and mucking chutes into a muck ring, which feeds the material onto the conveyor belt. Operation of the cutter head in the other direction to compensate rolling of the shield is theoretically possible, but the muck removal system in the cutter head only works in one direction to make the best use of the available space. Blockage or sticking of the muck chutes by excavated muck in combination with water ingress is the most common cause of stoppages. Depending on the properties of the ground, the choice is between a reduced number of muck chutes of open form, large and thus easily accessible, or alternatively a cutter head construction with as many muck chutes as possible, which evens out the material flow and nears the capacity limit of the subsequent conveyor belt (Figure 5-3).

5.2.2 Shield machines with pressure chamber

Pressurised support of the face now only requires the entire tunnel to be under pressure in exceptional cases, even with compressed air operation. Pressurisation is normally restricted to the excavation chamber. The transport of the muck through the pressure bulkhead behind the excavation chamber also has to cope with the pressure differential [271].

Dry transport

The normal solution is a conveyor belt or chain conveyor enclosed in an encapsulated housing, which feeds the material into a continuously or discontinuously operating pressure lock system (Figure 5-4).

Figure 5-2 Transfer of excavated material by loading excavator onto a conveyor belt in an open shield, Hagenholz Ost 1975/76

5 Muck removal

Figure 5-3 Cutterhead of Mixshield, Finne Tunnel (top): 6 buckets + main transport along the outer steel construction of the 6 main arms, 2008; cutter head of shield TBM, Zürich-Thalwil (bottom) with 16 perimeter buckets, 8 cone buckets and 4 back buckets, 1998 (Herrenknecht)

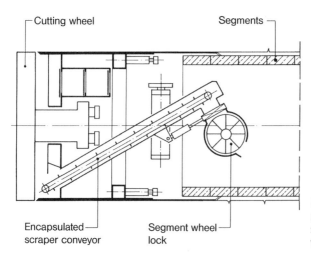

Figure 5-4 Transport out of the pressurised space with encapsulated scraper conveyor and segment wheel lock

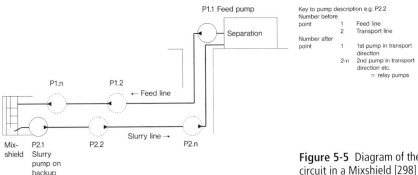

Figure 5-5 Diagram of the slurry circuit in a Mixshield [298]

The encapsulated pressure housing can be continued to the muck cars, which then have to be monitored as pressure vessels. Dry muck requires little preparation for removal apart from the limitations of grain size demanded by the conveyor and the lock.

Hydraulic transport

The removal of muck as a suspension in a fluid, which can be pumped by a centrifugal pump, is the most elegant and space-saving but also the most expensive transport solution to overcome pressure differences. The support medium also functions as transport medium. After the slurry has been pumped to a separation plant for the removal of suspended solids, it is pumped back to the shield machine in a circuit (Figure 5-5). If the pump capacity is exceeded due to the advance of the tunnel, then relay pumps can be installed in the shaft or the tunnel to raise the pressure.

The material excavated by the cutting wheel settles, at a speed depending on the grain size, into the fluid rotating slowly with the cutting wheel.

An intake screen protects the transport system from dangerously large grain sizes. The material held on the screen is broken to a suitable size for transport by crushers in the excavation chamber.

The effective tool for crushing stones and blocks has proved to be a jaw crusher mounted in front of the screen. The edge length of the stones produced by this crusher varies according to the type of crusher and the shield diameter between 400 and 1200 mm. Particular attention has to be paid to the fixing of the crusher, the distance between screen and crusher and the size of the opening to the screen past the crusher in order to avoid pressure fluctuations in the slurry line.

For tunnels in clay soil, where a continuous flow of material out of the excavation chamber is important to avoid unnecessary pressure fluctuations at the face, roller crushers are used. Two intake rollers with an active feed function are positioned in front of the suction intakes. These are also capable of breaking up any clay lumps to a size suitable for transport. The active feed function avoids clumping of clay lumps directly at the intake, which effectively prevents the vacuum cleaner blockage effect observed in shield tunnelling in clay without active feed.

In addition to the types of crusher already described, the cone crusher used in small diameters should also be mentioned. The excavated material is broken between welded crusher bars between the cone-shaped back of the cutting wheel and the tapered suction intake, before it can pass through the circular screen opening. This system has no active feed function and has to be fed from outside by an additional mechanical feeding device.

Various types of crushers are common depending on use (Figure 5-6).

Stone traps or crushers located in the slurry line have no significant effect in protecting against oversize material because the pipework and valves inside the shield machine cannot be made with much larger diameter than the following pipework in order to ensure that similar transport speeds are maintained.

The design of the cutting wheel and excavation chamber for the hydraulic transport from the face to the intake of the suction pipe requires a lot of experience. The low flow speed of the fluid in the excavation chamber in larger machines can lead to problems. Soil building up in dead corners has to be avoided as well as encrustation of cohesive soil on the surface of the cutting wheel or in front of the screen. Directional flushing jets fed by the suspension coming back from the separation plant are an aid to undisturbed operation. Whereas the classic design of machines for operation in sand features the almost complete feeding of the bentonite suspension at the back of the excavation chamber, machines for operation in clay are designed with swivelling bentonite jets and the feeding of part of the suspension through a rotating rotary coupling into the centre of the rotating cutting wheel.

The state of the technology for Mixshields being built today is to split the feed flow to the following outlets:

– flushing outlet in the upper part of the submerged wall,
– cutting wheel centre along the central ridge,
– cutting arm flushing,
– flushing of the cone in front of the submerged wall to the left and right of the opening,
– bentonite flushing jets left and right of the screen or the crusher,
– crusher flushing.

a)

b)

c)

Figure 5-6 Installation location of various types of crusher in the excavation chambers of slurry shields
a) Jaw crusher in front of the intake screen
b) Roller crusher
c) Cone crusher

Depending on the shield diameter and the specific requirements on a project, not all of these outlets will be used, as the required fittings and pipework take up space in the shield and make access forward more difficult. The various possible flushing locations are exploited by shield operators to accelerate the settling process of the excavated muck while taking the direction of rotation of the cutting wheel into account. Figure 5-7 shows an example of the basic distribution of the suspension feeds into the excavation and working chamber of a 10 m Mixshield.

The design of the excavation chamber in Mixshields should assist the transport process as far as possible. Flat surfaces of the shield blade and the submerged wall, the partial cladding of the cone in the invert, the submerged wall openings that are smaller nowadays, cone flushing openings next to the submerged wall opening and the appropriate arrangement of back buckets (Figure 5-8) on the back of the cutting wheel contour have all proven successful.

When the ground is excavated by a roadheader, the cutting head combines excavation and grading functions, so that no oversized pieces can get into the suction pipe ending at the cutter bit (Figure 4-17).

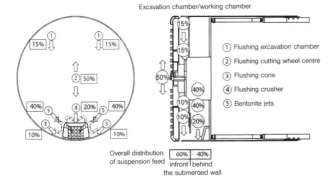

Figure 5-7 Distribution of the suspension feeds in the excavation and working chamber of a Mixshield [298]

Figure 5-8 Back buckets on the back of the cutting wheel of a Mixshield (Herrenknecht)

Figure 5-9 Screw conveyor (retracted into the outer pipe) of an EPB shield (Herrenknecht)

High density slurry transport

In the simple construction of a blind shield, the plug of soil punched out by the shield blade is extruded through one or more openings in the pressure bulkhead under the support pressure. The muck emerges as a continuous sausage and generally falls onto a conveyor belt for further transport (Figure 4-11).

The soil consistency and homogeneity demanded by a blind shield are created in a controlled manner in an EPB shield, with the possible addition of fluid and foam (Chapter 11).

The plastic remoulded soil must be transported by the support pressure to the intake to the transport system, as this cannot apply suction (Figure 11-6).

Screw conveyors (Figure 5-9) have proved successful for overcoming the pressure step into the tunnel. The diameter of these has grown with increased capacity up to 1,400 mm.

The revolution speed of the hydraulically driven screw conveyor drives can be controlled in order to influence the support pressure in the excavation chamber. When the pressure differential is large, two screw conveyors can be mounted in series in order to achieve secure sealing in the intermediate soil plug (Figure 11-17). Injection nozzles in the outer pipe for the addition of lubricants like bentonite or foam and the shut-off gate at the outlet are standard equipment. A number of maintenance hatches for the inspection of the flight, the fitting of additional hardened metal strips on the inside and even a split construction of the conveyor screw casing are further possibilities for specific projects.

Under low loading pressure, the front end must project into the excavation chamber by at least two turns. Retractable screws in telescopic outer pipes have proved useful for maintenance purposes. The intake opening in the pressure bulkhead can then be closed and the conveyor pipe opened without pressure using the conditioned soil in the excavation chamber for sealing.

The pitch is generally only about 50 % of the screw diameter. This increases the sealing effect of the soil in the screw conveyor. This gives for screw conveyors with a central shaft a maximum grain size that can pass through of about 40 % of the internal diameter.

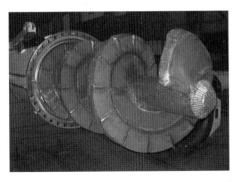 

Figure 5-10 Screw with and without central shaft

In order to increase this maximum size in soil containing a lot of stones, screw conveyors have been built without a central shaft (Figure 5-10). The construction of the screw is weaker but permits a change of direction of rotation in case of an exceptional blockage in the pumping direction, as the screw then immediately jams in the outer pipe. Due to the use of disc cutters in the cutting wheel to break boulders while they are still bedded into the face, screw conveyors are now almost always constructed with a central shaft. The integration of stone traps can also be omitted.

The slippage of the material being transported through the screw conveyor is very variable depending on the pressure difference, conveyor gradient, screw geometry, soil properties and lubrication. It can be about 50 % for conveyors at 30°, without pressure difference and a screw diameter of 1,100 mm, i.e. the soil moves half as fast as the actual screw motion. Larger diameter screws are worse in this respect.

As mentioned above, the use of a crusher in the full excavation chamber of an earth pressure shield is not possible, so the only remaining effective measure against stones and blocks is the fitting of the cutting wheel with disc cutters.

5.3 Transport along the tunnel and up shafts

The choice of equipment for transport in tunnels and shafts is determined by the method used to bring the muck out of the excavation chamber, but also by economic factors. The available machinery, traffic control in the tunnel, traffic safety, possible obstruction by construction work in the tunnel and the problem of ventilation all have to be considered.

5.3.1 Open transport

Transport along the tunnel to above ground is the same whether the muck is cleared from the face as dry material or high density slurry.

Transport along the tunnel from the machine conveyor or the discharge of the screw conveyor is normally on a conveyor belt to the shaft. Even in tunnels up to 6 km long and with curves, modern conveyor belt systems have proved capable of coping with curves and need just one drive station. The length is extended continually until the belt stored in the belt cassette at the tensioning and drive station is exhausted. Then a further section of belt is pulled in and the joints are vulcanised. The support construction in the tunnel is continually erected with the advance of the tunnel (Figure 5-11).

Only in exceptional cases is transport along the tunnel in wheeled muck dumpers or on rails (Figure 5-12).

These forms of transport in the tunnel are only needed today for the supply of tunnelling operations with segments, grout for filling the annular gap and pipes for the extension of supply pipework. The frequently encountered gradients of more than 4 % (the limit for rail transport) can be coped with without problems by wheeled vehicles, called Multi-Service Vehicles (MSVs, Figure 5-13). These vehicles also assist flexibility in the construction site facilities area, not requiring the extensive track layout and marshalling areas, which were formally typical on major tunnel projects.

When tunnel construction is accessed by a shaft at the end of the tunnel, dry material is normally transferred once again from a tipping point in the shaft and lifted by crane to an intermediate stockpile above ground or the transport skips are handled with portal crane and unloading traverse.

Instead of a crane with skips or grabber operation, continuously operating bucket conveyors are sometimes used, which cannot however be used for anything else on the construction site. There is also a danger, particularly with sticky muck, that the material remains in the buckets at the top and blocks the transport system. If the tunnel starts from a cutting, then transport can continue to an (intermediate) tip, which ensures effective shield tunnelling without disturbance from the rebuilding of the muck transport system. The objective of high-capacity tunnelling with conveyor belt transport is therefore either to erect an inclined band in the shaft, if the shaft is long enough, or in an additional small shaft or cutting, in both cases running to a tipping point (or landfill site) above ground.

5.3.2 Piped transport

Hydraulic transport (Figure 5-14) offers high capacity in a very small cross-section. When a tunnel is long or the cross-section is so small that muck trains cannot pass each other, this method has first enabled respectable advance rates. Hydraulic muck transport does, however, require a second means of transport for construction materials [121].

Figure 5-11 Installation of a conveyor belt support construction in the backup during tunnelling (left), horizontal belt cassette after the transfer to the tipping belt at the portal (right), Finne Tunnel 2009

Figure 5-12 Muck transport in the tunnel on mucking trains with loco (top), handling of the skips in the shaft with portal crane and unloading traverse (bottom and right), Wientalsammler 2005

The slurry is pumped by a revolution speed-controlled pump behind the shield through telescopic pipes (or flexible hoses) to the pipework installed in the tunnel.

Changes of direction at the shaft are simple to deal with, and it is also possible to continue for great distances above ground. Hydraulic transport does, however, have considerably higher energy consumption than dry transport, i.e. higher operating costs.

A second pipeline runs from the separation plant to carry the suspension, after cleaning and if required replenishment, back to the shield machine (Figure 5-5). If the transport distance exceeds the capacity of one pump, then relay pumps are installed. Centrifugal

Figure 5-13 Multi-service vehicles for the supply of tunnelling operations with segments, pipes, consumables etc. (Techni Metal), Socatop 2006

Figure 5-14 Centrifugal pump for hydraulic muck transport

Figure 5-15 Piston pump for muck transport in the tunnel, Emisor Oriente, Mexico, 2009

Figure 5-16 Muck emerging from the pipe in high density slurry transport

pumps with special impellers can pass stones up to about 60 % (in exceptional cases up to 80 %) of the diameter of their pressure-side connection without a significant reduction of efficiency. Slurries can be pumped reliably up to peak density values of 1.45 t/m^3, i.e. with a solid matter weight of approx. 720 kg/m^3 of suspension. Because of the sometimes very uneven feed from the excavation chamber, average solid matter transport has to be estimated with lower values (approx. 1.2 t/m³ corresponding to 325 kg/m³).

A flow speed of more than 3 m/s should be maintained in the slurry line [298], in order to avoid blockages caused by settling of the suspended muck. On the other hand, the flow speed should also be as low as possible to minimise wear in the pipework. Water as a transport medium requires higher flow speeds compared to, for example, a bentonite suspension with higher carrying capacity. The curves of pump or pipe performance required for the design of the pipeline can be calculated or taken from nomograms [161].

Higher densities are also sometimes pumped in tunnels and above ground, but only when the muck still has pronounced fluid properties [84]. In this case, piston pumps are used (Figure 5-15).

The material has to be prepared for pumping very carefully. Less transport medium has to be moved backwards and forwards than with hydraulic transport, but the expensive machinery required is only worth it if the flowing sludge, which is not suitable as backfill, can be pumped into a pit or similar without separation or further treatment (Figure 5-16).

5.4 Quantity determination and measuring equipment

The desire to determine the quantity of soil removed from the face by measuring the amount of muck (volume, weight and density) has led to the increasing installation of measuring equipment in transport systems.

Volume measurement (open transport):

The counting of the filled muck cars is a commonly used method in tunnelling with open shields or EPB shields. The cars are normally filled up to a line marked on the inside. The bulking factor (tipped bulk density), the effect of swelling and the conditioning of the soil in earth pressure machines all influence the results, but can be estimated through a careful analysis of the ring values in uniform geology. The result is first known at the end of the advance stroke.

When a piston pump is installed as a closure at the discharge from the screw conveyor, the required volume can be calculated from the number of plunger strokes during the advance, as long as the pumping capacity of the piston pump can be used completely for the transport process. The pumping of the pump is dependent on the supply from the screw conveyor, the soil properties and the pressure conditions.

When the discharge conveyor is uniformly filled, the surface of the loaded cross-section can be detected optically or acoustically and used to calculate the transported cross-section, which together with the belt speed gives the transported volume (Figure 11-7). Very variable surfaces, for example stones or clumping, give false results.

Weight measurement with a belt scales (open transport)

The transport belt in the tunnel (or the discharge belt after hydraulic transport) is run over belt scales and the weight (belt load) of the muck is determined as a mass flow. The material transported in open muck cars or skips can also be weighed; the residual muck in the wagon has to be taken into account.

Volume and/or mass balance by measuring the flow and density (with hydraulic transport)

When the muck is transported in a pipeline, the flow speed and density of the transported material can be measured by magnetic induction and radiometric density measurement. From these two values, the transported mass can be calculated. The mass of the material excavated is calculated as the difference between the measurements in the feed and slurry line (Figure 5-18).

Critical evaluation of measurement equipment

All methods of measurement initially offer the tempting possibility of detecting impermissible extra excavation (stealthy soil removal). Digital displays of the results with any number of figures after the point tend to increase trust in the measurement equipment.

Under more detailed consideration, however, this verdict should be put into perspective:

– The advance of the cutting wheel or the alteration of the overcut during a stroke alter the resulting excavated quantity. The consideration of these factors would require yet more complicated measurement equipment.
– The measurement tolerance of equipment robust enough for construction site use can be up to 5 %. This corresponds to 5 cm more or less advance per metre of tunnel. Constant attention should be paid to the calibration of equipment, as for example the data from belt scales is more or less dependent on the belt tension of the extending conveyor belt according to installation location.
– Errors due to the method of measurement should be taken into account and minimised if possible. For example, the correct choice of the arrangement and orientation of radiometric density measurement equipment can considerably improve the quality of the results (Figure 5-17) [299].
– The largest source of errors is the soil mechanics behaviour of the excavated muck. When the volume is to be determined, the bulking factor cannot be known in advance (measured values lie between 1.3 and 2.0), and when the weight and density are to be measured, the consolidation of the undisturbed soil can only be estimated in a range of about 10 % from the site investigation. When the soil is bedded in strata with variable transitions, this can increase errors still further. A sensible procedure is to determine a range for the theoretical material in the relevant section of the tunnel from the data in the site investigation report.

On account of the factors described above, values measured in the tunnel are in practice either calculated and compared with a theoretical design value [31] or compared with the average of the results from the last 10 rings. This procedure changes the viewpoint from the various error factors to the finding of early information. In addition, the shield operator is given a current "ideal" value based on the current state of the advance (= current thrust cylinder stroke).

5.5 Separation

Hydraulic transport always requires separation of the transported solids from the transport medium at the end of the pipeline. In shield tunnelling, this is the excavated muck and as the transport medium water or (bentonite) suspension. The quantities of the slurry to be separated are between about 50 m³/h for very small tunnels in fine-grained soil and up to 2,800 m³/h for very large diameter shields.

Transported solid materials, which under unfavourable conditions can exceed more than 30 % of the nominal advance rate, are possible and should be included in the dimensioning of screen areas as a buffer. Large variations of the quantities arriving, which at worst can result in the formation of a plug, should be taken into account in the design of the separation plant. The average density pumped from a shield will often reach values of about 1.3 t/m³. This value can be expected to vary downward and will sometimes inevitably be exceeded. Geological properties, like close grading, which could result in too little load or under some circumstances too much load affecting only certain components, also have to be considered.

It is now technically possible to separate any soil from the carrier fluid suspension or water on the construction site. But purchasing and operating costs prevent the complete separation of the transport medium from the soil and demand a compromise between the handling of the suspension loaded with fine material and disposal and tipping.

  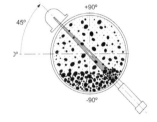

Figure 5-17 Suitable installation location of a vertical radiometric density-measuring device in a shaft and on the tunnelling machine (if horizontal, than tilted upwards at 45°) [299]

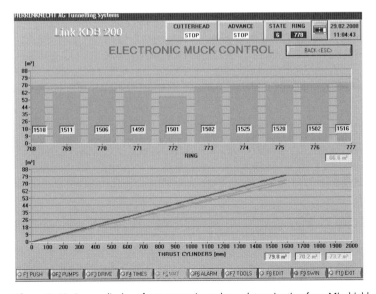

Figure 5-18 Screen display of a comparative volume determination for a Mixshield using flow measurement

5.5.1 Separating process

The practically applicable separation processes are sedimentation and filtration. Thermal separation by evaporating the fluid is too expensive and is not used.

Sedimentation

Sedimentation denotes the settling of the transported material in the transport fluid. The sinking speed of the transported material can be determined using the Stokes resistance formula [175]. The flow limit and viscosity of the transport medium on the one hand and the size and density of the soil particles on the other are the essential parameters, which have an influence. Increasing viscosity of the transport medium leads to reduced sink speeds and increasing particle size and density lead to increasing sink speed.

As the mineral grain density of soils only varies slightly, the sink speed shows very similar values in very different soils.

All soil pores are still filled with fluid after settling, so there are different water contents in the settled solids depending on grain shape and consolidation.

Filtration

The loaded suspension is forced through a filter with a defined size of passing grains. During this process, the passages through the filter are increasingly narrowed by accretion of suspended material. The grain size and shape determine the separation effect and both characteristics can vary widely in different soils. After the formation of a waterproof filter cake, the filter element has to be cleaned before further use. The filter performance and the residual water content of the cake can be influenced by the pressure applied.

5.5.2 Separating devices

Complete removal of the fine particles from the suspension is difficult but not necessary in most cases. In transport media, which also provide a support function at the face, a content of suspended particles capable of forming a filter in the suspension fed back into the tunnel is actually desirable. But the content of fine material, which essentially determines the flow limit of the suspension, should not increase too much in order not to impair the performance of the separating equipment [5], [218].

Settlement basins (Figure 5-19) are reliable and energy-saving facilities when pure water is used as transport medium and there is sufficient space. They are used in micro-tunnelling. Bentonite contents of only a few kg per m³ of fluid worsen the settlement process considerably. The residual drainage of the settled fine material can be troublesome when it is cleared. Winter operation is scarcely possible due to frost.

Regenerators

Regenerators are becoming increasingly popular in restricted construction site areas and for increasing fine grain content in the transport flow. They are offered as made-to-order compact plants and mostly mounted on the suspension basins needed for operation (Figure 5-20). They consist of the components:

5.5 Separation

Figure 5-19 Settlement basin in container format as intermediate tip in micro-tunnelling, pipe jacking, Witten

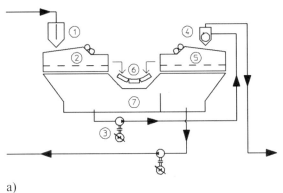

① Impact box
② Preliminary screen
③ Cyclone feed pump
④ Hydrocyclone ø 250 mm
⑤ Dewatering sieve
⑥ Discharge belt
⑦ Suspension collecting tank

a)

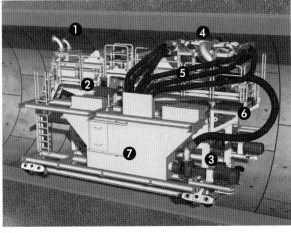

b)

Figure 5-20 Separating plant with preliminary screen and simple cyclone stage with drainage screen for 400 m3/h of slurry flow, mounted on a backup
a) Schematic diagram
b) Plant in situ, S-502 Lake Mead 2009

- preliminary screen,
- one- or two-stage cyclone plant,
- dewaterer.

The entire plant is normally installed in a housing if there is a risk of frost and to comply with noise emissions requirements. Construction in container format makes the plant easy to transport and assemble.

The **preliminary screen** as vibrating screen (also described as a coarse screen) serves to separate the very coarse material like coarse gravel or stones (grain sizes larger than 3 to 5 mm) and protect the subsequent equipment from damage. In general, a preliminary screen with a very large screen area is used, or a bar sizer in the impact box. The purpose of protective screening is to remove coarse material down to a largest grain size, which is small enough to avoid blockages in the subsequent coarse grain screen. The performance of the protective screen plant should therefore be dimensioned for the full volume flow of the slurry circuit.

Screen facilities are equipped with linear vibrating screens with double motors in order to keep the screens in action constantly. The oscillation of the vibrating screen therefore has to overcome the high viscosity of the solid material through the adhering bentonite content. Practical experience has shown that the removal of the solid material can be optimised at the smallest possible water content by higher energy inputs in the magnitude of 5 g (= five times acceleration due to gravity) and higher. The increase of the energy input and the screen discharge output is also associated with an increase of the transport speed of the screened material and thus a reduction of the stay time on the screen. Special polyurethane screens, synthetic coatings with wide screen openings or coatings with vertical drainage bars and screen catch basins in double-decker screen machines for the protection of the drainage screen underneath are some of the constructional possibilities to increase the size or performance of screening plants.

Curved screen areas offer a higher capacity per unit area for fine grain separation than flat screens. Cohesive clays are separated by the preliminary screen in the form of pellets. When the clay content is high and the screen area is barely adequate, the screen will tend to become blocked. The fluid then flows over the layer of clay and into the outlet. This effect can generally be countered with a simple stepped screen deck and possibly the alteration of the imbalance frequency or the screen coating.

Hydrocyclones work like centrifuges with a fixed rotor [284]. Due to the lack of a drainage run, they have a grading function. Since the centrifugal energy is provided by the pump, the fluid throughput is coupled closely to the cyclone diameter and the loading pressure. If single cyclones are connected in parallel, the total throughput can be adapted to the circulation quantity. Solid material tends to fall outwards in the gravity field and emerges much more concentrated at the bottom in the tailwater or the tailwater pocket, whereas the cleaned suspension is fed through the headwater.

The cyclones are fed by pumps from a collection basin, which is connected behind the protection screen. The pumps are thus independent of the volume flowing through the slurry circuit and can be dimensioned to suit the hydrocyclones.

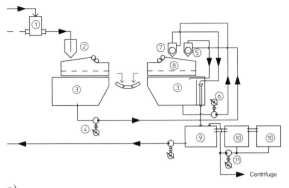

1: Flushing distributor
2: Impact box with preliminary screen
3: Tailwater collection basin
4: Coarse cyclone feed pumps
5: Coarse cyclone
6: Fine cyclone feed pumps
7: Fine cyclone
8: Drainage screen
9: Travelling basin
10: Fresh bentonite basin
11: Feed pump

a)

b)

Figure 5-21 Two-stage cyclone plant with two-stage preliminary and drainage screens
a) Schematic diagram
b) Plant in operation, Inntal crossing H3-4, Radfeld-Wiesing, 2007

Coarse cyclones separate according to exact type up to a threshold of **70 to 150 µm**. The threshold defines the minimum grain size, which the cyclone separates out of the cyclone tailwater.

Single-stage plants can still produce satisfactory results in medium sand.

Single-stage plants are increasingly being used to separate the solid material from the additional water discharge of the conveyor belt muck of open shields in solid rock. The conveyor belt coming from the cutter head in a trough is turned upward, and the water carried with the muck including solids spills over the sides into the drainage trough underneath. In a separation plant installed on the backup, the solids are separated and fed by an intermediate conveyor to the main conveyor (machine or tunnel conveyor). The separated water is drained into the tunnel drainage.

In two-stage plants, the threshold can be lowered into the coarse silt range.

For the second stage, small-diameter cyclones can be used, in which the higher radial acceleration reduces the threshold down to a range of 35 μm or less. The essential difference to a single-stage plant is however in the more even loading of solids in the second stage, since the cyclones largely even out fluctuations in the pumped density. Depending on the geology, fine stages are currently equipped with multi-cyclones with a diameter between 75 and 125 mm, arranged as a hydrocyclone manifold in a block. The inner liners of the cyclones can often be exchanged, so that changes to the fine cyclone stage can be made with little effort, if for example the expected geology is not encountered or wear increases or the cyclone stage is overloaded.

Figure 5-21 shows a two-stage cyclone plant with preliminary and drainage screens for 2,800 m3/h of suspension throughput. Each cyclone group has its own dewaterer.

Figure 5-22 shows the results of separation tests with suspension from the same soil and same density. The diagram shows the grading curves in the headwater of cyclones of various diameters compared to the grading curve of the incoming suspension and the grading curve of the suspension in the centre of a centrifuge. For the operation of the circulation, the residual solids in the headwater are significant. The term "threshold", which is normally based on d_{50}, i.e. the grain size, which makes up 50 % of the weight of solids in the grading curve, does not make sense for flat grading curves. This should rather be defined with d_{90}.

In the practice, the definition d_{50} is generally used, as fines with such a grading curve are either separated directly with the separation of the finest material or, if possible, have already been separated as closed clumps on the preliminary screen.

In the tailwater of the cyclone, a fluid with a high solid concentration emerges, reaching loads of up to about 1,000 g/l. With approximately single-grained sands, it may be possible to omit a further drainage screen if a vacuum-controlled pocket is installed in the tailwater.

The idea for tunnels in very coarse-grained geology, considered and sometimes applied, to switch off the first cyclone stage to maintain the solids as stopping grains cannot prevent the expensive addition of further agents in all geological strata, but can reduce their quantity and the associated expense.

Flocculation agents are ineffective in a cyclone due to the great turbulence of the fluid.

Dewaterers should be installed to separate the tailwater with its increased solids concentration from the excess water. The device consists of oscillating gap screen mats with gaps between 0.2 and 0.5 (occasionally 0.7) mm, which prevents the formation of a solid cake at the upper edge.

Until the fine sand limit, a soil-water mixture can be dewatered well in this manner. An increasing content of fine and very fine grains, particularly bentonite, quickly blocks the forming soil filter. The dewaterer overflows.

If the coarse grain skeleton has grain diameters less than the dewatering slots, then the screen mat cannot hold any solids and the soil remains in circulation.

In comparison to the good separation performance of the cyclone, the dewatering screen is the weak point in a system if the fines content is high.

5.5 Separation

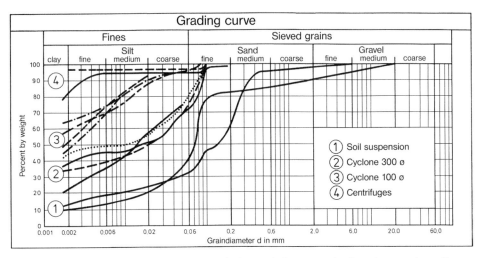

Figure 5-22 Grading curves of soil suspensions before and after separation in cyclones and centrifuges

Separation of very fine material

Various systems are in use for the very fine separation of silt contents, but they tend to have rather limited performance and be expensive with high energy consumption. Therefore these components are only operated in part of the flow. The main flow from the last cyclone stage is fed directly into the slurry circuit. Part of the flow is fed into the silt separation plant. These components are:

– the chamber filter press and
– the centrifuge.

The separation of the finest material is normally the part of the plant, which limits the advance rate of tunnels in ground with high fines content.

Filters

Filters in use today are almost exclusively chamber filter presses (Figure 5-23).

Figure 5-23 System MS chamber filter presses, overall view (left) and individual filter plates (right)

The chamber filter press is based on the system, that flushing from the fine cyclone stage is fed into the space between filter plates with filter mats. After the chamber filter is filled with flushings, it is pressurised and the solid phase is separated from the liquid phase over the filter element. The chamber pressure, the filter fabric and the duration determine the thickness of the filter cake and its solid content. The filter fabric normally provides a supporting function for the coarser solid particles, whereas the finest particles are held back by the growing filter cake. The chamber pressures of current plants are up to 16 bar. After the liquid phase has escaped, the single plates are pulled part and the solid particles are vibrated from the space by the application of compressed air to the filter plates.

Chamber filter presses work discontinuously and require a higher purchase price than continuous band filters. In comparison to band filter presses, they produce much drier filter cake. In comparison to centrifuges, the performance of chamber filter presses is slightly lower (5 to 7 t/h), but the quality of the product is much higher. When estimating a chamber filter press, conditioning with flocculation agent needs to be considered. Low filtration volumes, which is beneficial for the support function of the suspension, has the effect of greatly reducing the cleaning performance of filtration.

Centrifuges

Centrifuges work as a sedimentation process, with the difference that the solid particles do not settle slowly under gravity but are pressed into the outside of a drum under centrifugal force (Figure 5-24).

The drum, into which the suspension is fed, is accelerated to high revolutions (in operation: 1,500 min-1 to maximum 2,500 min-1). The solid particles are deposited on the outside, while the fluid phase, called the centrate, collects in the diameter of the drum. Because of the very great acceleration (2,500 to 4,000 g), sedimentation takes place much more quickly. In order that the solids do not fill up the drum, a screw conveyor transports the deposited solids to the end away from the feed. The drum is tapered to the solid discharge end in order that the centrate cannot be discharged. The fluid phase is discharged over an adjustable overflow weir.

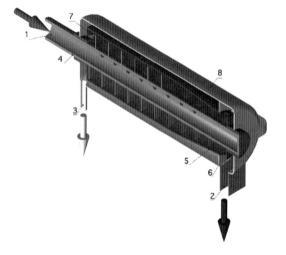

1: Intake pipe to feed the centrifuge with suspension
2: Ejection of solids
3: Discharge of centrate water
4: Rotating charging chamber
5: Rotating drum
6: Outlet for solids
7: Adjustable overflow edges (weir plates)
8: Centrifuge housing

Figure 5-24 Basic principle of a centrifuge

The separation capacity is adequate under construction site conditions to a grain size of about 5 μm. The correct setting of the operating conditions (revolution speed, flow) enables most of the bentonite contained in the suspension to be retained in the centrate. The addition of flocculation agent is unavoidable to achieve the discharge performance required from a centrifuge today (11 t/h).

For construction site use, the DC solid-bowl type has become accepted. Apart from the high purchase price of high-performance devices, the operating costs should also be considered.

5.6 Suitability of the muck for landfill

The problem of tipping excavated material mainly concerns the plastic muck from earth pressure balance shields and above all the products of separation. The residual water content ranges from about 20 % (dewatered medium sand) to 120 % (band filter cake with flocculation agent). Flocculation agents make the discharged soil seem drier than untreated material with the same or even lower water content, but this effect reduces in time.

The question of suitability for landfill is not precisely defined in soil mechanics. It is associated with ideas like

– whether heavy vehicles can drive on it,
– the formation of slopes,
– no bleeding of excess water etc.

An attempt to formulate a definition according to the parameters solid content and shear strength can be found in [194].

The grain fractions separated out from the hydraulic transport circulation in the preliminary screen and the cyclone stages are uncritical into the silt range.

The tipping of plastic muck material from EPB machines requires close control of the layer thickness.

In contrast, the material pumped as high density slurry and the product of band filter presses and centrifuges cannot be simply tipped. The material tends to flow plastically and even walking on the filled area can be dangerous. If the production of fines is low, it can be mixed together with the sand separated by the cyclone.

Consolidation through the addition of lime or storing in a permanently fenced pit are possible solutions, which would have to be investigated locally. The water-retaining effect of flocculants dissipates in a few weeks, which has to be borne in mind with regard to the crumbly appearance of filter cake. The product of chamber filter presses working at high pressure seems even drier, but the same consideration applies.

Suitability for tipping therefore has to be defined on a project-specific basis and investigated on site (Figure 5-25).

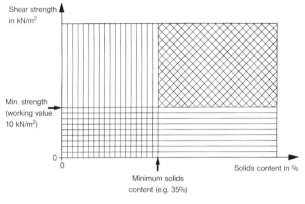

 Generally uncritical

 Critical due to solids content

 Critical zone
(not suitable for tipping)

Figure 5-25 Suitability for landfill, assessment based on the parameters solids content and shear strength [194]

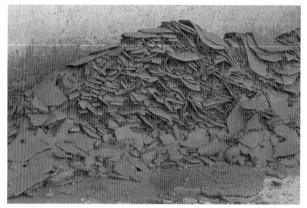

Figure 5-26 Filter cake produced by a chamber filter press, Lefortovo Tunnel

Figure 5-27 Product of fine soil separated by a centrifuge

CLEVER CONVEYING

Conveying powerful ideas.

H+E Logistik supplies customised conveyors and tunnel belt systems for all companies which have a lot to move. The smooth transport and distribution of excavated, raw and construction materials and of all types of goods demand flexible and reliable conveyor systems with seamless interfaces. Our customers receive economically and technically optimised systems designed to meet their specific requirements. Our engineering and our expertise have already proven effective in over 100 projects worldwide.

- **Curving and extendable belt conveyor systems**
  for tunnel boring machines, drill and blast applications, mining, construction materials industry, port management
- **Vertical and horizontal belt storage units**
  for optimal utilisation of construction site conditions and for rapid belt extension
- **Precision stackers**
  for the deposition of excavated material or bulk materials
- **Vertical conveyors** for use where space is limited

H+E Logistik GmbH
Josef-Baumann-Str. 18
D-44805 Bochum
Germany
Tel. +49 (0)234 I 950 23 60
Fax +49 (0)234 I 950 23 89
www.helogistik.de

Soil Dynamics with Applications in Vibration and Earthquake Protection

■ For numerous geotechnical applications soil dynamics are of special importance. In seismic engineering this affects the stability of dams, slopes, foundations, retaining walls and tunnels, while vibrations due to traffic and construction equipment represent a significant aspect in environmental protection. Foundations for mechanical equipment and cyclically loaded offshore structures are also part of the spectrum of application. This book covers the basics of soil dynamics and building thereon the practical applications in vibration protection and seismic engineering.

CHRISTOS VRETTOS
Soil Dynamics with Applications in Vibration and Earthquake Protection
2012.
approx. 200 pages,
approx. 90 fig.
Softcover
approx. € 55,–*
ISBN 978-3-433-02999-2
Date of publication:
II. Quarter 2012

Author:
Christos Vrettos is Professor and Director of the Division of Soil Mechanics and Foundation Engineering at the Technical University of Kaiserslautern.

Order Online: www.ernst-und-sohn.de

Ernst & Sohn
Verlag für Architektur und technische Wissenschaften GmbH & Co. KG

Customer Service: Wiley-VCH
Boschstraße 12
D-69469 Weinheim

Tel. +49 (0)6201 606-400
Fax +49 (0)6201 606-184
service@wiley-vch.de

*€Prices are valid in Germany, exclusively, and subject to alterations. Prices incl. VAT. Books excl. shipping. 0226300006_dp

---

Dramix®

better together

REINFORCING YOUR TUNNEL PROJECT

Dramix® - economic steel fibre reinforcement

- **MEET YOUR SPECS** with cost-effective fibres
- Count on **TIMELY DELIVERIES**
- Benefit from Bekaert's **WORLDWIDE PRESENCE**
- **WORK SAFELY** with stable quality
- Apply concrete solutions **TAILORED TO YOUR PROJECTS**

Bekaert GmbH
Otto-Hahn-Str. 20 · D-61381 Friedrichsdorf
Deutschland · building.germany@bekaert.com
T +49 6175 7970 137 · F +49 6175 7970 108
http://dramix.bekaert.com

Geotechnical Engineering Handbook – the English version of the German bestselling reference work! This handbook covers all aspects of the field, explaining the theory as well as describing practical applications. Editor: Ulrich Smoltczyk.

■ Volume 1 covers the basics necessary for any construction activity in foundation engineering. This systematic introduction to the assessment of soil and rock properties provides an insight into the requirements of Eurocode 7.

Volume 1: Fundamentals

2002. 808 pages. 616 fig. 82 tab. Hardcover.
€ 205,–*
ISBN: 978-3-433-01449-3

■ Volume 2 of the Handbook covers the geotechnical procedures used in manufacturing anchors and piles as well as for improving or underpinning foundations, securing existing constructions, controlling ground water, excavating rocks and earth works. It also treats such specialist areas as the use of geotextiles and seeding.

Volume 2: Procedures

2002. 701 pages. 558 fig. 67 tab. Hardcover.
€ 205,–
ISBN: 978-3-433-01450-9

■ Volume 3 of this Handbook deals with foundations. It presents spread foundations starting with basic designs right up the necessary proofs.

Volume 3: Elements and Structures

2003. 666 pages. 500 fig. 58 tab. Hardcover.
€ 205,–
ISBN: 978-3-433-01451-6

Package-Price for Volumes 1–3
€ 550,–*
ISBN: 978-3-433-01452-3

Online-Order: www.ernst-und-sohn.de

Ernst & Sohn
Verlag für Architektur und technische
Wissenschaften GmbH & Co. KG

Customer Service: Wiley-VCH
Boschstraße 12
D-69469 Weinheim

Tel. +49 (0)6201 606-400
Fax +49 (0)6201 606-184
service@wiley-vch.de

* € Prices are valid in Germany, exclusively, and subject to alterations. Prices incl. VAT. Books excl. shipping. Journals incl. shipping. 0233100006_dp

6 The tunnel lining

6.1 General

The tunnel lining has to permanently guarantee structural safety, durability and serviceability for the entire duration of the use of the tunnel. It supports the interior against the surrounding ground, forms a barrier against the penetration or emergence of water, transfers the internal loads from installations and traffic and, depending on the design of the TBM, may serve as an abutment for the thrust cylinders. The design and constructional detailing of the lining has to meet the requirements of the tunnel use, the acting loads and the conditions imposed by the construction process.

Temporary support with shotcrete and rock bolts is only an option with a gripper TBM. This type of support is dealt with in Maidl: "Hardrock Tunnel Boring Machines" [203].

The boring process normally demands a circular cross-section. The internal radius is determined by the use and the resulting requirements, such as a structure gauge that has to be maintained in transport tunnels or the required hydraulic cross-section in water or ventilation tunnels.

The dimensioning of the lining is decided depending on the structural actions, which mainly come from ground and water pressure. On account of the circular cross-section of the tunnel and the loading, which is not normally radially symmetrical, the system and support lines rarely coincide. This results in a transverse bending action on the tunnel lining in addition to compression, which can lead to the installation of structural reinforcement. The water pressure acting on the lining is influenced both by natural conditions and by the waterproofing concept. In tunnel systems subjected to water under pressure, the natural water conditions are restored after the completion of tunnelling. In this case, the lining must be designed to resist the entire water pressure. In tunnel systems not subjected to water under pressure, the water is drained so that no water pressure can build up.

The machinery used has a great influence on the possible types of lining. If a open gripper TBM is used, temporary support can be provided with shotcrete and an inner lining of in-situ concrete can be concreted later. If a shield is used normally, an immediately load-bearing segment lining will have to be used.

6.2 Construction principles for the tunnel lining

6.2.1 Single-layer and Double-layer construction

The lining of tunnels can consist of one layer (single-pass) or two or more layered construction [185], [189], [196].

In double-layer construction, the individual layers are constitutionally and functionally separate. The outer lining is installed as the tunnel advances and is designed to provide immediate support for the excavated cavity against the expected ground pressure. There are normally no serviceability requirements and thus no waterproofing requirements. These are complied with for the lifetime of the tunnel by the second inner layer installed as a permanent lining. For tunnels subject to water under pressure, the inner lining is designed to resist the applied water pressure. The inner lining also has to permanently support the ground pressure if the structural stability of the outer lining cannot be guaranteed for the lifetime of the tunnel. This is the case, for example, when aggressive groundwater, which can damage concrete, causes rotting of the outer lining. The inner and outer layers of construction are normally constructionally separated, for example by foil, in order to keep the inner lining free from indirect actions. In waterproofed tunnels, the waterproofing membrane undertakes this function.

Single-layer constructions can either be real single-layer solutions fulfilled by one construction system or composite solutions, in which two or more layers undertake the requirements on the lining with the individual layers contributing to the resistance of external loading as a composite (Figure 6-1). The first category includes, for example, tunnels with a single-layer segmental or extruded concrete lining and also tunnels in stable rock, which require no support for structural purposes but are provided with a smooth internal finish such as in-situ concrete for aesthetic or operational purposes. The second category includes, for example, tunnels whose lining consists of a composite of shotcrete applied for temporary support and a subsequently concreted bonded inner lining to complete the tunnel support system.

For both variants, the single-layer or composite cross-section must ensure structural safety and serviceability, particularly concerning waterproofing of the tunnel tube, for the entire period of use.

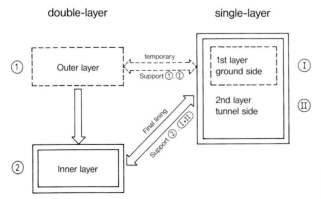

Figure 6-1 Definition of single-layer tunnel construction [185]

As the installation of precast segments and the application of shotcrete are both possible for a TBM-driven tunnel, the whole range of single and multi-layer tunnel construction types is available. There are also new developments using steel fibre reinforced concrete. Table 6-1 gives a schematic overview of the already used application possibilities.

Table 6-1 Matrix of possible lining construction types

		Inner layer or single-layer construction			
		Shotcrete	In-situ concrete	Segments	Extruded concrete
Outer layer	Shotcrete	✓	✓		
	Segments		✓		
	Extruded concrete		✓	✓	
Single-layer construction		✓	✓	✓	✓

The question as to whether single or double-layer construction is better is primarily determined by economic aspects. But the risks for the construction of a waterproof tunnel also have to be considered. The construction of a segmental lining as the final tunnel support and lining is already standard worldwide for pressurised shield tunnelling. But for a hard rock TBM tunnel, the matter is not so clear-cut. A single-layer lining is only the economic option under certain conditions.

Comparisons of tenders for a number of major transport projects in Switzerland have shown that double-layer construction is still 5 % to 10 % cheaper because the production costs are lower due to thinner segments, less reinforcement and less stringent precision requirements. Better advance rates can be achieved due to the shorter ring building time and the annular gap can be filled by stowing pea gravel instead of relatively expensive grouting, which is required to compress the gaskets in the longitudinal joints of the segments, among other reasons.

6.2.2 Watertight and water draining construction

In order to deal with the presence of water in the ground, there are two basic possibilities: the groundwater can be collected by drainage and led away, or the tunnel can be made watertight all round and resist water under pressure. Figure 6-2 shows the functional principles of these two methods through the example of a single-layer segment lining (left) and a double-layer cross-section with in-situ inner lining (right).

In drained tunnels, the tunnel vault is made watertight to protect the interior from water penetration. This can be achieved by a plastic waterproofing membrane hung in front of the inner lining or the construction of the vault as a waterproof concrete structure. The groundwater flows around the technically waterproof tunnel lining into the drains at the sides, where it is collected and carried away. This system is also known as drainage pipes with umbrella waterproofing and is suitable for percolating water or water under pressure. As no hydrostatic pressure can build up around a drained tunnel, the tunnel lining only has to be designed to resist ground pressure.

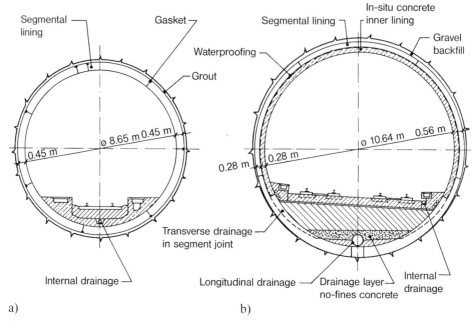

Figure 6-2 Construction principle of watertight and drained tunnel linings
a) Botlek Tunnel, internal diameter 8.65 m, single-layer, pressure-resistant
b) Murgenthal Tunnel, internal diameter 10.64 m, double-layer, drained

A disadvantage of tunnel systems that drain water away is the sometimes heavy expense for the maintenance of the drainage system due to encrustation and sintering, resulting in high operating costs. In order to maintain their function, drainage pipes should be flushed regularly. Recent investigations have shown that the use of low-elution shotcrete, constructional alterations to individual components of the drainage system and alterations to the overall system, like for example the exclusion of air by backing up the water, can reduce the problem of sintering [198].

Tunnel systems that drain water have an impact on groundwater conditions and thus the ecological system by lowering the groundwater table [184]. It may therefore be a requirement for environmental protection reasons to construct a waterproof tunnel without any effect on groundwater conditions after the completion of the tunnel.

Since the 1960s, the use of improved materials (plastic waterproofing membranes, plastic waterstops, waterproof concrete, new grouts etc.) have enabled the construction of tunnels with all-round waterproofing against water under pressure [258].

While single-layer, watertight segmental lining are now standard technology for shield tunnels, the proportion of the total number of tunnels with in-situ lining constructed to resist water under pressure is very low. For example, a survey of experience with tunnel waterproofing in Switzerland investigated altogether 239 projects, of which 233 were designed to resist seepage water and only 6 against water under pressure [310], [258].

In a study of the experience with waterproofing systems in German road and rail tunnels, altogether 9 tunnels with waterproofing against water under pressure were analysed. In summary, the results showed that the contractually required waterproofing classes had not been achieved straight away without additional measures. In systems with plastic waterproofing membranes, localised leaks were often caused by damage during the installation of reinforcement. This was, however, often only detected in the form of visible damp patches after the first exposure to water under pressure, which made subsequent repair measures necessary, often extensive and made difficult by the continued water pressure. With a single-layer segment linings, waterproofing can be impaired by defective installation of segments resulting in damage in the form of heavy cracking or spalling.

Experience shows that not even two-layer waterproofing systems make it possible to ensure the required waterproofing without additional measures. A concept should therefore be produced for tunnels intended to resists water under pressure at the design stage to enable the subsequent improvement and repair of the waterproofing system with some prospect of success, even under water pressure.

6.3 Segmental lining

6.3.1 General

Segments are precast elements, which are installed in a ring to serve as tunnel lining. The particular feature of a segment lining is the high degree of jointing, in addition to the segments themselves. The joints can be differentiated into longitudinal joints between the segments in a ring and ring joints between the rings.

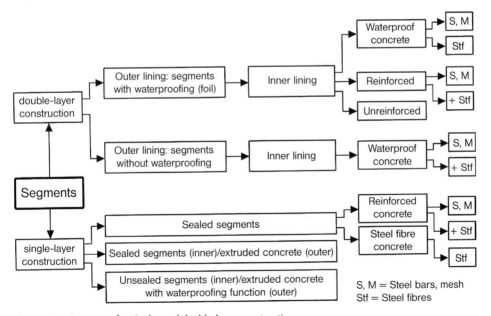

Figure 6-3 Segments for single- and double-layer construction

The use of segments is essential in TBM tunnelling if the gripping of the machine into the rock mass in order to produce the thrust forces is not possible due to insufficient rock strength. In such cases, the thrust forces are resisted by the already installed lining, which then works as an abutment in the direction of the tunnel axis. This requires immediately available load-bearing capacity, which cannot be provided by a shotcrete or in-situ concrete lining.

Figure 6-3 shows the spectrum of construction possibilities for segments for single- and double-layer construction in tunnelling.

Segments are usually installed using an erector in the protection of the tailskin of the TBM or braced directly against the rock mass behind the shield. In a subsequent working step, the annular gap remaining between the segment ring and the sides of the excavation is filled or grouted with suitable material through appropriate openings in the segments or through the tailskin. This limits the loosening of the surrounding ground, enables continuous transfer of the external ground pressure into the lining and provides the bedding required for the stability and structural safety of the tunnel tube.

Concrete segments are standard today and have mostly superseded steel and cast iron segments or tubbings for cost reasons. For further information about the use of steel and cast iron for tunnel linings, see [83], [143], [183], [196].

Specifications for segmental linings, which ensue from the local geological and hydrological conditions and also process-related and economic aspects, have led to numerous different construction variants for concrete segments.

The thickness of segments is determined according to structural and constructional criteria. The minimum thickness is mostly determined by the need to transfer the thrust cylinder forces and the resulting load-bearing area of the thrust pads. The usual thickness is 20 to 50 cm. Larger tunnel cross-sections also require thicker segments, for example segments 60 cm thick were required for the fourth tube of the Elbe Tunnel.

The width of concrete segments varies between 1.0 and about 2.0 m, with a current tendency to wider segments due to production developments in formwork technology and in the technology of transporting and installing segments. This enables quicker tunnelling and the reduction of joint lengths. But the increasing width of segments also concentrates more secondary loading in the joints resulting from production and installation tolerances and thus leads to a worsening of the cracking and spalling problem. Increased width of segments also reduces the "margin" for driving round curves and increases the necessary stroke of the thrust cylinders.

Segments should be reinforced around the ring to resist bending from external loading. Nominal reinforcement in both directions is also recommended to ensure serviceability. The splitting tension resulting from the thrust cylinder loading the ring joints should also be covered with appropriate reinforcement. This also applies for the transfer of eccentrically acting compression forces in the longitudinal joint.

6.3.2 Constructional variants

6.3.2.1 Block segments with rectangular plan

This variant is the most commonly used form, with a ring being built of five to eight single segments and a keystone. The rectangular geometry results in flat ring joints, and each ring alone is stable and load-bearing. The wedge-shaped keystone is generally smaller than the other segments and is installed last. This allows the closing of the ring by inserting the keystone along the tunnel direction. The spreading action can produce a pressurising of the segment lining around the ring.

It is advantageous for structural reasons to produce all the segments with as similar size as possible, i.e. the keystone has about the same opening angle as the other segments. But for ring installation, a smaller, easily handled keystone is better. It should be investigated on each project, which system is more advantageous. A large keystone is becoming ever more accepted.

The use of block segments with a rectangular plan has become the norm for single-layer, watertight tunnels. Waterproofing is ensured by an all-round gasket profile, which fits into a groove intended for it in the longitudinal and ring joints. In addition to the provision of sealing gaskets, reliable waterproofing requires high quality installation of the ring while minimising secondary actions or excessive stiffness of the segment tube. The latter is normally effected by offsetting the longitudinal joints of adjacent rings (Figure 6-4). This reduces on the one hand the deformation capability of the tube by interlocking to mechanically bond adjacent rings (Section 6.3.3.2) and on the other hand eases the problem of waterproofing crossing joints.

A quite different procedure for keystone installation is the five-piece segment lining with keystone at the bottom (Figure 6-2 b), which is usual in Switzerland. The joints between the segments are not waterproofed, but are supplemented by an inner lining with waterproofing foil in-between. The ring installation sequence is as shown in Figure 6-5: first, the two invert segments are installed and then the side segments. These have to be laid on slewing carrier rollers, because the segments are not bolted to each other and no holding force is available from the thrust cylinders during ring installation. Then the crown segment is installed from below with the erector. The erector stays in place to maintain the position. Then the invert joint is widened using dowels inserted into the openings provided in the invert segments and the straight keystone is inserted into the gap at the bottom. Finally, the widening of the invert segments is slackened and the no longer required erector and the support rollers are removed, causing the crown segment to settle slightly. The segment ring is now in its final position and the thrust reaction ring for the thrust cylinders can be moved forward again.

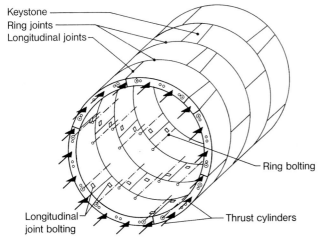

Figure 6-4 Block segments with keystone and longitudinally offset joints

124 6 The tunnel lining

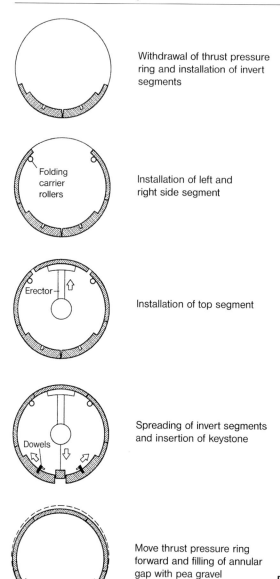

Withdrawal of thrust pressure ring and installation of invert segments

Installation of left and right side segment

Installation of top segment

Spreading of invert segments and insertion of keystone

Move thrust pressure ring forward and filling of annular gap with pea gravel

Figure 6-5 Ring installation procedure with five-piece segmental lining with keystone at the bottom

The advantages of this method are the avoidance of excessive secondary loading due to violent insertion of the wedge-shaped keystone and the short time needed for ring building of 15 to 20 minutes. The disadvantages are the lower installation precision and the resulting spalling of the edges of the segments.

In general, a tunnel alignment is a three-dimensional curve, which the TBM has to follow as precisely as possible. The ring being installed has to follow the direction taken by the TBM without damage occurring due to contact of the shield skin on the lining [7]. This would result in a one-sided opening of the ring joint in the curve. This is, however, not permissible for a single-layer tunnel with waterproofing requirements, as it could cause the gaskets to lose their function. In order to enable driving around curves without opening the ring joints, rings are formed with a taper on one side. Two basic systems can be used:

– When a tapered universal ring is used, any angle can be obtained through the appropriate rotation of the ring. The advantage is the lower formwork and logistics effort through the use of only one ring type, which is a decisive advantage in segment production. When offset longitudinal joints are specified, however, it should be noted that a compromise is often necessary regarding the rotation of the ring for the optimal steering of the shield. A further disadvantage of the universal ring is that the keystone position varies. If the keystone is in the invert, then ring assembly has to begin with the segment in the crown, which in this case can only be held in position by applying the thrust cylinders. This has to be taken into account in the design of the TBM, as additional measures are necessary to protect the crew in the erector area.
– The production of a number of ring types as right-, left-turning and parallel rings can maintain uniform geometry of the longitudinal joints with a defined keystone position. The disadvantage is the relatively high logistical complication and the resulting higher segment production costs.

The use of block segments with conventional shield machines normally demands the coupling of the excavation and ring assembly operations. This means that excavation can only continue after the installation of a ring is complete. This limitation results primarily from the limited travel of the thrust cylinders.

The desire for a continuous installation sequence has led to the development of the spiral segment, which has sometimes been used in Berlin and Stuttgart (Figure 6-6). This variant has not, however, become established.

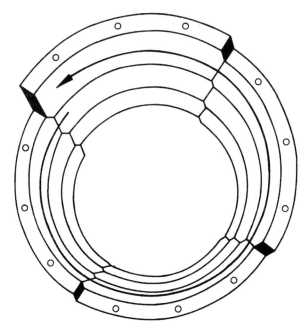

Figure 6-6 Spiral segment, Berlin underground, 1965/66 [196]

6.3.2.2 Hexagonal segments

Hexagonal segments were first used as long ago as 1961 for the construction of the Happurg water tunnel [196]. This type of segment is being increasingly used today as the outer layer of a two-layer tunnel system, but also for single-layer solutions, particularly when a double shield TBM is used [113], [182], [288], [287]. Because of the hexagonal shape of the individual segments, there is no continuous ring joint in this system, as this is offset by a half segment width between adjacent segments. The mechanical interlocking of the ring joints also results in an altogether stiffer tube compared to rectangular block segments without offset longitudinal joints. A ring built of hexagonal segments can consist of altogether four segments in smaller diameters, with the invert and top segments and the side segments opposite each other (Figure 6-7).

A significant economic advantage of this segment system is that ring construction only requires one type of segment. This leads to considerable cost savings in segment production compared to the use of rectangular block segments. Only the invert segment is often constructed with differences from the standard segment for operational reasons.

Disadvantages of this variant result from the size of the segments with increasing diameter and the resulting transport and assembly problems. Hexagonal segments are therefore preferably used for smaller diameters up to 4.50 m, although larger diameters are possible, as demonstrated by the Plave II and Doblar II pressure water tunnels in Slovenia, which are both of 6.98.m diameter [287].

6.3.2.3 Rhomboidal and trapezoidal segment systems

The desire to use mostly automated processes for segment installation has led to the development of segment systems with jointing dowels and guiding rods [294]. This is intended to reduce secondary loading during installation by optimising the segment geometry and by installing guiding rods in the longitudinal joints as well as centring dowels in the ring joints to achieve a high precision of assembly and ring quality.

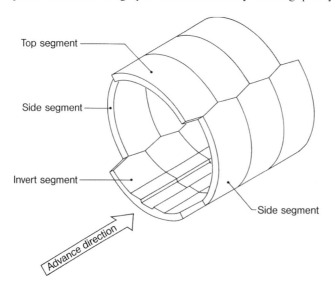

Figure 6-7 Schematic construction of a tunnel tube of hexagonal segments [288]

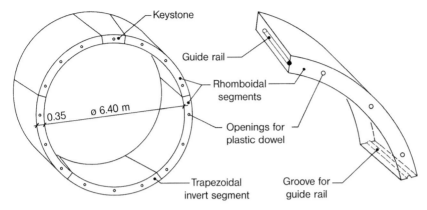

Figure 6-8 Tunnel cross-section, EOLE Paris, internal diameter 6.40 m

Figure 6-8 shows the segment system used for the construction of a single-track rail tunnel in Paris [294]. The segment ring consists of a trapezoidal invert segment, four rhomboidal segments and a trapezoidal keystone. The angled longitudinal joints of the segments means that the sealing gaskets of the segments only come into contact in the last few centimetres of installation movement. However, angled joints lead to kinematic lack of fit and stresses when a ring is deformed in its plane by a curve.

The ring joints are connected with three Conex plastic dowels per element (Figure 6-9) and one per keystone. Precise installation is ensured by 5 cm wide guidance grooves in the longitudinal joints. This system was also used successfully on a rail tunnel in Milan.

6.3.2.4 Expanding segments

In stable and relatively dry ground, ring construction can also be undertaken behind the shield. The ring is then expanded against the excavated ground and assumes a stable position. No further filling or grouting of the annular gap is required as long as the tunnel can be excavated with a uniformly round excavation profile.

a)　　　　　　　　　　b)

Figure 6-9 Conex Lining System, dowelled segments, Passante Milan (Mayreder) [196]
a) Plastic dowel
b) Dowels prefitted in a segment

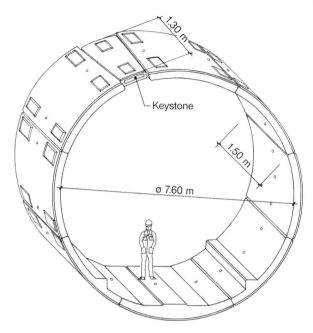

Figure 6-10 Typical cross-section of the Eurotunnel with expanding segments on the British side [96]

Expanding segments were originally developed for use in London Underground construction in clay with a long stand-up time [162]. Figure 6-10 shows a typical cross-section, which was installed in large numbers on the British side of the Channel Tunnel. The spreading of the segment ring is determined by how far the wedge-shaped keystone is inserted. This has a smaller width then the remaining segments [16].

This lining system enables short ring building times and thus high advance rates. One major disadvantage is that each ring assumes a different geometry, which can be associated with corresponding offsets at the joints. As a result of the relatively large twisting of the longitudinal joints, expanding segments are not normally suitable for single-layer, waterproof construction.

6.3.2.5 Yielding lining systems

According to the model representation used with shotcrete construction, the convergence of the excavated cavity leads to the formation of a so-called rock mass ring, which contributes to bearing the ground pressure. The time of installation of the support is therefore of particular significance, as this determines the share of the load falling on the lining. The earlier the support can be installed, i.e. the less time the rock mass can relax through deformation, the larger is the load the support has to resist. It should, however, be noted that depending on the geological conditions, it is also possible that the ground pressure can increase further with increasing deformation and softening of the rock mass.

A segmental lining is immediately load-bearing and relatively stiff in bending, which provides immediate resistance to deformation of the rock mass after the passage of the shield machine and the filling of the annular gap. This can lead to extremely high loading under high overburden or in squeezing rock, which in extreme cases could lead to the failure of the lining.

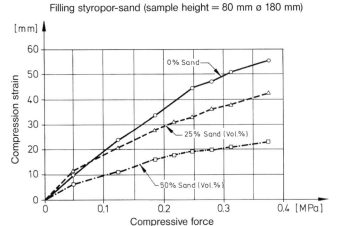

Figure 6-11 Trials of yielding stowed fill with high deformation potential [255]

For this reason, various attempts have been undertaken to develop yielding lining systems, which permit controlled convergence of the sides of the excavation. The measures can be differentiated into methods concerning the filling of the annular gap and methods concerning the longitudinal joints.

The use of a special grout for the filling of the annular gap can provide yielding support in combination with a segment lining. This can be a mixture of Styropor beads and sand. Particularly suitable could be the use of Styropor beads surrounded by cement paste, which would solve the problem of separation. Figure 6-11 shows the deformation potential of a Styropor-sand mix.

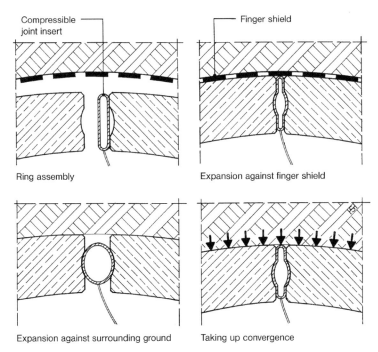

Figure 6-12 Yielding construction of a longitudinal joint using plastified steel tube [248]

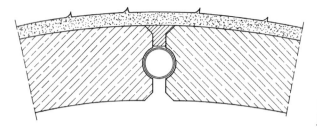

Figure 6-13 Proposed yielding joint detail from Thompson [282]

The yielding of a segmental lining can also be accomplished by the use of compressible inserts in the form of steel or plastic profiles. Figure 6-12 shows diagrammatically how convergence can be permitted by the plastic deformation of a steel tube inserted along the joint. The compression force in the lining is also limited by the transverse compression that can be resisted by the pipe in the plastic state.

A continuation and improvement of this principle is shown in the detail of a yielding joint construction (Figure 6-13). The yielding element in this detail is a plastic hose filled with water under pressure, which is grouted after the deformation of the rock mass has declined. Overloading of the segments can be prevented by valves in the water outlet pipe, which open at defined water pressures. The compression of the segment ring is enabled by the reduction of the volume of the hose due to water discharge.

Another proposal is for the use of a closed, plastically deformable plastic element (Figure 6-14), which also meets the requirements for waterproofing. This element, which takes up almost the entire face of the joint, consists of chambers filed with compressible air-entrained concrete. The load-deflection behaviour of the system can be adjusted by changing the details of the chamber system.

The use of compressible air-entrained concrete for filling the annular gap should also be able to yield with a set deflection.

In a shaft at the Ibbenbüren coal mine, a segmental lining was constructed with Meypo yielding elements in the longitudinal joints. For the construction of the shaft, the effective overburden corresponded to a depth of 1,650 m, which increased in operation due to further coal removal to be equivalent to a depth of about 2,000 m.

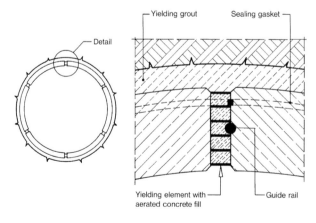

Figure 6-14 Proposal for a yielding plastic element [276]

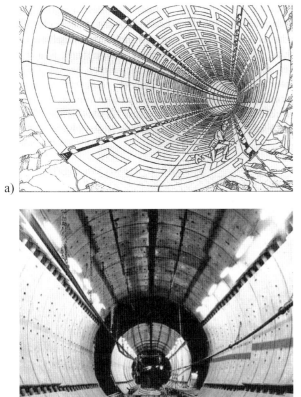

Figure 6-15 Yielding segmental lining at the Ibbenbüren shaft landing [40]
a) schematic illustration
b) completed segmental lining

The tunnel lining shown in Figure 6-15 consists of eight segments with eight longitudinal joints whereas four are detailed as yielding joints. The cross-section has a clear diameter of 8.5 to 9.5 m and a support resistance of over 1,000 kN/m^2. The yielding around the circumference is 30 cm per yielding element, corresponding to about 4 % of the total circumference.

The yielding mechanism of the Meypo yielding elements is shown in Figure 6-16. The essential components are the shear ring with a hardened shear pin to ensure the transfer of shear forces and the yielding plunger, which folds up in waves under compression loading.

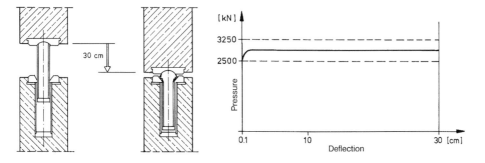

Figure 6-16 Detail and working curve of the Meypo yielding element [40]

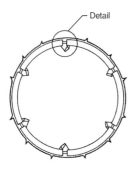
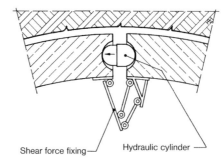

Figure 6-17 Proposal for a reusable yielding lining element [17]

The working curve of the yielding elements shows nearly ideal plastic behaviour. Up to about 85 % of the maximum load, the element is stiff, with the load resistance being maintained with further plastic deformation until the last few millimetres of travel.

The effectiveness of the yielding lining system has been demonstrated with deflections of up to 17 cm [17]. The first application did, however, show that the cost of the process is too high for normal tunnelling. A further proposal is therefore the inclusion of a reusable plunger, which can be removed after the convergence has died down and the longitudinal joints have been concreted and used again for ring installation behind the face (Figure 6-17). Hydraulic cylinders are used as yielding elements. The transfer of shear forces is ensured by a trussed articulation system, which is also reusable. This system has not yet been tried in practice.

6.3.3 Joint details

Because of the construction of each ring out of segments and the construction of the lining out of rings, the degree of joints in the tunnel tube is relatively high. The joints are either longitudinal joints running parallel to the tunnel axis or radial ring joints, and these have different functions and construction. The suitability of the chosen joint detail with regard to load-bearing, risk of spalling and waterproofing should if possible be demonstrated in appropriate test series [29], [52], [261], [262].

6.3.3.1 Longitudinal joints

The longitudinal joints transfer axial ring forces, bending moments from eccentric axial forces and shear forces from external and sometimes also internal loading. This occurs mostly through the contact at the contact surfaces, and in some cases also through the bolting of the longitudinal segment joints. In the usual precast concrete segment system, the longitudinal joints are hinges or partial hinges (concrete hinges) from the structural point of view with a limited capacity to transfer bending moments.

The engineer responsible for detailing can essentially consider three different groups of joint types for the detailing of the longitudinal joints. Longitudinal joints can have

– two flat contact surfaces
– two convex contact surfaces
– convex/concave contact surfaces.

A further variant is the tongue and grooved detail, which however is normally constructed as a hinge with flat contact surfaces.

Longitudinal joints with flat contact surfaces

A detail with flat contact surfaces as shown in Figure 6-18 geometrically prevents free rotation of the segments. Thus the longitudinal joints can transfer not only axial compression force and shear force (through friction) but also bending moments, which reduces the moments in the segments.

The rotation of the longitudinal joints takes place at the contact surfaces through elastic and plastic compression strains. Figure 6-19 illustrates the basis for the determination of resistance to rotation. Because only compressive stresses can be transferred, equilibrium can only occur when the resultant of the external forces R acts inside the cross-section.

In order to prevent the introduction of compressive stresses at the outer edge of the cross-section, outside the core surrounded by reinforcement, the contact surface is normally reduced. The width b of the reduced thickness at the joint is normally 1/3 to 1/2 of the total segment thickness. The splitting tension resulting from the introduction of concentrated compressive stresses has to be taken into account and covered by sufficient reinforcement.

The reduction of the thickness of the reduced section leads to a higher rotation capability of the longitudinal joints. This is of particular significance for single-layer waterproof tunnel linings, as the rotation capability has to be limited to prevent the sealing gaskets "breathing".

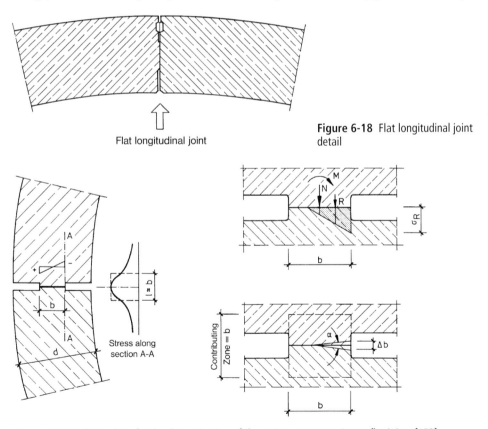

Figure 6-18 Flat longitudinal joint detail

Figure 6-19 Relationships for the determination of the resistance to rotation at flat joints [153]

134   6 The tunnel lining

The construction of the longitudinal joints with flat detail offers advantages during segment installation in that the offsets resulting from the inevitable imprecision in installation do not normally lead to concrete spalling.

Longitudinal joints with two convex contact surfaces

With flat contact surfaces, the splitting tension loading increases with increasing rotation due to the narrowed contact surfaces. If the axial compressive forces are large, there is a danger of concrete spalling at the outer edges, which could extend into the gasket area in single-layer waterproof linings. If the compressive forces and the angle of rotation are very large, the convex joint detail shown in the diagram in Figure 6-20 is recommended, in which the width of the contact area is not dependent on rotation angle [16].

Figure 6-21 shows the scope of application for flat and convex joints depending on the compressive force and the angle of rotation. The diagram was produced by evaluating tests.

The radius of curvature of the convex surfaces depends on segment thickness, the magnitude of the loads acting and the permissible angle of rotation. The detailing of the curvature radii must consider various aspects. If the radius is too small, the area through which the loads are transferred is reduced and the splitting tension is worsened. If the radius is too large, the rotation capability of the segments is limited.

During ring building, this system is not stable because there is no ring compressive force and no resistance to rotation. Suitable measures (e.g. bolting) should be undertaken to ensure that the ring does not collapse during construction.

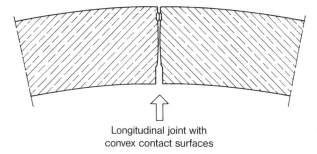

Longitudinal joint with convex contact surfaces

Figure 6-20 Longitudinal joint with convex contact surfaces in the Great Belt Tunnel [16]

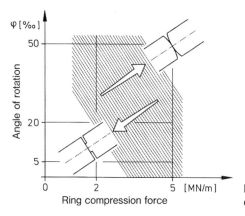

Figure 6-21 Scope of application for flat and convex longitudinal joints [16]

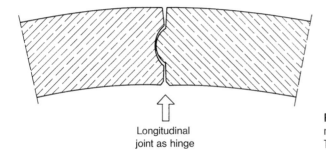

Longitudinal
joint as hinge

Figure 6-22 Detail of a longitudinal joint as hinged joint, Quarten Tunnel [163]

Longitudinal joints with convex-concave contact surfaces

Longitudinal joints with convex-concave contact surfaces as shown in Figure 6-22, which are also referred to as hinged segments, generally have a high rotation capability of the segments. In order to reduce the friction resistance and to provide more play during the installation of the segments, the radius of curvature of the concave surface is normally detailed larger.

For constructional detailing, a large angle of rotation makes it necessary to ensure that unwanted contact surfaces at the edges of the joint are avoided, which could result in spalling due to the concentrated introduction of compressive stresses. The edges of the concave side of the joint are particularly at risk because it is not practical to provide enough reinforcement at this location. In order to avoid damage to the edges, the central part of the longitudinal joint surface is normally detailed as a hinge.

The convex-concave joint detail leads to greater stability of the ring during installation, which is why this detail is preferred for use with expanding segments [16]. The centring effect represents an aid to assembly.

Hinged segments are mostly used for two-layered tunnels, as the waterproofing of the joints in single-layer waterproof construction has not been successfully solved [163].

Longitudinal joints with tongue and groove

Joints with tongue and groove (Figure 6-25) offer good guidance for assembly and with flat contact surfaces transfer axial forces, moments and shear forces. Because the sides of the groove cannot be reinforced effectively, there is a risk of concrete spalling if the available play is only slightly exceeded.

A special form of tongue and groove joint is the one-sided tongue/groove, i.e the inside-face spur. This is normally only used for small keystones in order to prevent them slipping out.

6.3.3.2 Ring joints

The plane of the ring joints lies at right angles to the tunnel axis. The loading on it comes mostly from the introduction and transfer of the thrust cylinder forces during construction. If the deformation patterns of two adjacent rings are different and the deformation is hindered, additional coupling forces (shear forces) occur in the ring joints.

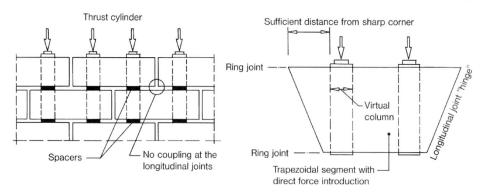

Figure 6-23 Transfer of the thrust forces into the ring joints [234]

The load from the thrust cylinders is introduced into the face of the ring plane through thrust pads to increase the support area and transferred through the segment into the next ring joint. Because flat contact in the ring joint cannot be assumed due to the unavoidable inaccuracies in segment assembly, there are normally contact zones at points, which load the segment as a deep beam. In order to avoid impermissibly high stresses and the resulting formation of longitudinal cracks, intermediate spacers are inserted to define the load transfer in the ring joint. These are glued centrally in the ring joint and should ideally be located exactly one behind the other in order to enable a straight flow of force in the longitudinal direction (Figure 6-23). The material used is rubber bitumen or hard fibreboard. Any deviations in the location of the load distribution boards should be covered by additional radial reinforcement.

As the height of the load introduction surface does not extend over the whole thickness of the segment, the loading from the thrust cylinders results in splitting tension, which should be covered by additional reinforcement.

Flat ring joint detail

The most simple ring joint detail is the flat, level joint as shown in Figure 6-24. Each ring is load-bearing on its own and there is no interaction between the rings, at least as a design assumption. The coupling is solely through friction, and the time-dependent decrease of the elastically stored prestress caused by the thrust cylinder forces, for example due to concrete creep, should be taken into consideration.

The flat ring joint provides no assistance for the assembly of the segments. This can be provided, for example, by the provision of plastic dowels as centring cones (see Figure 6-9). This is not normally intended to provide a mechanical coupling for shear transfer. This can be implemented with permanent bolting, as was used for example in the Great Belt Tunnel (Figure 6-24).

Tongue and groove systems

Various joint details are possible in order to simplify ring assembly and also to provide mechanical coupling of adjacent rings. Particularly for single-layer watertight segment construction, the deformation of the joints and offset joints should be kept as small as possible.

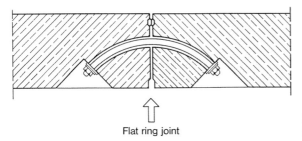

Flat ring joint

Figure 6-24 Flat ring joint in the Great Belt Tunnel [16]

One method is to construct the ring joint as a tongue and groove system. The tongue typically occupies more than half of the segment thickness and has a height of 10 to 25 mm. In order to enable the defined transfer of the coupling forces, coupling strips of rubber bitumen can be provided in addition to the boards to distribute loading from the thrust cylinders. This is illustrated in Figure 6-25.

Reinforcement of the tongued and grooved element to resist the coupling forces is difficult to implement while keeping the required concrete cover. Assembly inaccuracies and the resulting secondary loading can rapidly lead to damage to the ring joint. In order to minimise such damage, the groove is made larger than the tongue. The available play is normally only a few millimetres and is quickly taken up by production and installation tolerances.

In addition to the detail with tongue and groove, convex-concave detail of ring joints is also known, as shown in Figure 6-26. In this case also, the edges of the concave surface are at risk of damage during installation.

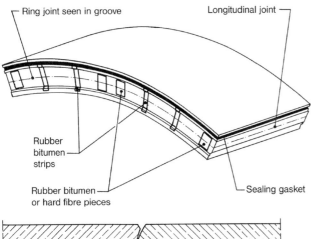

Figure 6-25 Segment detail with tongue and groove ring joint [52]

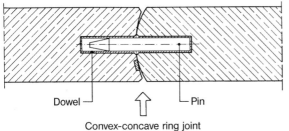

Convex-concave ring joint

Figure 6-26 Convex-concave ring joint detail [263]

Pin and socket systems

In contrast to tongue and groove, pin and socket systems provide coupling at points, e.g. at the quarter points of the segment (Figure 6-27). This can limit secondary loading resulting from assembly tolerance in the ring joint to the location of the pin. The coupling forces are, however, locally concentrated, which would be distributed along the joint in a tongue and groove detail. According to local conditions, the coupling locations can be heavily loaded in operation.

As the pin is normally thicker than a tongue would be, it can be reinforced to a certain degree. The dimensioning of the height of the pin should be planned to ensure that in case of failure, the pin would shear off and not the edge of the socket, which ensures continued waterproofing.

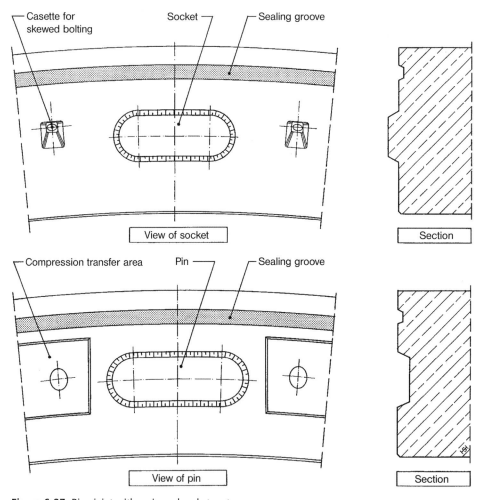

Figure 6-27 Ring joint with a pin and socket system

6.3.4 Steel fibre concrete segments

The external actions on a tunnel lining often lead to compression loading with only slight eccentricity, which would not require reinforcement for structural reasons. But the provision of nominal reinforcement is normally specified and necessary. In such cases, steel fibre concrete offers a suitable and economic alternative to conventional reinforcement [196].

Steel fibre concrete is a relatively ductile construction material with a high working capacity. The crack-distributing effect of the fibres in concrete is ideal for the required waterproofing of a single-layer segment lining. The positive effect on the cracking behaviour under the typical bending and compressive loading on a tunnel lining is particularly noteworthy [69], [99], [224], [230].

A further important advantage of the use of steel fibre concrete is the strengthening of the fragile corners and edges of the segments, which cannot be adequately reinforced conventionally due to the required cover [51]. The extent of damage at such locations can be significantly reduced.

Due to the industrial production methods used, higher fibre contents can be provided than in in-situ concrete. A typical fibre content would be 60 to 80 kg/m^3 with fibre lengths of up to 60 mm.

6.3.5 Filling of the annular gap

The process of shield tunnelling with segmental lining leaves a gap between the excavated ground and the lining, termed the annular gap. This has to be filled with a suitable material in order to provide the appropriate bedding for the segment tube and to ensure a uniformly distributed transfer of the loading from ground pressure and also to counter any loosening of the surrounding ground.

6.3.5.1 Filling with gravel

On shield tunnelling in hard rock, the annular gap is normally filled with pea and closely graded gravel (Figure 6-28), which is blown in (stowed), for example, by a normal dry shotcrete machine [286]. The filling of the annular gap should be undertaken as soon as possible after ring closure. This requires the appropriate openings to be provided in the segments for the hoses.

The pores in the gravel fill can then be grouted with fluid mortar in a subsequent working stage, in order to prevent the permeable fill acting as a drain.

6.3.5.2 Mortar grouting

In tunnels in ground with less stability, the annular gap is grouted. The necessary grouting pressure is matched to the prevailing ground and water pressure. As the grout is not normally taken into account in the verifications of structural safety of the tunnel support, no particular requirements are placed on its strength. But the stiffness must reach at least that of the surrounding ground in order to justify the assumed modulus of subgrade reaction.

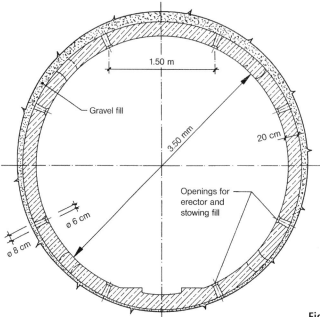

Figure 6-28 Gravel stowing [263]

In order to be pumped successfully, the grout must be capable of flowing sufficiently. When the grout is injected into the annular gap, a part (depending on the ground properties) of the mixing water is lost into the surrounding ground, which activates the grain skeleton of the grout as a support medium. At the same time, the loss of filter water represents a volume loss for the fill material, the quantity of which should be limited by using a mix with the lowest possible water content and high solid material content.

Good compaction can be achieved by adding fine material such as fly ash. Cement is normally used as binder to ensure rapid stabilisation of the grain skeleton. The hardening behaviour and strength gain have to be selected so that the grout in the hoses can be injected without problems even after long stoppages, in order to limit the requirement for flushing and cleaning work to a minimum.

The grout can be injected through openings in the segments or through the grout lines in the tailskin. For grouting through the segments, the lockable openings are fitted with threaded connections for the grouting hoses. A further possibility is the provision of plastic backflow valves integrated in the segments.

As a precondition for tunnelling with little settlement, grouting should be undertaken as near behind the shield as possible. Particularly when the ground has little or no stand-up time, no voids should be permitted to remain behind the shield tail for collapsing ground. The development of high-performance shield tail sealing systems, such as modern plastic seals or steel brush seals, has enabled the grouting of the annular gap directly through the shield tail. This means that the annular gap created with the advance of the shield can be filled immediately and the ingress of soil prevented.

6.3.6 Measures to waterproof tunnels with segment linings

Single-layer tunnel linings are mostly subject to waterproofing requirements. Transport tunnels have to be secured against the penetration of seepage water or formation water under pressure, while water tunnels have to limit the loss of water. As a segmental lining has a high degree of joints, the effort required to ensure waterproofing is relatively large.

6.3.6.1 Gaskets

In most single-layer transport tunnels with segmental linings in hard rock and loose ground, the penetration of formation water or groundwater is prevented by the provision of continuous gaskets in the longitudinal and ring joints. The closing of the segments compresses the butting gaskets and seals the joint (Figure 6-29). The contact pressure of the sealing frames has to exceed the water pressure acting on one side. The concrete surface of the groove must be free of air holes, in order to prevent water bypassing the gasket.

The compression behaviour of the profile and the detail of the groove must be matched to each other to prevent concrete spalling behind the groove due to splitting tension. The relationships between joint opening and compression force and between testing pressure and joint opening are shown in Figure 6-29 through the example of a profile from the manufacturer Dätwyler.

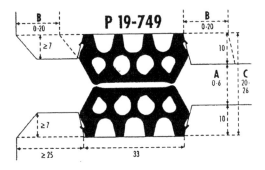

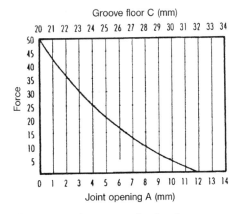
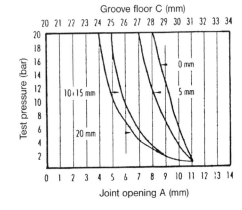

Figure 6-29 Elastomer sealing band

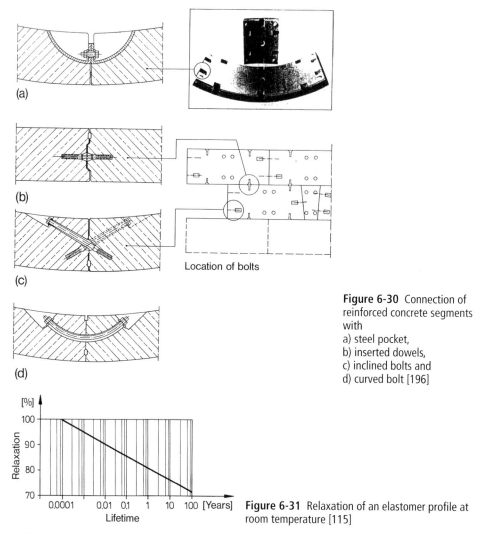

Figure 6-30 Connection of reinforced concrete segments with
a) steel pocket,
b) inserted dowels,
c) inclined bolts and
d) curved bolt [196]

Figure 6-31 Relaxation of an elastomer profile at room temperature [115]

In the ring joints, the required contact pressure is applied by the introduction of the forces from the thrust cylinders and is stored elastically. In the longitudinal joints, the sealing profiles are compressed by the axial compression force from ground and water pressure. "Breathing" of the sealing frame during construction is prevented by temporary bolting, and Figure 6-30 shows some possible variants. Near the portals and at crosscuts, the bolting is permanent.

Sealing materials used at the moment are natural rubber, plastics, elastomers, neoprene, silicon and swelling rubber (Hydrotite etc.), which are subject to stringent durability requirements. Considering the possible period of use of the tunnel, the functionality of the seal must be preserved for 100 years, including the consideration of any relaxation and ageing effects. As shown in Figure 6-31, the compression is reduced to about 70 % of its original value through relaxation, depending on the composition of the sealing material [115].

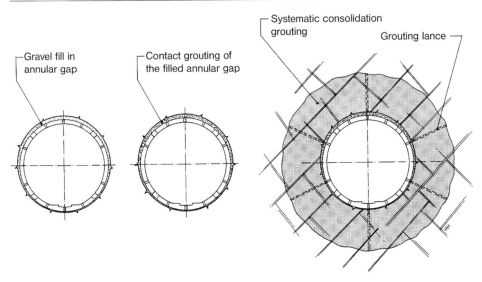

Figure 6-32 Grouting programme at the Evinos Tunnel [285]

6.3.6.2 Grouting

Another way of ensuring the required waterproofing is to reduce the flow of incoming formation water or the outflow of water from the tunnel by filling the joints in the surrounding ground by grouting.

For the construction of the Evinos Tunnel, a water tunnel with a single-layer lining of hexagonal segments (Figure 6-32), this method was successfully used to limit the water loss to the quantity specified in the contract. A grouting plan was designed with contact grouting of the fine-grained gravel fill in the annular gap followed by systematic consolidation grouting of the surrounding ground.

The joints have to be adequately sealed so the required grouting pressure can be applied. For this purpose, the longitudinal and ring joint are mortared over. As the grout is injected, the filtering out of the excess water from the grout presses this into the joints so that the full grouting pressure can be applied and the joints are completely filled and load can be transferred over the full area [288].

6.3.7 Production

Segments are normally produced in specialised precast concrete works. On larger tunnel projects, it can be more economical to set up a production facility on site, as for example at the Channel Tunnel and the Belt Tunnel, and the areas required for this need to be taken into account in the planning of the site facilities. Particularly on large projects, reliable supply of segments requires detailed planning and logistics.

Steel formwork is the only option in order to comply with the dimensional requirements for segments. The precision specifications for the production of segments are determined by the intended waterproofing function and to minimise secondary constraint loading. Progress in formwork technology has made possible formwork tolerances similar to mechanical engineering. For rail tunnels in Germany, the stringent dimensional specifications of the DS 853 [62] apply, as illustrated by excerpts in Table 6-2. This specification has often been included in contract documents on international projects.

Table 6-2 Dimensional tolerances for segments [62]

Segment width	± 0.5 mm
Segment thickness	± 2.0 mm
Segment arc length	± 0.6 mm
Longitudinal joint flatness	± 0.3 mm
Ring joint flatness	± 0.3 mm
Twist angle in the longitudinal joints	± 0.04°
Taper angle of the longitudinal joints	± 0.01°

The maintenance of tolerances should be checked regularly at the shortest possible intervals in order to be able to spot warping of the formwork during production as soon as possible. Exceeded tolerances could result in unplanned constraints to the segments during assembly and in service. These can reach a magnitude, which cannot be resisted by the concrete or reinforcement in the concrete. If it proves impossible to maintain the tolerances in practice, the effects on the segments should be investigated.

The taper of the longitudinal joints can be mentioned as an example. With an increasing angle of taper, the uniform introduction of the load over the width of the segment assumed in the structural calculations can no longer be guaranteed. This leads to eccentric load introduction into the joint, which could have to be compensated with additional reinforcement.

Quality assurance in precast production of segments should be verified in the production phase through a suitable quality assurance programme with a quality assurance handbook and associated working instructions.

6.3.8 Damage

Most damage to segments is caused during the construction phase. This can result in the formation of single cracks or even large-scale spalling. The cause of such damage is mostly the occurrence of impermissibly large contact stresses resulting from high thrust loads combined with production and installation tolerances in the segments resulting in a geometrical lack of fit. The segment design should be investigated for unwanted contact joints through geometrical and kinematic studies in the design phase.

Figure 6-33 shows an example from the El-Salaam siphon, a drinking water tunnel under the Suez Canal, with spalling of the longitudinal joint of the adjacent segment to the keystone, which was to some extent due to the unsuitable constructional detailing of the longitudinal joint with a spur between the keystone and the adjacent segment.

If segments are only used as excavation support behind the TBM and as the outer lining of a tunnel, damage can be regarded as uncritical unless the structural stability of the segment is endangered or the resistance and transfer of the thrust forces is impaired. In such cases, repair of even large-scale damage can be omitted.

Figure 6-33 Spalling of a segment in the El Salaam Siphon under the Suez Canal

When the lining is a single layer of segments with waterproofing function, damage is however critical if it impairs the serviceability or the waterproofing. In such cases, repair of the damage is essential, often requiring extensive work, the cost of which should not be underestimated.

There follows a description of some causes of damage to single-layer waterproof segment systems during segment assembly, on application of the thrust forces, in the shield tail and after the segment has left the shield with appropriate repair measures.

6.3.8.1 Damage during ring building

Clumsy assembly of rings with the erector is the most frequent source of damage to the concrete surface and gaskets during ring building. Damage is often also caused as the keystone is inserted due to the limited space available. The longitudinal joints of the new ring being assembled are not normally fully pressed together by the erector and the connection bolts so that the ring appears to be too large and the normally tapered keystone can only be inserted with great force. If the tolerances are inadequate or if the ring assembly is imprecise, this can cause wedging and unwanted contact at the sides, leading to concrete spalling. This needs to be borne in mind in design and construction.

6.3.8.2 Damage while advancing the machine

After the ring has been assembled, the thrust cylinders again apply the required thrust force. As the last ring to be assembled is not completely within the shield skin, i.e. is not yet bedded, each individual segment has to bear the axial force from the thrust cylinders independently, as there is only a limited load-bearing action of the ring. This effect can be worsened by one-sided thrust forces resulting from steering curves or correction radii.

Excessive thrust forces can sometimes also cause longitudinal cracks in the central third of the segment. Such cracks are often only noticed due to water ingress or damp patches two to three rings behind the shield but are have mostly been caused immediately after the first application of thrust force. Such damage occurs more often when segments are supported in a statically indeterminate system with more than two load transfer points per segment than with a statically determinate system. This is due to the fact that just slight assembly inaccuracies lead to partial ineffectiveness of a support and the segment without abutment is thus loaded in bending by the thrust cylinders. Even heavy reinforcement against tension resulting from bending cannot support the high thrust cylinder forces if a support point is ineffective.

This effect can be avoided by meticulous ring building with intentional compression of the ring joints of the last ring to be assembled. While the ring is being assembled, the thrust cylinders are retracted. At the same time, however, sufficient axial force must be maintained in the ring joint by the bolts to create enough friction to prevent the segments moving against each other. Only load transfer plates placed exactly in the line of the thrust cylinder forces can transfer the force from segment to segment without damage, as shown in the "column model" (Figure 6-23).

6.3.8.3 Damage in the shield tail seal

If the segment ring is not centred in the shield tail, the contact bar of the tailskin in front of the shield tail can come into contact and apply concentrated loading to the lining and attempt to push the segment ring into a central location in the shield tail. If a ring is installed eccentrically to the shield tail, then it will be forced into the central position by the seal as the shield tail seal slides past it. If this movement is prevented by the geometry of the concrete, for example with a pin and socket or tongue and groove joint detail, then damage can be caused to the segment in the shield tail.

This type of damage can be minimised if, for example, tapered segment rings are used and always installed so that the ring centreline is centred as exactly as possible in the shield tail. A frequent problem in practice is that a ring is incorrectly installed and the eccentricity to the shield tail is actually magnified.

It is essential to constantly monitor the clearance in the shield tail and control the ring assembly programme in order to prevent any dangerous increase of the clearance. It has to be pointed out that it is still the usual practice to measure the clearance in the shield tail manually. Local conditions and pressure of construction schedules can often lead to this not being checked often enough, which is only human. The only solution is the use of automatic measuring systems. These often prove unsuitable in practice due to local conditions, but the equipment is constantly being improved (serviceability). Manual measurement should, however, be history.

Continuous monitoring and control and careful analysis is the only way of avoiding damage.

6.3.8.4 Damage after leaving the shield

After leaving the tailskin, the segment ring is loaded by the pressure of annular gap grouting and the prevailing ground and water pressure. The resulting deformation can show a type of "trumpet effect". The loads acting on the ring in this construction state are quantitatively the highest loads in the entire sequence of temporary and permanent loading. A further unfavourable circumstance is that the back part of the ring is still largely unbedded in the protection of the shield tail and is only completely loaded as the machine advances.

Extreme rotation of the segment ring can occur as it emerges from the tailskin, which can lead to the failure of corners of the contact surfaces.

Such rotation can be limited by careful ring assembly with planned compression of the longitudinal joints while they are still in the shield tail. In addition to the constructional details, balanced and simultaneous grouting of the annular gap is an absolute precondition for the avoidance of this problem. The data from annular gap grouting should be recorded and evaluated continuously. Control and analysis of possible damage can be used to optimise the construction process.

It is also possible that the segment ring in the area of unhardened grout becomes oval under flotation, which can lead to impermissible deformation with concrete-to-concrete contact and spalling. The properties of the grout should be selected to avoid this. Continuous control of segment deformation can be used to determine the cause of damage.

After leaving the shield tail, the segment rings are also loaded by the concentrated load from the wheel sets of the first backup in the area where the grout is still soft. If the location of the joints is unfavourable or the grouting of the annular gap is defective, the load of the backup has to be carried by the joints, which can lead to displacement and severe damage to the joints. The load from the first backup carriage should therefore be spread as much as possible and be as far from the shield tail as possible.

6.3.8.5 Repair of damage

As long as the stability of the structure is not endangered, damage to the internal surfaces of the lining can be repaired relatively easily. Repair mixes based on artificial resin have proved successful for concrete repairs. Large areas should be filled in with shotcrete after careful removal of the affected or damaged concrete layers. The affected concrete can best be removed by high-pressure water jetting. If the damage extends through the entire depth of the layer and could have damaged the external waterproofing or its supporting concrete, the lining will have to be broken out completely.

Waterproofing defects are the most common source of damage, and can also lead to expensive long-term damage. When elastic joint sealing compounds are used, grouting through the lining to the external surface has proved successful. If the damage is restricted to the joint, which more often affects the keystone, then a thin injection lance can be pushed between the gaskets. The selection of grout depends on local conditions. Cement paste, sodium silicate and plastics have been used successfully.

Swelling (hydrophilic) rubber is also used for the repair of waterproofing defects. Swelling rubbers expand their volume many times on contact with water, resulting in the contact pressure required for waterproofing when the expansion is restricted. The provision of a second inner waterproofing groove for the insertion of swelling rubber profiles can provide for this eventuality.

6.4 In-situ concrete lining

6.4.1 General

In-situ concrete lining in TBM tunnelling are mostly used as the inner layer of a two-layer construction in combination with an outer support lining of shotcrete or segments. The construction and function is essentially the same as in-situ concrete linings in conventional rock tunnelling [183], [197].

6.4.2 Construction

Tunnel inner linings can be concreted with or without reinforcement. The requirement for bending reinforcement results from the local loading conditions. The necessity or otherwise of nominal reinforcement is determined by applicable national concrete codes or the requirements of the client. Prefabricated sheets of mesh are normally used.

The construction thickness of an in-situ inner lining should be at least 30 cm, with reinforcement 35 cm.

The concrete grade of the inner lining is determined by structural requirements, but it should also be noted that increasing concrete strength also leads to more heat of hydration and a considerable reduction of the work capacity, so concrete strength should not be unnecessarily high in order to limit the width of cracks in service.

For rail and road tunnels constructed of waterproof cement in Germany, a maximum crack width of $w_{k,cal} = 0.2$ mm [62] or $w_{k,cal} = 0.15$ mm has to be verified [42]. In addition, a maximum water penetration depth of 30 mm is specified.

Steel fibre reinforced concrete represents a sensible alternative to conventional reinforcement. The positive effect of the steel fibres on the load-bearing, deflection and cracking behaviour of concrete is particularly well suited for the typical loading on a tunnel lining, with high axial compression and relatively low bending moments. But the technical process and constructional advantages of steel fibre reinforcement have also been demonstrated. As an example, underground and urban rail projects in Dortmund showed economic advantages. For the use and design of steel fibre reinforced concrete in tunnelling, see: [64], [65], [69], [99], [189], [224], [230].

6.4.3 Concreting

Concrete, reinforced concrete and steel fibre concrete are cast in blocks using mobile formwork units with hydraulic striking, with typical block lengths of between 8 and 12 m (Figure 6-34). For small cross-sections, a full-round unit is used with the invert and vault being concreted in one pour without a working joint. For larger cross-sections, where the flotation force of the wet concrete would be too large, the invert is concreted in advance followed by the vault with a suitable formwork unit. This variant is also used with invert segments.

The minimum concrete strength required for striking should be determined in a structural calculation. Careful curing of the concrete is important.

The joints between the blocks can be detailed as contact joints or as expansion joints. Waterproofing can be provided by the use of appropriate expansion joint waterstops, the width of which is selected according to the active water pressure, but should not be less than 30 cm.

The wet concrete is transferred to a concrete pump in the tunnel and pumped into the formwork unit through a distributor ring. The distributor ring ensures uniform distribution of the wet concrete into the formwork. When concreting the vault, the permissible concrete level difference resulting from the construction and anchoring of the formwork must be complied with. Compaction is provided by external vibrators. The construction of the formwork unit must ensure that it can resist the loading of concreting and compaction without impermissible deformation of the profile. An adequate number of concreting windows should also be provided in order to control the concreting process.

Figure 6-34 Formwork for the concreting of an in-situ inner lining with umbrella waterproofing over the carriageway area, Murgenthal Tunnel

When steel fibres are used, the negative effect of the fibres on the workability of the concrete should be noted. In this case, the introduction of a quality assurance system specially adapted for the application of steel fibres in tunnel construction is recommended [188].

6.5 Injected concrete, Extruded concrete

After the introduction of single-layer segment linings in the early 1970s, intentions turned to the development of a shield with continuous concreting of the lining. Tunnellers have long hoped for the introduction of a tunnelling method with a slipformed lining, and the large number of patent applications from that time reflected attempts to introduce this very economical method. Lucke presented a slip form using steel fibre concrete in 1975 [180]. The process is illustrated in Figure 6-35.

A slipform with liquid-filled rubber membrane to simplify steering is towed behind a tunnelling machine. Lucke pointed out the problems of placing the concrete with pumps and supporting the shield forces. Lucke described this process at the time as "extruded tunnel lining". Hochtief took up the idea and developed the "extrusion process" to a stage ready for application. But the process was not actually used until 1978 for the construction of a sewer [181]. This was not a true slipform, as for example used to construct chimneys, which has not been achieved yet.

The term "extruding", with the related terms extruded concrete and extrusion process, is actually an imprecise description of the process; "extrusion" normally denotes the manufacture of plastic parts through an extruder. The term "injected concrete" is a better description of the process and is therefore used at the head of this section. In the practice and in the literature, however, the term "extruded concrete" used by Hochtief has become the norm and will therefore be used from now on. The concrete produced by this process is called "extruded concrete".

The extrusion process in its current form basically corresponds to the grouting of the shield track dealt with in Chapter 7, although in this case the requirements for material quality (waterproofing, strength) are higher, since the hardened lining will have to provide load-bearing support. Steel fibre concrete offers many advantages in this case, particularly where reinforcement is required for strength, but also for waterproofing. The installation of bar or mesh reinforcement is not compatible with this process. If the inner lining does not consist of permanently installed segments, a travelling steel formwork has to be used, which the shield pushes against to advance.

Construction process

The extrusion process is characterised by the pumping of concrete from a number of pipes through the travelling face formwork in the shield tail into the annular gap between shield and surrounding ground. This circular space is bounded externally by the surrounding ground, internally by a steel formwork, which is moved forwards in steps and at the front by the elastically supported face formwork (Figure 6-36). The steel formwork normally consists of 1.20 m long rings, which can be struck down into single segments (segment formwork), which is moved forwards using a special transport device and erected again in the protection of the shield tail [35].

The total length of the moving formwork depends on the strength development of the concrete and the forecast maximum advance rate and is about 15 m.

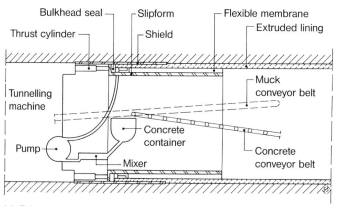

(a) Principle

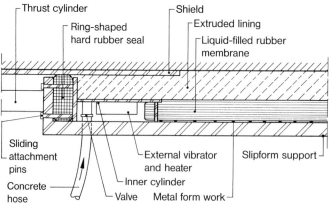

(b) Detail

Figure 6-35 Concept for a system with a slipform for extruded tunnel lining [180]

Workability properties of the concrete

The required workability properties of the concrete result form the planned process technology.

The concrete being pumped through a number of openings (Figure 6-37) in the face formwork must distribute itself evenly in the annular space and fill up all cavities. The pressure must be controllable within close limits so that equilibrium can be maintained against the prevailing ground and water pressure. These requirements must even be fulfilled during stoppages of the concreting process of limited duration, in order to ensure the effectiveness of the tunnelling machine even in case of short-term operational interruptions.

This demands a concrete with special properties, which even remains in a fluid state after emerging from the concreting pipe within a sufficiently extended flow zone behind the face formwork, perhaps for some time.

The concrete, which may be enhanced with steel fibres, must fulfil partially contradictory requirements regarding its mixing capability, workability and strength development.

Good workability properties of the wet concrete can be achieved through compliance with the general guidelines for the production of flowing and pumping concrete and the addition of special additives for this application.

Care should also be taken that the flow properties can change, not only through the normal setting process of the concrete but also through the loss of filtrate water into the surrounding ground (consolidation) as a result of the applied pressure (Figure 6-38).

In a fresh flowing concrete, the internal shear strength is very low. Triaxial shear tests with preconsolidated concrete samples have shown that the shear strength of the wet concrete containing steel fibres can increase to an effective friction angle of up to 44° after only a short consolidation time. In contrast, the shear strength of an unconsolidated sample is only about 0.03 N/mm² [35].

The selection of cements with higher grinding fineness, the use of appropriate concrete admixtures and the maintenance of certain minimum fines contents make it possible to sufficiently stabilise the workability properties of the wet concrete and meet the practical demands on construction sites.

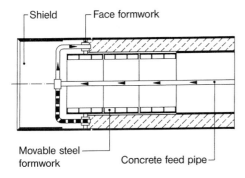

Figure 6-36 Schematic diagram of the extrusion process [35]

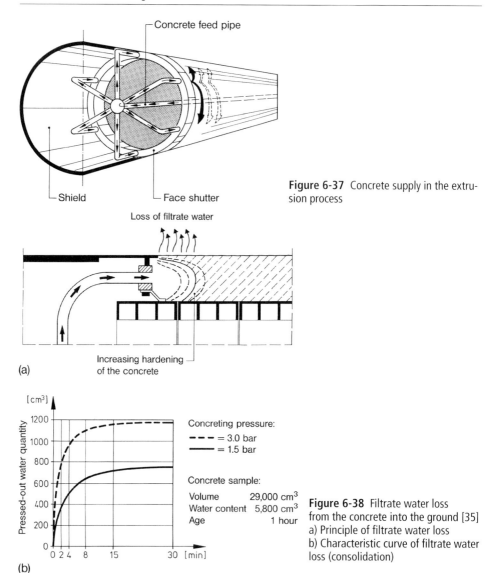

Figure 6-37 Concrete supply in the extrusion process

Figure 6-38 Filtrate water loss from the concrete into the ground [35]
a) Principle of filtrate water loss
b) Characteristic curve of filtrate water loss (consolidation)

Properties of the extruded concrete lining

The mixes used fulfil high requirements regarding early strength development (Figure 6-39), final strength and water impermeability without requiring particular measures to compact the concrete. The early strength after 12 h is 12 N/mm², so that striking is already possible this early. The 28-day strength is more than 35 N/mm². Various methods are available for the subsequent sealing of the working joints depending on the use of the tunnel and the required degree of waterproofing of the tunnel lining. In heavily water-bearing ground, as with the Métro in Lyon, a second in-situ concrete inner lining is concreted conventionally [35].

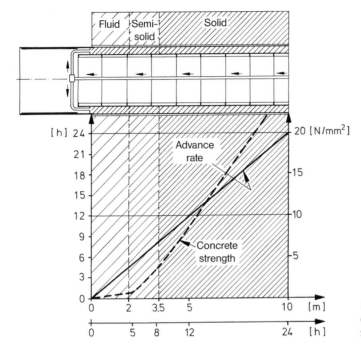

Figure 6-39 Development of concrete strength [35]

Mechanical equipment

A **telescoping follow-up shield** has an articulated connection to the tunnelling machine and is equipped with a telescoping joint (Figure 6-40). This enables for relative movement between the first and follow-up shields, typically one to one and a half formwork widths. In normal operation, the telescopic joint is in the central position so that the two dependent operating modes, advance and lining assembly, can be decoupled. When the tunnel advance is interrupted, the concreting process is slowed.

In the shield tail is the movable, elastically supported **face formwork** with inner and outer seals. The elastic support of the face formwork consists of hydraulic cylinders, whose closed hydraulic system can be preloaded by a gas buffer with a selected adjustable pressure. The gas volume is designed so that pressure fluctuations during the movements of the formwork in operation are very small. This ensures on one hand that the pressure applied to the concrete by the formwork and vice versa remain almost constant, and on the other hand that variations of the volume flow supplied by the concrete pump can be compensated without significant pressure variations.

The **concreting system** consists of a concrete pump located in the shield backup and the pipework to the face formwork.

Through a distribution valve located in the shield, the concrete is pumped through the face formwork in preset quantities and in any sequence through up to eight filling openings (Figure 6-37), which are arranged at equal spacing around the face formwork. This permits uniform volume distribution with short flow routes and avoids the formation of clods (dead zones) behind the face formwork.

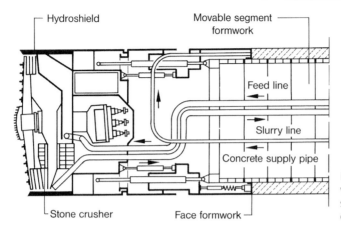

Figure 6-40 Hydroshield with telescoping follow-up shield for the extrusion process, Métro Lyon, 1993 [35]

Applications until now

Table 6-3 shows the applications of the extrusion process.

Improvements regarding quality control of the lining and the formation of working joints enable the expectation of further developments in this construction process, which is primarily interesting for economic reasons.

Table 6-3 Applications until now of the extrusion process

Year	Client	Project	Excavation quantity [m^2]	Length [m]	Geology	Lining	Remarks
1978–1979	City and state of Hamburg, Construction office	Sewer in Hamburg-Harburg, contract 1	10.2	1,169	Fine sand, medium sand	Extruded steel fibre concrete, single layer	Hydroshield, slurry circuit
1980–1982	City of Frankfurt, urban railway, Construction office	Underground railway Frankfurt, contract 36	37.0	1,650	Clay, limestone, gravel sand	Extruded steel fibre concrete, reinforced concrete inner lining	Blade shield, telescopic excavator
1984–1985	Ste d'Economie Mixte du Metro de l'Agglomeration Lyonaise	Métro Lyon, Line D, contract 1, France	33.2	2,484	Gravel sand, weathered granite	Extruded steel fibre concrete, reinforced concrete inner lining	Hydroshield, slurry circuit

Year	Client	Project	Ex-cavation quantity [m²]	Length [m]	Geology	Lining	Remarks
1984–1986	German Railways Head office Karlsruhe	Investigation heading for the Freuden-steinTunnel, contract 2	21.2	1,800	Marl, anhydrite	Extruded concrete, single layer	Expandable blade shield, full-face excavation
1987–1989	East Japan Railway Company	Water tunnel, Shinano-gawa River, Japan	55.4	3,100	Silty rock, sandstone, gravel	Extruded concrete, single layer	Open shield, cutting arm
1988–1990	City of Essen, Underground railway construction office	Essen underground railway, contract 32/33	39.8	3,044	Marl, silt, limestone banks	Extruded concrete, reinforced concrete inner lining	Earth pressure shield, muck removal by rolling stock
1990–1993	Japan Railway Construction Public Corporation	Akima Tunnel, Hokuriku, Shinkansen line, Japan	88.6	4,000	Sandstone, tuff	Extruded concrete, single layer	Open shield, cutting arm
1992–1994	Regione Lombardia Commune di Milano	Passante Ferroviario Milan, Italy	50.6	3,853	Sand, gravel	Reinforced concrete segments with extruded bedding	Earth pressure balance shield, muck removal by rolling stock

6.6 Shotcrete layers as the final lining

Intensive research in recent years has led to the situation that shotcrete can now be produced as structural concrete with the appropriate quality requirements [185]. Regarding the strength, this means that strength class B 25 according to DIN 1045 can now be achieved without problems. An important aspect of this is the uniformity of the material properties. One example of the problems facing the use of shotcrete is the tolerance allowance of about 5 N/mm² more than formed concrete in the initial test. It is therefore currently assumed that the range of variation of the compression strength of shotcrete is twice that of concrete placed in formwork. The uniformity can be considerably improved by appropriate technical measures.

When shotcrete is used as the final lining with waterproofing requirements, it should be born in mind that complete coverage is very difficult, if not impossible, when steel ribs and mesh are used and this can cause leaks. The possible spraying shadow behind these inserted items means that the use of a waterproof shotcrete cannot be assumed to have a waterproofing effect for the overall construction [190]. The use of steel fibre reinforced shotcrete certainly offers advantages in this case [201].

6.7 Structural calculations

The design of tunnel structures has a special status in structural engineering. Starting with the load assumptions, then the model formation and calculation methods and on to dimensioning and suitable safety factors, the normal practice in structural engineering has to be adapted to the practice in geotechnical engineering.

The special status of tunnel structures is mostly due to the particular properties of the ground around the tunnel. This is at the same time a loading and a load-bearing element. The loading from the ground can only be determined with certain statistical uncertainties, which can be kept within bounds through the experience of tunnelling engineers. The ability of the ground to bear load can also be determined with the same degree of certainty, although in this case the difference between the load-bearing capacity of rock and rock mass (or between soil and ground) is of particular significance. Working from the results of investigation at a few points, the design engineer then has to force the complex characteristics of the ground into a model for processing using the available calculation methods. This process has to make numerous assumptions, which should be considered in the interpretation of the calculation results [197].

The methods of performing structural investigations into the stability of tunnel linings are varied. These can be differentiated into analytical and numerical processes, with the latter becoming ever more significant.

Constantly refined constitutive laws and calculation algorithms enable the complex load-bearing behaviour of rock as a construction material to be represented ever more realistically. The step from the geological model to the mechanical-mathematical model for the assessment of the performance of a construction method or to dimension a tunnel lining has gained transparency with the improvement of the latter. The imperfections of the geological model as a basis for the design and the resulting encounters with unexpected conditions during the construction phase still have to be considered as empirical engineering for tunnelling, requiring constant observation and the employment of experienced personnel.

As the discussion of basics and methodology for the structural design of tunnels alone would be sufficient to fill this book, reference should be made here to the literature. An overview of calculation models and procedures can be found in [184] and [197], where references to further literature can also be found.

7 Shield tail sealing, grouting works

Shield tail seals and grouting equipment are important components of all shields working with simultaneous installation of a segmental lining, particularly those with active face support, the slurry and earth pressure balance shields. Shield tail seals seal the back end of the shield against groundwater and soil, prevent bypassing of support fluid or grout and are an essential part of the measures for the support of the face according to the principles already described. Another important part of these measures is the grouting equipment for the pressurised filling of the annular gap (Chapter 2). Equipment for grouting the surrounding ground has also become essential equipment for many shields. This is an important aid in tunnelling through problematic geological conditions, to overcome unexpected events and for localised preparation of the ground for the type of shield being used.

7.1 Shield tail seals

When precast segments are used for support, the advance of the shield results in the creation of an annular cavity behind the shield tail, which is bounded externally by the surrounding ground and internally by the segments.

The shield tunnelling machine is separated from this cavity by the shield tail seal, whose constructional thickness determines the width of the annular cavity in combination with the design taper and the intended overcut (Chapter 2). The width of the annular gap for the types of seal in common use today is generally between 115 and 140 mm (85 mm – minimum, 185 mm – maximum) depending on the shield diameter. The constructional width of the sealing gap is between 60 and 95 mm with a working range of the seals of 30 to 50 mm (ring building clearance).

Larger gap widths of more than 250 mm have been implemented when using the extrusion process. This is a concrete layer, which can undertake structural and waterproofing functions (Chapter 6).

The shield tail seal should reliably close the gap between the shield tail and the tunnel lining extrados. It has to resist considerable pressure, which depends on the prevailing ground and water pressure and the resulting grouting pressure. At the Eurotunnel, for example, routine grouting pressures on the French side were up to 12 bar [66].

At one time, various flexible materials like cleaning rags, hessian, wood wool, string etc. were pushed in by hand to resist the heavy loss of air (under compressed air operation) or the leakage of grout into the shield. Such sealing methods have long been replaced with elastic plastic and later steel brush seals.

When plastic and steel brush seals are used, a fixed mounting of the seals on the shield is disadvantageous. As the shield advances, the new cavity created in the annular gap has to be simultaneously filled with grout in order to prevent penetration of soil and water into the annular gap in unstable ground or groundwater. Particularly when the grout is injected through the segments, it is more likely that the volume differences resulting from the advance of the shield will lead to a pressure loss in the grout immediately behind the shield tail seal, with the result that soil and water enter the annular gap. Even if the ingress of soil and water compensates the pressure loss in the grout, this still results in an impairment of the bedding of the segment rings and a disturbance of the natural structure of the ground. Even when the grout is injected through the shield tail, constant grouting pressure can only be achieved to a limited extent.

7.1.1 Plastic seals

Plastic seals are often described as rubber seals. One of the first forms of plastic seal introduced consists of elastic foam elements, which are supported and protected by individual spring metal plates. In the specialist world, this type of seal construction is known as Lynacell sealing and is used in open shield TBMs suck a thrust ring, with which the annular gap in the invert is grouted with cement mortar and the sides and crown are filled with pea gravel (Figure 7-1). The spring elements made of thin sheet metal strips slide over the outer face of the segments. Each spring plate can be exchanged individually if repair is necessary. This type of sealing requires uniform ring assembly, as is the case when using the five-piece Swiss segment system with keystone at the bottom in shift bars provided in the shield tail of the open shield TBM described in Chapter 6, if heavy wear is to be avoided. This type of seal is not capable of sealing against considerable grouting or water pressure, but has the advantages of simple and cheap basic construction, rapid exchangeability of the components and the omission of secondary consumables.

As part of the development of the Hydroshields, a type of seal was developed, which can also undertake the sealing of the shield against the support fluid, which can penetrate inside the shield if bypass routes are open. After the intermediate form of the giraffe-neck seal, the S1 seal (Figure 7-2) consisting of individual rubber segments was developed, which are supported by ball-jointed rods from the end of the shield tail. The seal consists of closed neoprene profiles with a sealing lip sliding on the back of the segments. In case the main seal fails, these seals are fitted with back-up seals, which are mounted further forward and can be inflated if required.

The S1 seal enables grouting through the shield tail and has proved successful on various projects under water pressures of up to 3.5 bar. In practice, this type of seal has proved to be a complex overall system together with the shield driving, the segment system and its ring assembly and the grouting of the annular gap, which demands a detailed understanding of the relationships from all parties. It reacts sensitively if the ring building clearance is used up, to larger offsets in the longitudinal segment joints and to localised grouting pressures, and requires segments with even, flat backs and a clean shield tail in order to prevent the ingress of soil and water or grout into the shield.

With increasingly stringent project requirements concerning the length of advance and the water pressures to be overcome, steel brushes have increasingly been used as the shield tail seal system for shields with active face support.

7.1 Shield tail seals   159

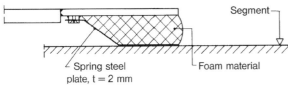

Figure 7-1 Lynacell seal

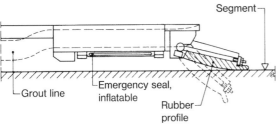

Figure 7-2 S1 seal

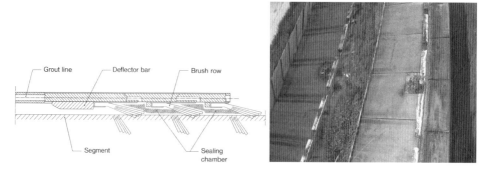

Figure 7-3 Brush seals

7.1.2 Steel brush seals

Steel brush seals (Figure 7-3), which were originally developed in Japan, are also fixed to the shield tail. Up to four rows of brushes in succession form individual sealing chambers together with the outer shield tail skin and the segments. In the chambers between the individual rows, a sealing compound is injected as sacrificial lubrication and maintained at a certain pressure level, which is about 2 bar higher than that of the grout [10]. The increased pressure in the sealing compound has the effect that water, soil and grout cannot penetrate into the sealed area.

Steel brush sealing has the advantage over plastic seals that penetration of soil and water into the annular gap is prevented when the pressure in the grout falls, as the pressurised sealing compound flows into the annular gap. Therefore care has to be taken that the sealing compound is environmentally acceptable. Steel brush seals are advantageous in case of steps at the longitudinal segment joints. Together with the sealing compound in the individual chambers, steel brush seals are capable of closing the resulting gap reliably. The multiple rows of brushes covered with spring plates are also more robust than plastic seals. Increasing wear as the tunnel advances can be compensated comparatively simply by increasing the supply of sealing compound. Another advantage of this type of seal is

Figure 7-4 Outer shield tail seal as tiled spring plate construction

the accessibility of the inner rows through the sequential installation of the segments in the last ring or by extending the thrust cylinder stroke to change the brushes. In case a problem occurs, special "chewing gum" sealing substances, which react with water, are available to be pumped in as an immediate measure.

7.1.3 Outer shield tail seals

With open shields with segmental lining and annular gap filling, there is no active reaction to the grouting pressure from the face support medium. Some assistance is offered in this case by outer shield tail seals, which are intended to prevent the grout running into the steering gap or forwards to the cutting wheel (Figure 7-4).

Outer shield tail seals, often described as excluders, are constructed as packets of spring steel plates like roof tiles, with sealing being provided by these sitting as closely as possible onto a flat outer surface of the surrounding ground. The overcut, the ground properties and the behaviour of the ground under excavation have a considerable influence on the sealing effect, in addition to the rough usage in the annular gap. An additional row of outer brushes with provision for the supply of sealing compound into the resulting chamber indeed offers the advantage of redundancy in the system, but also loads the shield tail with the resulting sealing chamber pressure and makes the starting process more difficult as the starting seal is passed.

7.1.4 Elastically supported face formwork for the extrusion process

The principle of the elastically supported face formwork used with the extrusion process (Figure 7-5) conforms to tunnel support with extruded concrete (Chapter 6). Instead of formwork units being moved forward, the bounding of the annular gap is provided by the installation of concrete segments.

The elastic support of the face formwork ensures a constant grouting pressure in the shield tail joint, even if the injected grout volume is not as large as the volume opened by the shield skin, and therefore achieves optimal bedding of the segment rings. If the grout mix has the appropriate composition, this can achieve an additional improvement of sealing. The elastic support of the face formwork consists of hydraulic cylinders, whose closed hydraulic system can be pre-pressurised by a gas puffer with a selected, adjustable pressure. The gas volume is designed to keep pressure fluctuations during the movements of the seal as low as possible in operation. This ensures on one hand that the pressure applied to the grouting material by the seal and vice-versa are nearly constant and on the other hand that fluctuations of the quantity delivered by the pumps can be compensated with as little alteration of pressure as possible.

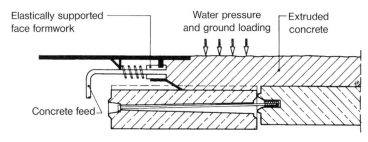

Figure 7-5 Elastically supported face formwork in the extrusion process [10]

One disadvantage of elastically supported face formwork is the required construction width of the annular gap, which leads to an increase of the required volume of grout. The last application of an elastically supported shield tail seal was the Passante Ferroviaria (Milan underground railway), where a 250 cm thick grouted layer was installed behind 300 mm thick reinforced concrete segments.

7.2 Grouting process

7.2.1 Requirements

The gap created between ground and segmental lining by the advance of the shield should be filled with grout under pressure behind the shield tail seal. This has the following objectives and tasks:

- **Bedding of the segment ring consisting of single segments:**
  Purely jointed rings – with more than three joints – are kinematically dependent on the support of the surrounding ground to maintain their shape. But semi-stiff rings with bolted and offset longitudinal joints also need bedding in order to keep bending moments and deflections small. The filling of the gap also has to achieve a fixed length of the segment tube in the ground to support the loads from the shield, particularly in curves.
- **Maintenance of the natural stress state of the ground and minimisation of settlement:**
  Less stress redistribution in the ground means less movement of the ground, which could show as surface settlement above the tunnel. Particularly with shields with slurry or earth pressure balance support, the contribution of stress relief in the face to settlement has been reduced to such a degree that complete, pressure-controlled grouting of the shield gap is of increased significance.
- **Isolation of the segments from immediate contact with soils aggressive to concrete:**
  The correct design of the constructional width of the shield seal can have an influence on the thickness of the surround created by grouting. In special cases, the construction of a layer of resistant material by grouting can be economically practical.
- **Improvement of the waterproofing of single-layer support:**
  The addition of additives to the grout capable of accepting deformation and bridging cracks can also help to waterproof defective locations in the segment seals. Such defects cannot be ruled out even if the ring building is meticulous.

In order to achieve complete, pressure-controlled filling of the annular gap around the lining, the annular gap must therefore offer a space closed on all sides for grouting. Any loss of filtrate water into the ground or the excavation chamber will have an effect on this system; however, correctly constructed grouted layers do normally achieved the required uniformity, as has been confirmed by many excavations of completed tunnels. Even small localised defects have no significance for the bedding of the lining rings and little significance for the maintenance of natural stress conditions and the improvement of waterproofing. Figure 7-6 shows the grouted space with segmental lining in a launching shaft.

Figure 7-6 Grouted annular gap, transition of a tunnel tube at a station (construction state), Cologne underground, contract M1, 1993

7.2.2 Conception

Two main groups of grouting systems can be differentiated depending on the type of machine and the segment system and its application, and these differ in the composition of the grouting material (Table 7-1):

Table 7-1 Annular gap filling with shield tunnelling machines with segmental lining

Application	Machine type	Segmental lining	Grouting system
Loose ground	shield with active face support	single layer, waterproofed	grouting of the annular gap with mortar
Hard rock	open shields	single layer	grouting of the annular gap with mortar
			grouting of the annular gap with mortar in the invert and pea gravel filling in the sides and crown
		double layer	grouting of the annular gap with mortar in the invert and pea gravel filling in the sides and crown

Systems with annular gap mortar are used with shield machines with and without active face support and single-layer segment linings. Pea gravel backfilling is limited to open shield machines and segmental lining in hard rock.

Regarding the sequence of the grouting process, primary and secondary grouting can be differentiated [234]. The aim of primary grouting is to enable immediate bedding of the rings in order to provide the preconditions for low-settlement tunnelling. The aim of secondary grouting is to fill cavities remaining around the tunnel after the passing of the machine and associated processes, e.g. settling of the primary grout.

Grouting processes can be differentiated into grouting through the shield tail and grouting through openings in the segments.

Grouting through the shield tail

Grouting through the shield tail is currently standard practice for shield machines with active face support. It has the advantage of immediately filling the shield track in soils, which immediately settle, such as non-cohesive soils. Grouting at the same time as the shield is advancing can reduce the ingress of soil into the annular gap and enable the bedding of the segment rings.

Grouting through openings in the segments

Openings in the segments are currently mostly used with shield TBMs with segmental lining and pea gravel backfilling, and for secondary grouting and quick fixing of the segments with two-part grout with single-layer segmental lining. Openings fitted with threaded connections are provided in the segments for the injection of the grout. These remain closed while the tunnel ring is being assembled. Another possibility is the use of plastic non-return (flapper) valves cast into the segments.

7.2.3 Grouting systems

For the grouting of the annular gap with mortar, there are two different grouting systems used today depending on the grout mix:

– classic wet sand mortar with and without cement,
– two-part grout.

Both systems require the appropriate machinery and equipment. The following section first describes the components for classic wet mortar. The deviations or specialities of two-part grout are described in a separate section.

Mortar handling and grouting pumps

Due to the required quantities of many m^3 per ring, the grout is almost always mixed above ground. This can be done with paddle mixers for semi-wet mixes or continuous mixers for oven-dried premixed products from silos. The grout is preferably pumped into the tunnel if the tunnel is relatively short, or else transported to the machine in mortar skips with mixing apparatus on a supply train. The grout is then transferred by a grout transfer pump into the container provided on the machine.

Double-piston pumps controlled by plate valve units pump the grout from the supply container. The provision of a number of pumps for direct supply to the grouting locations enables simultaneous grouting of a number of locations around the perimeter of the annular gap. The grouting is controlled according to volume and pressure. The quantity is adjusted through the number of strokes according to the advance rate. The grouting pressure in each grout line can be set according to the relevant prevailing ground and water pressure, with upper and lower limits being set for the grout supply to be stopped and restarted (Figure 7-7). According to the grout mix recipe, peristaltic (hose squeezing) or screw pumps are also sometimes used.

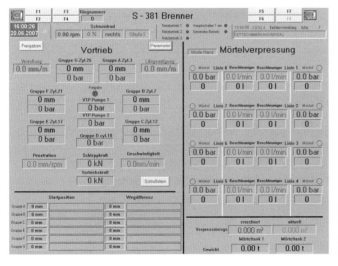

Figure 7-7 Grout pumps in the backup (top); operating panel (middle); display at the operators cabin (bottom)

Grout lines

The grout can either be pumped through openings in the segments or through grout lines integrated into the tailskin (Figure 7-8 left). The latter method means that the annular gap can be grouted immediately as it is created behind the shield tail as the shield advances. In order to prevent flow currents around the ring, the grout must emerge from a ring line through a slot facing backwards. This ideal condition is approximately achieved by arranging a number of injection connections around the perimeter of the annular gap.

Grout lines in the shield tail, which are often called injection points to differentiate them from the grout supply hoses in shield and backup, should preferably be integrated into the steelwork (Figure 7-8 left). External grout lines (Figure 7-8 right) can exceed the sealing capacity of the starting seals and also cause indentation of the shield tail when steering movements result in contact with local obstructions, which not even the scrapers mounted further forward on the shield skin in the direction of advance can completely avoid.

The cross-section of integrated grout lines is, depending on the shield, between 50 and 65 mm diameter in order to avoid blockages from the maximum grain size in the grout mix being used. A reliable guideline value for the maximum grain size is one third of the available passage cross-section. Oval and round sections are used. The extremely oval slot cross-section of external injection lines is primarily intended for use with two-part grouts and is no longer suitable for cleaning with flushing nozzles often called "mice".

The number of lines depends on the shield diameter (Table 7-2). A spacing of 4 to 5 m between adjacent lines can be stated as a guideline. Reserve lines, as are often demanded in specifications, weaken the strength of the shield tail without being of additional use, as in practical operation only the actively used grout lines are kept open continuously.

Table 7-2 Number of grout lines in the shield tail depending on shield diameter

Shield diameter	Number of grout lines
4 to 6 m	4
8 to 10 m	6
12 to 13 m	8

Figure 7-8 Grout lines integrated into the shield tail as injection points (left) or as external construction on a Japanese shield (right)

Two-part systems

Two part systems can be differentiated into genuine two-component wet mix grouting with the addition of accelerator either before entering the grout line or through an additional accelerator line facing backwards in the shield tail.

In two-component systems, a thin cement solution (component A) with a density of about 1.2 t/m^3 and water content of 80 % is pumped through the grout lines into the shield tail. The required accelerator (component B), usually with a mixing ratio of 5 to 10 %, is added either through separate mixing nozzles arranged at the back of the grouting ducts in the shield tail (Figure 7-9) or at the discharge from the shield tail. The latter variant requires active switching and flushing of the duct in the shield tail immediately after the end of pumping (Figure 7-10).

Both fluids are pumped with electrically powered screw or liquid pumps. The motors are equipped with frequency controllers, which enables the adjustment of the quantity pumped. Each line has a flow meter and manometer.

The injector principle has the advantages of less constructional thickness, leading to thinner practical annular gap widths, the lack of moving parts, the shorter mixing distance and more simple handling.

The grouting system should keep the grout under a constant pressure with continuous flow to the location of the gap as it is created behind the shield tail. Grouting systems today are therefore not generally controlled just by volume, but also pressure-controlled. The grouting pressure should ideally be measured in the cavity being grouted, but for practical reasons the pressure is normally measured at the grouting connections in the shield or the segment as applicable.

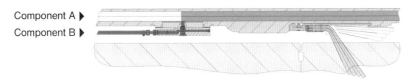

Figure 7-9 Addition of accelerator through feed openings at the back end of the grout line (injector principle)

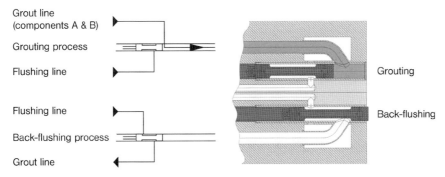

Figure 7-10 Supply of accelerated grout through the grout port poppets in the shield tail (premix variant or back-flushing principle)

7.2.4 Grout

The stiffening behaviour and the grain structure of the grout must primarily enable it to flow around the segments and fix their position. The ready-mixed grout should have very good flowing properties if it is a wet mix with high solids content, but also with relatively low filtrate water loss. Good workability until placing, even if this is delayed by machinery problems (unplanned stoppages), good pumping properties (avoidance of blockages) and rapid hardening in the annular gap are all desirable. The control of the consistency and stiffening behaviour of the grout, depending on the geological and hydrogeological and other current local conditions, is therefore of particular importance. The grout must, for example, be capable of flowing into place after a long standing time in the grout lines in order to minimise cleaning work as far as possible. The hardening and stiffening behaviour must be designed so that steering movements of the machine are still possible after long stoppages.

Philipp [234] gives examples of two mix recipes for grout, which are shown as examples in Table 7-3. A third mix is also shown (mix 3), which was used in gravelly soil for the construction of the underground railway in Cologne, contract M1.

Table 7-3 Examples of mix recipes for grout

Component	Percent by weight
Mix 1*	
Cement	14
Rock flour	17
Fine sand	58
Water	10
Mix 2*	
Cement	24
Pock flour	48
Silo lime	12
Water	17
Mix 3**	
Cement	4
Sand	58
Filler	25
Bentonite	1
Water	12

* [234],
** Cologne underground, contract M1

The use of coarser aggregates up to 8 mm is currently typical for grouting the annular gap, often with a content of 4/8 mm of nearly 50 % (which should strictly be no longer be described as mortar), sometimes high W/C ratios depending on the mix and the geology and a high content of fillers like fly ash and limestone dust. While mixes without cement are normally only used in Germany for weekend mixes before planned stoppages, mixes without any cement are regularly used on sites in France.

Although good flowing mortar first hardens after 5 to 7 hours, the immediate supporting effect has been demonstrated. The final strength of the mortar based grout is set at 3 to 7 N/mm^2 [234].

7.3 Grouting for ground improvement

The performance and range of applications and the degree of utilisation of shields are primarily determined by the geological and hydrogeological properties of the ground, or by the optimal adaptation of the machine to the prevailing conditions. Shield machines in recent years have increasingly been fitted with equipment for grouting, in order to be able to tunnel through problematic geological formations and unexpected faults, or to reduce water ingress. Experience shows that the development of universal shields has not been successful. Various types of shield are therefore combined and geological formations, which are outside the range they can cope with, are treated by grouting to comply with the requirements of the type of shield being used. Grouting in this case means the injection of flowing materials into the ground to be improved in order to consolidate or waterproof it. The cavities can be cracks and joints in solid rock, or pores in loose ground.

Grouting is therefore a suitable way of increasing the flexibility of the shield tunnelling construction process. Some other possible reasons for grouting are to enable entry into the excavation chamber, for example for maintenance or to remove obstructions, to tunnel through fault zones or reduce water inflow through special preparation of the ground.

Grouting can enable the freeing of the shield tail seal after damage has occurred, and has also been used successfully to remedy defects in the tunnel lining. If, for example, leakages appear in the completed lining, these can often be remedied by waterproofing the surrounding ground.

In addition to the geological and hydrogeological conditions, the grouting technology and grouting scheme for a tunnel project will also be influenced by operational aspects, as the grouting is undertaken by injecting into the face through the cutting wheel/cutter head and pressure bulkhead, or more radially through the shield skin or the segments. The obstruction of tunnelling operations and the associated processes caused by the grouting work has to be kept to an absolute minimum.

7.3.1 Machinery and equipment

The equipment required for grouting must be included in the planning and construction of the tunnelling machine, particularly when changeable geological formations are present. This can include:

– openings in the pressure bulkhead and/or in the shield skin with mechanical locking devices and rotating BOP seals,
– drills, preferably permanently installed in the shield in order to save time for installation,
– facilities for the production and injection of grout; when these are intended for frequent use; for longer tunnels, these should be provided in the backup.

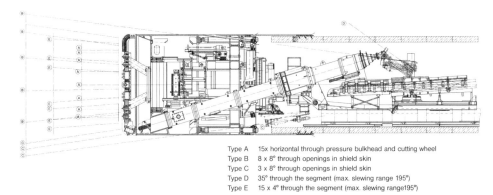

Type A	15x horizontal through pressure bulkhead and cutting wheel
Type B	8 x 8° through openings in shield skin
Type C	3 x 8° through openings in shield skin
Type D	35° through the segment (max. slewing range 195°)
Type E	15 x 4° through the segment (max. slewing range195°)

Figure 7-11 Arrangement of grouting openings in the shield and backup, open shield at Arrowhead, 2007

The number and arrangement of openings in the shield skin and in the pressure bulkhead have to take into account the requirements of the construction process, i.e. the grouting scheme, and also the geological and hydrological conditions. The drilling equipment required competes for space with the other equipment required and can therefore sometimes lead to a lengthening of the shield. Figure 7-11 shows an arrangement of grouting openings in the shield and backup areas. This arrangement permits the production of a grouted canopy, i.e. the creation of an umbrella of improved ground over the intended path of the tunnel, and also the improvement of the ground properties along the alignment through the production of a grouted block immediately in front of the shield.

The drills should if possible be mounted permanently in the shield in order to avoid time lost installing them. This intention is, however, often not practical due to the limited space and some disturbance of the main work of tunnelling is inevitable.

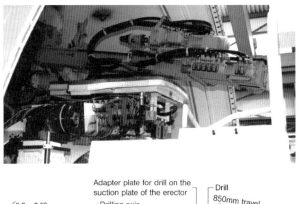

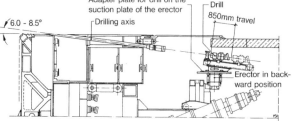

Figure 7-12 Mounting of a drill on the segment erector with an adapter, Duisburg TA7/8a 1995

Figure 7-13 Drill for grouting holes mounted on a bridge between shield and backup, (open) shield for the CLEM7 road tunnel, Brisbane, 2008

Alternatively, adapters can be provided to permit the rapid installation of the drill on the suction plate of the erector. The actual drill can be stored in the backup. Figure 7-12 shows a possibility for mounting a drill on the segment erector using an appropriate adapter.

Figure 7-13 shows a drill for grouting holes, which works from a working platform between shield and backup.

The production plant for the grout should be provided in the backup in longer tunnels. These facilities essentially consist of silos, mixers, stock containers and pumps. During grouting work, the grouting pressure and quantity (absolute and relative) should be measured and monitored continuously.

7.3.2 Grout

The requirements for the grout are determined by the hydrological and geological conditions. For technical and economic reasons, multi-phase grouting may be more suitable, with less expensive material being used for the filling of large volumes in a first phase followed by higher quality material for the filling of fine cracks in a second phase of grouting.

The grout must be capable of penetrating into the cavities as well as possible. The following requirements are placed on grouting material [208]:

– stability against sedimentation when suspensions are used,
– stability against separation due to hydrostatic pressure when suspensions are used,
– good flow properties, which can be adapted to the ground conditions,
– adequate strength/resistance against any aggressive water,
– no unpleasant odour development in the tunnel,
– no contamination of the lining with impairment of its serviceability,
– low production costs,
– environmental acceptability (groundwater).

Cement suspensions are often used for grouting, in this case mostly blast furnace cement, or bentonite-cement suspensions, which are ideal for the grouting of larger volumes. The bentonite reduces the tendency of the grout to sediment, with the addition of just a few per cent of bentonite leading to the desired effect [208].

Table 7-4 shows some examples of mixes for bentonite-cement suspensions, which were used for the service tunnel at the Channel Tunnel on the French side (Chapter 8).

Table 7-4 Mix recipes for bentonite-cement suspensions for the service tunnel on the French side of the Channel Tunnel [208]

Suspension	w/c [-]	Cement [kg]	Bentonite [kg]	Plasticiser [l]	Water [l]
Bentonite/cement	2.00	300	26	–	600
Bentonite/cement	2.94	300	40	–	880
Bentonite/cement/plasticiser	2.00	300	26	0.75	600
Bentonite/cement/plasticiser	2.94	300	40	0.75	880

The addition of plasticiser improves the flow properties of the suspension considerably, but also worsens some other properties like the filter stability and strength. The hardening time of the suspensions shown in Table 7-3 is about 4 to 5 h and the achievable compressive strength is 1.5 N/mm^2 after 7 days and 3 to 4 N/mm^2 after 28 days [208]. The decisive factor for the propagation capacity of the suspension in rock is the W/C ratio, which determines the viscosity. Excessive W/C ratios should be avoided as the suspensions then tend to bleed, i.e. lose water from the mix (Syneresis). In areas where the cavities are small, this can cause thickening of the grout with a danger of blockages. The unavoidable increase of grouting pressure can also fracture the ground.

For the grouting of finer cavities, the construction materials industry offers special grouting mixes based on chemicals (epoxy resin, polyurethane), which offer good workability properties. The cost of such products is often a multiple of that of cement or bentonite-cement suspensions, and their environmental impact should be investigated before use. A summary of currently used grouting materials is shown in Table 7-5.

Table 7-5 Grout used in the Arrowhead Tunnel, 2007 [103]

Item	Description	Pre-Excavation grouting			Annular Grouting	
		Water	Ground	Voids	Stage 1	Stage 2
A	Cementitious Products					
1	Portland Cement – Type II	●	●	●	●	●
2	Portland Cement – Type III	●	●	●		
3	Rapidset Cement	●	●		●	
4	Micro-Fine Cement	●	●	●		
5	Flyash		●		●	

7.3 Grouting for ground improvement

Item	Description	Pre-Excavation grouting			Annular Grouting	
B	**Chemical Products grouts**					
1	Polyurethane grout	●	●		○	
2	Sodium silicate	◆			◆	
3	Colloidal silica				○	
4	Minova Celbex 802	●			◆	

● Primary grouting material
◆ Secondary grouting material
○ Trial and test grouting material

Micro-Fine Cements
Used for pre-excavation grouting (Stage I) in most rock types for the reduction of water ingress and ground improvement for tunnelling. Various mixes and pressures (Stage I).

Type II and III cement
Used for pre-excavation grouting (Stage II) in most rock types for the reduction of water ingress and ground improvement for tunnelling, following the pre-excavation grouting with micro-fine cement and for filling the annular gap.

Silicate gel
Used in aquiferous joints, which represent problems for the penetration of micro-fine cement – also at thin concentrations. Successful for the reduction of water ingress.

Polyurethane grouting
Used in tunnelling with limited success for the control of water ingress and to close joints. Not successful under high water pressures when injected into drilled holes.

Chemical grouting
Water glass is used in the tunnel and at the portals for the consolidation of loose breccia and to control the water inflow. Limited success in dense rock formations and in clay.

7.3.3 Grouting work at the Channel Tunnel

Extensive grouting works were undertaken for the tunnelling of the service tunnel on the French side of the Channel Tunnel. This was necessitated by extensive zones of faulting and folding, which could lead on the one hand to strong one- and two-dimensional loading of the chalk formations and on the other hand, particularly in the upper strata, to closely spaced, irregular jointing systems [208].

Grouting measures were necessary at the Channel Tunnel in the following cases:

– in areas, in which the tunnel alignment passed through the grey chalk and poor rock conditions were expected (Figure 8-7),
– in areas, in which the permeability of the blue chalk was over 10 Lugeon and heavy water ingress had to be expected,
– passing through fault zones,
– during longer interruptions to tunnelling to seal the ground around the shield.

The grouting materials used were bentonite-cement suspensions (Table 7-4) and silicate-based materials.

In the three-bore tunnel system (Figure 8-6), the central service tunnel was bored in advance as an investigation tunnel. The section of the service tunnel already lined with the final segments could therefore be used for grouting works for the following main tunnels without any disruption to tunnelling. The shield machines for the main tunnels were almost free of space taken up for grouting equipment and most stoppages for advance grouting from the main tunnel shields could be avoided.

The shield for the service tunnel had two permanently installed drills, which enabled the drilling of investigation holes up to 70 m deep through the cutting wheel. In case grouting through the cutting wheel or the shield skin was necessary, additional drills could be installed, which were kept ready during tunnelling on the grout production car behind the backup. Their maximum drilling depth was 25 m.

Particular attention was paid to the specification of grouting pressure near the tunnel. The grouting pressure had to be set to ensure the correct propagation of the grout into the ground even under the full water pressure (up to 10 bar). Pressure losses in the transport hoses and packers had to be taken into account, but on the other hand exceeding of the maximum permissible pressure had to be avoided in order to rule out damage to the segment lining.

Figure 7-14 shows the grouting scheme for a section through a fault zone. The purpose was to produce the widest possible zone of waterproofing around the fault. Stoppage of tunnelling work for a time was unavoidable.

In order to avoid uncontrolled grout flows, grouting was performed in three stages. After localisation, the shield was advanced to about 25 m from the fault zone and a 15 m long grouted cylinder (Stage 1) was produced extending to immediately behind the fault zone. The cylinder served in particular to protect the face from ingress of water. Then the shield was advanced 15 m and the grouting process was repeated (Stage 2). After the completion of tunnelling work, the drill for the drilling of the radial grouting holes was mounted on the erector.

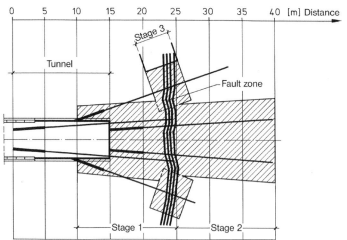

Figure 7-14 Grouting scheme for sections through fault zones, Channel Tunnel [202]

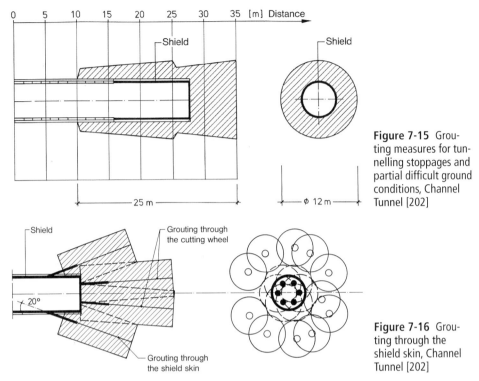

Figure 7-15 Grouting measures for tunnelling stoppages and partial difficult ground conditions, Channel Tunnel [202]

Figure 7-16 Grouting through the shield skin, Channel Tunnel [202]

Figure 7-15 shows the grouted cylinder prepared for the service tunnel shield machine for longer stoppages in rock of high permeability. The maximum angle of the holes drilled through the cutting wheel of 4° and the drilling depth of 25 m made possible the achievement of an effective radius of 12 m.

Generally, the grouting measures shown in Figure 7-15 proved sufficient. Under particularly poor conditions, however, it was necessary to increase the range of the grouted cylinder to include zones further away from the shield area (Figure 7-16). In addition to grouting through the cutting wheel, holes were drilled from the shield skin at an angle of 20° to the tunnel axis.

Figure 7-17 shows the grouting scheme for the advance grouting for the outer main tunnels, carried out from the working platforms in the service tunnel. In Phase 1, investigation holes (E) were drilled with a spacing of 5.60 m (four segment rings). The water quantities were measured after 60 cm and then every 5 m. If a measured value exceeded an equivalent permeability value of 20 Lugeon, two additional holes were drilled for grouting (I) for the production of the grouted vault shown in the section (Phase 2). In Phase 3, control holes (C) were drilled into the zone between to check the results. If a measured value exceeded the threshold of 20 Lugeon, the zones between were grouted in Phase 4.

Figure 7-18 shows the grouting scheme for the protection canopy to protect the pressure relief works undertaken by miners. In the investigation phase, the borehole water quantities were measured behind the segments (0.30 to 0.60 m) after the first 5.00 m of drilling, after every further 5 m of drilling and at the end of the hole.

If the arithmetic average of the measurements was less than 10 Lugeon and no single measurement exceeded 20 Lugeon, then single local grouting holes were sufficient and a complete grouting of the area (Figure 7-18) could be omitted.

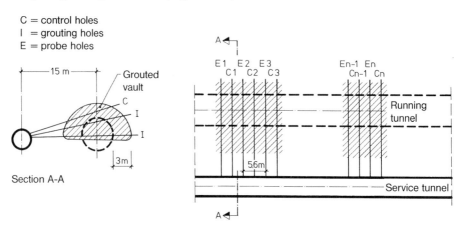

Figure 7-17 Advance grouting for the main tunnel from the service tunnel, Channel Tunnel [202]

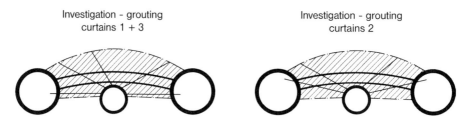

Figure 7-18 Example of a grouting scheme for the pressure relief connections, Channel Tunnel [202]

8 Open shields

The term open shields describes tunnelling machines without any closed system to balance the pressure at the face in order to provide soil support or hold back water. These include TBM-S for hard rock and SM-V1, SM-V2, SM-T1 and SM-T2 according to DAUB (Chapter 1.2). To the extent that the natural standup time of the ground is insufficient, this is achieved mechanically. This means that the tunnel cannot operate near the groundwater table; either it has to be lowered by means of advance measures like wells, or if the tunnel is below the groundwater table, the ground has to be predominantly permeable or prepared with measures like grouting or ground freezing.

In case small quantities of seepage water or localised water lenses are encountered, open dewatering can be undertaken at the face of open shields. Injection wells and lances are suitable measures. Almost all support types as already described (Chapter 6) can be used.

If the cross-section is large, the face is divided and, except for shields with closed cutter heads, remains accessible. Open shields are at a relatively low level of technology concerning the methods of face support. The good level of flexibility, particularly with hand or partially mechanised open shields, enables tunnelling through all types of loose ground and rock, even if the face consists partially or completely of rock or contains boulders. Other advantages are the lack of breakdowns and the possibility of excavating non-circular cross-sections with hand or partially mechanised shields. The low level of investment in machinery compared with other types of shields enables economic solutions, even for short tunnel lengths.

Open shields are often used for small cross-sections, such as the excavation of drainage interceptors. In addition to conventional shield tunnelling, open shields are also used for relatively short pipe jacking stretches (Chapters 13 and 14).

8.1 Shield construction

Open shields can be categorised according to the type of excavation into:

- hand shields (now only used in low-wage countries),
- part-face excavation,
- full-face excavation.

8.1.1 Hand shields

Hand shield denotes shields with the ground mostly being dug by hand. Until the introduction of compressed air in shield tunnelling by Greathead in 1886 (Section 1.3), all shields were open and advanced by hand digging. Low advance rates and high wage costs now limit the use of hand shields to the advance of short stretches and tunnels in low-wage countries, where the use of mechanised or partially mechanised shields exceeds the bounds of cost-effectiveness. In technologically undeveloped countries with low wage costs, hand shields are sometimes used for longer tunnels. For example, an open shield from the 1950s was still in operation at the end of 2000 for the construction of the underground railway in Charkov (Ukraine).

In Germany, an open shield with hand digging was used for the construction of the underground Line 1 in Nuremberg from August 1973 to June 1976 (Figure 8-1). The ground to be excavated varied between stable Keuper sandstone and loose sands. An open shield was used in non-cohesive sands for a length of about 2,000 m [222]. The external diameter of the hand shield used was 6.04 m and the length 5.77 m. The stability of the face was limited to the non-cohesive sand running out onto the platforms at its angle of repose. Daily advances of 8 to 10 m were achieved after familiarisation. In order to limit settlement, intermediate or working platforms were arranged in the shield, with the middle and upper platforms being hydraulically extendable. These platforms divided the face and thus the sloping material into individual sections.

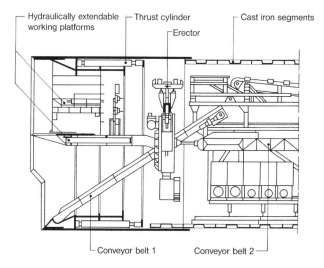

Figure 8-1 Hand shield, Nuremberg underground Line 1, 1973/76 (Westfalia Lünen)

Figure 8-2 Hands shields with rectangular or horseshoe-shaped cross-section (Mitsubishi)

Hand shields are inherently flexible and can have non-circular cross-sections without additional technological problems (Figure 8-2).

8.1.2 Part-face excavation

Since hand digging only enables slow advance rates and often results in unacceptable conditions for the miners, shields are often equipped with backhoes (Figure 8-4) or road headers (Figure 8-3) to excavate the ground (Chapter 4).

Worth mentioning are systems, in which the backhoe or cutting arm is part of a modular system and can be exchanged in the tunnelling machine according to the conditions encountered.

The field of application of partially mechanised open shields is generally limited to smaller cross-sections (interceptors etc.); sometimes, however, they are also used for short stretches of tunnel of more than 10 m diameter.

Example: Hofoldinger Tunnel

Project data
Tunnel length: approx. 17.5 km,
Tunnel diameter: 3.48 m,
Geology: loose gravel, clay, nagelfluh
Lining: reinforced concrete segments enclosing steel pipes,

Machine data
Part face machine MHSM1, ⌀ 3.48 m, weight 99 t, length 10.77 m.

The 17.5 km long Hofoldinger tunnel is the fourth and last construction section of the new drinking water supply tunnel 30 km long from the Mangfall Valley.

Tunnelling work at the Hofoldinger Tunnel started in March 2000 with a SM-V2, but had to be given up without success due to technical difficulties. Münchener Stadtwerke (Munich City Utility) tendered the work again. This time, a SM-T2 was to be used.

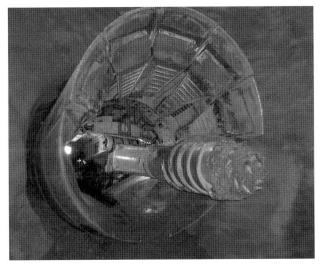

Figure 8-3 Open shield type SM-T2 with breasting plates and road header for the construction of the Hofoldinger Tunnel, 2002 to 2007

Figure 8-4 Open shield with integrated backhoe arm, external diameter 7 m, London Docklands Railway City Extension 1988, (Herrenknecht)

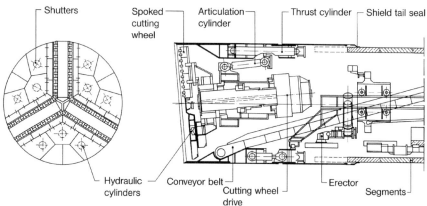

Figure 8-5 Open shield with spoked cutting wheel and shutters, Hanover Underground, Marienstraße, 1985 to 1987 [291]

A part face machine with combined excavator and breaker unit was designed in collaboration with Herrenknecht, which could be rebuilt to road header operation after 100 m underground in order to increase the advance rate (Figure 8-3). The "high-tech" machine then achieved very good advance rates through gravels of all types and even managed up to 20 m per day through very hard nagelfluh.

Partially mechanised shields are also often fitted with hydraulically operated breasting plates, which provide additional mechanical support in individual areas and around the perimeter of the shield. These can be seen retracted in the crown above the excavators in Figs. 8-3 and 8-4.

8.1.3 Full-face excavation

Open shields with fully mechanical excavation equipment are used in order to enable tunnelling in harder ground formations or increase the advance rate in loose ground. This type of machine already has a relatively high degree of technical complexity.

The full-face excavation machinery can be constructed as a cutting wheel or cutter head and fitted with the appropriate tools. The arrangement of shutters between the spokes of a cutting wheel, as shown in Figure 8-5, achieves a reduction of settlement and controlled entry of soil into the shield. This type of shutter is used in low- to non-cohesive soils. No further uses of this type of shield are known.

8.2 Projects

8.2.1 Example: Eurotunnel – under the English Channel, 1988 to 1991

The most prominent example of all for the use of shield technology is the crossing of the English Channel, where long stretches were driven with open shields. After the previous unsuccessful attempt at a crossing in 1975 had been abandoned for financial reasons [156], [202], private investors started to investigate the crossing of the Channel again at the start of the 1980s. In 1986, a French/English consortium consisting of the concessionaires Channel Tunnel Group and France Manche (CTG/FM), were awarded a contract to construct the link consisting of two single-track railway running tunnels and a central service tunnel.

Project

The running tunnels with a spacing of 30 m have an internal diameter of 7.6 m and the service tunnel in the middle has an internal diameter of 4.8 m (Figure 8-6). The route of

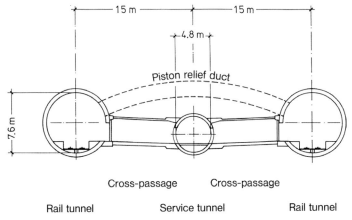

Figure 8-6 Tunnel cross-section of the Eurotunnel at a cross-passage [202]

the tunnel crosses the English Channel at its narrowest point. The distance between the portals at Castle Hill north of Folkestone (England) and near Coquelles (France) is about 50.5 km (Figure 8-7). About 37 km of the length of the tunnel runs directly below the English Channel at depths of up to 126 m below sea level. The shallowest depth below the seabed is 17 m not far from the French coast and the greatest depth in the middle of the Channel is 60 m. The tunnel connecting east of the launching shaft in Sangatte reach the station at Coquelles a further 3.7 km inland. On the English side, three tunnels about 9.8 km long under land connect the starting cavern with side access at Shakespeare Cliff and the station portal of Castle Hill [165].

Every 375 m are cross-passages with a diameter of 3.3 m connecting the three tunnel. They serve as emergency exits and for maintenance. The cross-passages are closed with water- and fire-tight doors, so the service tunnel is available as an independent escape system. At a spacing of 250 m, the running tunnels are connected by piston relief ducts with a diameter of 2.0 m in order to relieve the piston effect of the passing trains.

At about the third points of the tunnel centreline, crossover caverns are provided to enable trains to change tunnels in case of breakdowns. While these, like the piston relief ducts and the cross-passages, were mined conventionally, all the tunnels were bored with altogether 11 shield machines.

① Upper chalk ⑥ English channel
② Middle chalk ⑦ Tunnel
③ Lower chalk ⑧ Crossover
④ Chalk marl ⑨ Shaft in Sangatte
⑤ Clay ⑩ Tunnel portal

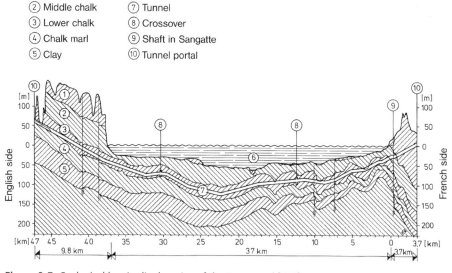

Figure 8-7 Geological longitudinal section of the Eurotunnel [202]

Geological and hydrogeological conditions

Extensive geotechnical investigations were undertaken along the entire route of the tunnel as part of the preparatory work for the project. The "lower chalk" formation is a mixture of chalk and clays and ideal for tunnelling regarding excavation and water impermeability. It is very uniform, plastic and bedded without faults, has a water content of 11 % and an average strength of 2 to 10 MN/m². Independently of water content, the lower chalk breaks easily, without tending to smear; this factor is a construction criterion for the cutting tools of a shield machine. The water permeability of the formation is about 2.5×10^{-7} m/s, as was observed on site in the tunnel bored by Beaumont in 1882/83.

The chalk marl layer underneath is indeed also impermeable to water, but much softer and shows strongly plastic behaviour under loading, which brings considerable problems regarding processing, loading and settlement.

In order to exploit the favourable geological conditions and avoid cutting through unfavourable zones as far as possible, the design intended the advance of about 90 per cent of the tunnel through the lower chalk. The vertical alignment of the drives was selected so that the thickness of this layer above the tunnel would be at least 10 m. Difficult ground conditions nevertheless had to be expected near the French coast, where boreholes has discovered faulting and fractured zones, which demanded appropriate tunnelling processes [202].

Tunnelling and machine concept

The tunnel drives were driven by altogether eleven shield machines, five on the French side and six on the English side, which all started from the respective starting shafts on the French side at Sangatte and the English side at Shakespeare Cliff. The entire route of the tunnel was therefore divided into four sections: French side under land, French side under the sea, and English side under land and sea. While the six drives from the English side were each driven by one machine, the two running tunnels under land on the French side were driven by only one shield due to the comparatively short overall length of just 6,542 m (Figure 8-8, Table 8-1). The drives of the sections of the service tunnel under the sea always stayed some kilometres ahead in order to gain additional information about the geological situation before driving the running tunnels and also to permit ground improvement by grouting (Chapter 7.3.3).

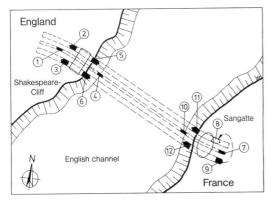

① English side, under land, service tunnel
② English side, under land, north running tunnel
③ English side, under land, south running tunnel
④ English side, under sea, service tunnel
⑤ English side, under sea, north running tunnel
⑥ English side, under sea, south running tunnel
⑦ French side, under land, service tunnel
⑧ French side, under land, north running tunnel
⑨ French side, under land, south running tunnel
⑩ French side, under sea, service tunnel
⑪ French side, under sea, north running tunnel
⑫ French side, under sea, south running tunnel

Figure 8-8 Tunnel sections of the Channel Tunnel [11]

Table 8-1 Details of the tunnel sections of the Channel Tunnel [53], [100]

Tunnel section	English side						French side					
	service tunnel under land (1)	running tunnel under land, north (2)	running tunnel under land, south (3)	service tunnel under the sea (4)	running tunnel under the sea, north (5)	running tunnel under the sea, south (6)	service tunnel under land (7)	running tunnel under land, north (8)	running tunnel under land, south (9)	service tunnel under the sea (10)	running tunnel under the sea, north (11)	running tunnel under the sea, south (12)
Tunnel length [km]	8.2	8.1	8.2	22.3	18.5	17.8	3.3	3.3	3.3	15.6	19.4	20.1
Manufacturer	Howden	Howden	Howden	Howden	Robbins/Markham	Robbins/Markham	Mitsubishi	Mitsubishi		Robbins	Robbins/Kawasaki	Robbins/Kawasaki
External diameter of machine [m]	5.76	8.72	8.72	5.76	8.36	8.36	5.61	8.62		5.77	8.72	8.72
Internal tunnel diameter [m]	4.8	7.6	7.6	4.8	7.6	7.6	4.8	7.6		4.8	7.6	7.6

Starting from the French side, faulted zones were forecast until reaching the lower chalk. The shields on the French side were therefore designed to be able to work as EPB shields in the faulted area and later work as open shields in the lower chalk formation [11], and should therefore be categorised as convertible shields (Chapter 12). In order to complete the description of the Channel Tunnel project here, details of the shields (Figure 8-9) are now given.

The single-layer lining on the French side consists of 1.4 m (service tunnel) and 1.6 m (running tunnel) wide, bolted segment rings of precast reinforced concrete segments installed in the protection of the shield. The segments were fitted with neoprene sealing gaskets.

On the English side, the more favourable conditions in the lower chalk meant that only open shields were used. The single-layer lining consists of 1.5 m wide reinforced concrete segment rings, which were only dowelled and expanded against the surrounding ground with a wedge-shaped keystone behind the shield. Areas where water ingress was later discovered were waterproofed by grouting.

In order to achieve the best possible advance rates, the shields were designed to allow simultaneous installation of the lining whenever the geological conditions permitted.

All machines on the English side were equipped with walking mechanisms. Hydraulic grippers transferred the necessary thrust forces radially into the surrounding ground. The cutting wheels were then pushed forwards by hydraulic cylinders from a stationary shield. The installation of the lining took place simultaneously behind the shield (Double shield mode). Only under unfavourable ground conditions, which prevented the transfer of the thrust forces into the surrounding ground by the walking mechanism, did the thrust forces have to be pushed against the lining by hydraulic cylinders. In this case, simultaneous installation of the lining was not possible.

On the French side, the thrust forces were mostly transferred to the lining behind the machine. The shield tail was, however, constructed so large that two rings could be assembled at the same time in its protection. The shields under the sea on the French side additionally had a walking mechanism, in order to enable simultaneous excavation and lining in areas of suitable geological formations. The excavated muck was removed from the excavation chamber of the machines on the French side by screw conveyors. The three shields under the sea were provided with a second opening in the pressure bulkhead to enable the installation of a conveyor belt for use when the shields were operated in open mode in favourable ground conditions.

English side

Sections 1 and 4

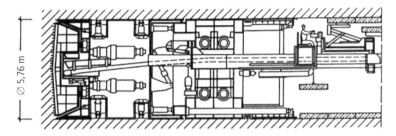

Sections 5 and 6

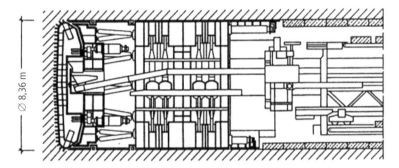

Sections 2 and 3

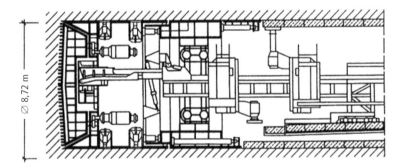

Figure 8-9 Shield machines for the Channel Tunnel [11]
English side, p. 186
French side, p. 187

French side
Section 10

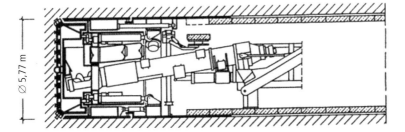

Sections 11 and 12

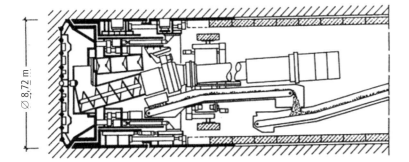

Section 7

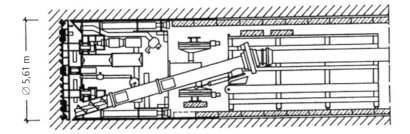

Sections 8 and 9

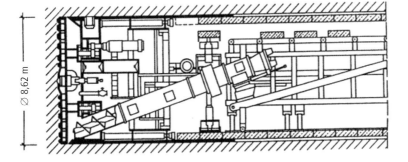

While the shields on the French side (Figure 8-10) were provided with suitable equipment for tunnelling through faulted ground and were capable of being converted to an EPB shield with closed mode, additional equipment proved necessary on the English side (Figure 8-11) to cope with unexpected water ingress. In normal operating mode, the excavated muck was removed from the excavation chamber on conveyor belts. The conveyor belt could be pulled back out of the excavation chamber in a few minutes and the pressure bulkhead automatically closed. At the same time, the steering gap behind the cutting wheel was sealed around the perimeter of the shield. The gap between the last segment ring installed behind the shield and the shield tail could also be sealed by a hydraulically operated closing system.

On the French side, grouting behind the shield was necessary in order to bed the segment rings, which had been assembled in the shield tail. On the English side, grouting was only necessary in zones with water, as the segment rings had already been expanded directly against the surrounding ground.

As the tunnelling work progressed, extensive grouting was undertaken from the service tunnel, particularly on the French side. The equipment and devices used have already been described in Chapter 7.

Progress of the project

The enormous investment costs for the project and the resulting enormous financing costs, which already represented a healthy 39 % of the total volume at the planning stage, justified great investments in increasing the advance rate of the machines [37]. The cost of all eleven shield machines was about 300 Mio DM, only about 10 % of the financing costs (as of 1991). The 150 km of tunnel required were driven in only 3½ years [109]. Figure 8-12 shows the performance curves for all eleven shield machines. The maximum advance of a shield in one month was 1,071 m (24 h/d, 7 d/week, one 8-hour maintenance shift per week). The average performance of the individual shields over the entire course of the project lay between a good 800 m/month and scarcely 300 m/month. The shields under the sea always achieved better advances due to the longer tunnel lengths [37].

Figure 8-10 Shield machine for the under sea running tunnels on the French side (11/12), external diameter 8.72 m, 1987 (Robbins/Kawasaki/FCB)

Figure 8-11 Shield machine for the running tunnels under sea on the English side (5/6), external diameter 8.36 m, driving through a crossover (Markham/Robbins)

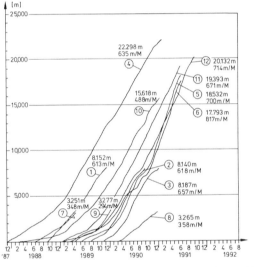

① English side, under land, service tunnel
② English side, under land, north running tunnel
③ English side, under land, south running tunnel
④ English side, under sea, service tunnel
⑤ English side, under sea, north running tunnel
⑥ English side, under sea, south running tunnel
⑦ French side, under land, service tunnel
⑧ French side, under land, north running tunnel
⑨ French side, under land, south running tunnel
⑩ French side, under sea, service tunnel
⑪ French side, under sea, north running tunnel
⑫ French side, under sea, south running tunnel

Figure 8-12 Performance curves at the Channel Tunnel [37]

Consideration of the early advance rates of the shields in comparison to the record values achieved as the project progressed show the cause of the impressive average performance. With one exception, none of the shields achieved an advance of 1,000 m in the first 18 weeks. As a comparison, the Beaumont machine used in 1881 had already achieved similar rates (1,000 m of tunnel in the first 18 weeks) [109]. Even considering that a machine length of 280 m means that the final assembly is only possible as the drive progresses, and the crews naturally require a certain familiarisation time, the decisive reasons for the subsequently achieved increase of performance can be explained by the introduction of specialised logistics management [109] and the technical optimisation of the shields, partially during the drive, which according to [37] took a year. The investment in logistics and the long technical optimisation phase on this project are only possible for corresponding tunnel lengths, so such levels of performance have only been achieved on similarly large projects. The comparatively problem-free geological conditions should also be considered, which scarcely produced any surprises. The shields driving the service tunnel also operated a few kilometres ahead and provided additional information about the geological conditions [109] and also enabled extensive ground improvement works to be carried out (Chapter 7) [202].

The relatively benevolent geological conditions on the English side enabled the extensive use of reinforced concrete segments expanded directly against the surrounding ground. This process has a long tradition in England and normally produces high advance rates [37].

In addition to the high advance rates, the exact steering of the machines should be emphasised. The shields in the service tunnel met with a deviation of only 350 mm horizontally and 60 mm vertically [295].

Despite the successfully technical progress on the project, some problems during the tunnelling work must be mentioned.

After about 1,000 m of advance from the English side, a zone of heavily jointed blue chalk was encountered. Rockfall of more than 1,5 m^3/m occurred continuously in the crown behind the shield tail. The water inflow had already increased to 400 l/min during the advance of the service tunnel.

It was largely thanks to the motivation of the Irish miners that the tunnel was successfully bored through this section despite these difficulties. Some rebuilding of the machine was also required. The service tunnel machine was fitted underground with a finger shield arranged as a hood in the shield tail in order to prevent collapses of the crown in the area where rings were being assembled. The upper segment transporter was also replaced by a working platform in order to enable the necessary grouting work shortly after the shield. From this platform, aluminium stiffening arches were installed to stabilise the shape and to ensure the circular shape of the rings as they were expanded despite the absence of reaction from the ground in the crown. The same modification was built into the machines for the running tunnels at the works.

Götz [109] describes a problem with one of the shield machines in the service tunnel. Damage was discovered to the front cutting wheel bearing. This meant that the cutting wheel could no longer be extended, and the machine continued to the end of the tunnel in this condition.

On one of the machines for the running tunnels under the sea on the French side, deformation of up to 75 mm was discovered in the centre of the cutting wheel, which had led to cracks in the weld seams of the spoke to the hub. The causes of this were the large central hub with relatively small intake openings and the flat form of the cutting wheel with a large span between supports of 5.30 m. The retroactive fitting of a centre cutter in the hub proved to be the solution to the difficulty through a reduction of the contact pressure by 40 %. This measure led to a performance improvement of 100 %.

There were also defects in the construction of the lining in the form of large offsets between neighbouring rings, which could not be avoided. Heavy water inflows through the radial joints had to be sealed with a lot of extra work [37]. The lining on the French side also gave problems with the transfer of the thrust forces into the lining. With an annular gap width of 190 mm and immediate grouting, the load was acting on segment rings, which were still practically without any bedding. This resulted in cracks, which had to be laboriously made good, as it was no longer possible to replace the segments.

The tunnelling works were complete in the middle of 1991 and the first shuttle made a test trip through the tunnel in summer 1993.

Figure 8-13 Overview of the Arrowhead Tunnel project

8.2.2 Arrowhead Tunnel

As part of a 70 km long drinking water supply project for the city of Los Angeles, two tunnels were to be driven with a total length of 13 km with an internal diameter of 4.87 m and external diameter of 5.53 m. The length of the individual tunnel drives was 9.7 km (east) and 6.1 km (west) (Figure 8-13). The primary lining consisted of 330 mm thick reinforced concrete segments. For the later use as a water tunnel, either a reinforced concrete inner lining or in fault zones a welded steel inner pipe was installed.

Tunnelling works had already been started in 1997 with open hard rock shield machines and a temporary segment lining without waterproofing. After 2,440 m of the east tunnel had been driven, the works had to be stopped due to high water inflows of 100 l/s into the tunnel. After the tunnelling machine had been dismantled, the already bored part of the east tunnel was mostly grouted to waterproof it and a steel liner was installed.

In 2001, the remaining tunnelling works were tendered once more and contracts were awarded in 2002 for mechanical tunnelling of 6,765 m (east) and 6,023 m (west) after a prequalification process for possible machine manufacturers.

Geological and hydrogeological conditions

The tunnels of the Arrowhead project lie parallel to and within a few kilometres of the trough of the well-known San Andreas fault and cross many active outliers of this fault zone. The overburden along the two tunnels ranges from 15 to a maximum of 640 m. The geology along the route of the tunnels is characterised by relatively fresh to slightly weathered metamorphic and crystalline igneous rocks. The metamorphic rock mainly consists of gneiss with thin layers of marble. The granite contains quartz-monzonite with slight contents of quartz-diorite and granodiorite. While tunnelling through heterogeneous rock, blocky rock at the face also had to be expected. Near the portal, in the fault zones and up to 100 m below ground level, weathered rock had to be expected. In the fault zones, water ingress with pressures up to 30 bar were possible.

Tunnelling and machine concept

The tunnelling and machine concept specified the following minimum requirements:

- Two shielded hard rock machines as open (single) shields,
- the installation of a waterproof segmental lining with an internal diameter of 4,877 mm,
- extensive provision for pre-excavation grouting from the machine (Figure 8-19),
- provision for sealing the machines tight against pressures of up to 10 bar during stoppages,
- cutter head displacement,
- maintenance of tunnel driving with a water inflow of 32 l/s.

Based on these minimum requirements, a machine concept was developed for both machines intended to maintain tunnelling advance even with high water ingress. If the manageable water quantity should be exceeded, it was intended to use permanently installed drilling equipment to grout ahead of the machine to reduce the water ingress to manageable quantities (Figure 7-11, types A to D). Drilling and grouting campaigns could be performed in both open and closed mode (against standing groundwater pressure). The creation of static groundwater conditions while boring and for grouting was one of the key points of the machine and tunnelling concept.

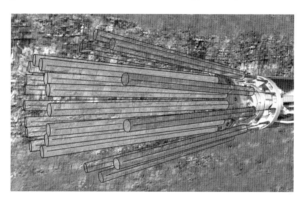

Figure 8-14 Diagram of an pre-excavation grouting body [inner circle by means of drilling through cutting wheel and face (pale) and outer peripheral circle by drilling through the shield skin (dark)]

Figure 8-15 Cutting wheels of the open shields for Arrowhead

Provision was also made, in case longitudinal drainage occurred in the annular gap behind the shield, to stop the flow of water and the washing out of the annular gap filling and regrout under static conditions. This was done by creating a sealing flange between segment and rock mass.

The shield construction with articulated shield tail and all sealing systems were designed for 10 bar of static water pressure. For the cutting wheel, a massive construction similar to a cutter head was used with 39 disc cutters $\varnothing$ 17" and one working direction of rotation (Figure 8-15). The primary mucking was by a screw conveyor $\varnothing$ 700 mm with exchangeable wearing elements discharging onto a following conveyor belt. The drainage at the machine was constructed as a multi-stage system with 575 m^3/h flow quantity and storage tanks to avoid sanding of the drainage pipework.

Progress of the project

The east machine started first from the Strawberry Creek portal at the end of August 2003, and the west machine started 1½ months later in the middle of October at the Waterman Canyon portal, only to be stopped after 40 m by a fault zone and the grouting work that became necessary. A mudslide during the Christmas break of 2003, which had been made more likely by previous bush fires near the construction site and flooded the portal area and the TBM, prevented the immediate resumption of tunnelling work and demanded the drying out and repair of the tunnelling machinery lasting into early 2004.

The advance rates of both shield TBMs proved, as expected, to be strongly dependent on the rock and groundwater conditions. While the advance rate in hard, competent granite could reach 244 m/month, the advance was severely disrupted in aquiferous rock with an instable face and in fault zones, as these demanded time spent on pre-excavation drilling and grouting. The tunnel drainage works and the excessive flow of fine material also hindered tunnelling work. An overview of the performance data is shown in Table 8-2.

Table 8-2 Tunnelling advance at the Arrowhead Tunnel, west and east

Advance rates	East tunnel	West tunnel
Tunnel length	6,840 m	6,062 m
TBM tunnel length	6,765 m	6,023 m
Maximum monthly advance	361 m	250 m
Maximum weekly advance	108 m	67 m
Maximum daily advance	29 m	24 m
Average monthly advance	118 m	90 m
Average weekly advance	30 m	22.5 m
Average daily advance	5.9 m	4.5 m

In order to control the groundwater ingress to the permissible quantity of 32 l/s for more than a week, an pre-excavation grouting body was maintained at least 6 to 9 m deeper than the shield with a minimum overlap of 6.5 m and constantly adapted to suit the current rock conditions. This entailed the performance of the drilling and grouting works listed in Table 8-3. With the intention of reducing the water inflow inside the 30 m interval to 16 l/s and the uncertainty whether the real water inflow had been discovered by the probe drilling, a water ingress of 0.06 l/s/m was specified as the threshold value for grouting work.

Table 8-3 Pre-excavation drilling and grouting works at Arrowhead Tunnel [103]

Length of the advancing grouted block ahead of the TBM:	6 to 12 m, depending on the rock mass conditions (minimum overlap of the investigation and grouting holes: 6.5 m) (max. effective length of grouting holes: 30 m)
Length of investigation holes into the face:	15 to 30 m, depending on the rock mass conditions (max. effective length: 45 m)
Number of investigation holes per boring cycle:	2 to 6 holes drilled, depending on the rock mass conditions
Number of water pressure tests in the hole for the determination of water inflow per boring cycle:	2 each, depending on the quantity and the certainty of the water inflow
Minimum number of grouting holes drilled per boring cycle:	2 at an angle of 4°
Maximum number of grouting holes per boring cycle:	15 holes at 1.5° through the face and 19 holes at 4° and 11 holes at 8° through the shield

In addition, the following additional measures were carried out (see also Table 7-5) [103]:
– Advancing waterproofing grouting,
– ground improvement grouting (if required, in many stages),
– filling of any voids encountered by grouting if required and along the shield and the segment lining in order to avoid fine material flowing from the face),
– installation of a waterproof segmental lining that provided full ground support with backfill grouting,
– installation of inflatable segment collars in the annular gap to the segment lining (to prevent water bypassing along the annular gap fill),
– drilling of drainage holes (at 20° through the segments, when necessary to limit the water pressure and to improve the rock conditions).

Due to the tunnelling experience, the advance characteristics of the machines were changed with a reduction of maximum revolution speed (from 9.75 to 7.5 ^{rpm}) and increase of the available torque (from 2,000 kNm to 3,520 kNm) to give characteristics more similar to an "earth pressure balance mode". In order to free the shield skin in squeezing rock, the maximum thrust force was doubled to 58,540 kN with a conversion to a two-stage 700 bar hydraulic pressure with a corresponding change to the hydraulic pumps and power units. On occasions, 11 additional temporary cylinders were used, raising the total thrust force to 114 MN, four times the original force of 29,270 kN. Further measures to overcome the friction on the shield skin were the enlargement of the overcut and lubrication of the skin with bentonite.

An additional drilling rig and an additional drilling level E (Figure 7-11, type E) in the face were also used occasionally. The tunnel drainage had to be adapted to cope with the fines content (up to 15 % of the volume, with peaks up to 50 %) and was converted to closed circulation with water as the transport medium and integrated fines separation with a nominal flow capacity of 150 m^3/h using screens and centrifuges.

Both drives were completed successfully with the breakthrough of the east machine at the end of April 2008 and the west machine in the middle of August 2008. The example of this project shows that a tunnel drive can be completed successfully despite high water inflows and pressures with open shields and the implementation of thorough drilling and grouting campaigns.

8.3 Double shields [203]

8.3.1 Development

Next to the development of the single shield TBM in Switzerland, another new machine concept was developed to combine a shield with a gripper TBM: the double shield TBM. This type of machine was developed by Carlo Grandori in 1972 [111] and was intended to achieve high advance rates even in poor or highly changeable geological conditions. As with the single shield TBM, installation of support is decoupled from the advance to permit continuous advance. Further advantages over the gripper TBM are better steering control in soft rock formations and the possibility of the variable installation of prefabricated support elements. In contrast to the single shield TBM, the double shield has alternative possibilities for thrust and gripping, since it can either operate using the gripper bracing of an open gripper TBM or also use its telescoping and thrust cylinders aligned along the direction of the tunnel.

Shield machines of this type are currently used very successfully to drive water tunnels with diameters from 3.8 to 7 m. In particular, the development of specialised segment systems for the double shield tunnelling of water tunnels, like hexagonal segments (Chapter 6), has made them more successful. The short construction schedules demanded for continuously lined pressure tunnels can only be achieved by using double shield TBMs, even in geological conditions ideal for a gripper TBM, because they enable boring and lining to be undertaken simultaneously.

Under ideal conditions, double shield machines in the diameter range of 5 to 7 m can achieve average advance rates of 35 to 70 m/d. The cycle time for the boring and installation of a segment ring (hexagonal segment, 1.3 m length) can be estimated at about 15 minutes for an excavation diameter of 5 m. This includes relocating the TBM twice (1½ minutes each and the assembly of a segment in about 3 minutes [286].

The development of a machine capable of continuous tunnelling in loose ground under the groundwater table has been a goal for many years. One example is the use of a machine with elongated shield tail, to permit the installation of two segment rings, for the Sophia Tunnel in the Netherlands (Section 8.3.3.3, Examples).

8.3.2 Functional principle

A double shield TBM consists of the front shield with cutter head, main bearing and drive, and a gripper shield with gripper shoes, shield tail and auxiliary thrust cylinders. The two shield sections are connected, as described in the section about telescopic shields, by the main thrust cylinders acting as telescoping cylinders (Figure 8-16). Comparable systems are also manufactured by Lovat and Herrenknecht.

The basic principle is that the machine is braced by the grippers of the gripper section acting radially against the perimeter of the excavated cavity, while boring and installation of the segmental lining take place simultaneously. Cutterhead and front shield are being pushed forwards by the telescopic cylinders. The auxiliary thrust cylinders in the shield tail only serve to hold the position of the segments after they have been installed. When the end of the full stroke of the telescoping cylinders is reached, the grippers are retracted and the gripper shield is pushed up to the front shield, with the extension of the auxiliary thrust cylinders holding the last segment ring in position. The machine is supported while the gripper shield is being regripped by the stabilising shoes, the shield skin of the front shield and the auxiliary thrust cylinders. It is possible to brace horizontally with the grippers but they are normally applied at an angle of 45° upwards in order to press the gripper shield downwards. This variable bracing is made possible by the arrangement of the grippers as a rocker construction, and achieves stable bracing to enable reaction to the vertical forces from the front shield.

If radial bracing using the grippers is not possible, the necessary thrust forces can also be applied by the telescoping cylinders (with the gripper shield remaining static). In this operating mode, the telescoping cylinders extend fully while the auxiliary cylinders only transfer the thrust to the last segment ring. Alternatively, in a further possible mode described as "single shield mode", the front and gripper shields form a fixed unit, the telescopic joint is fully closed with its cylinders retracted, and the necessary thrust force is provided by the auxiliary cylinders. This means that simultaneous excavation and ring installation is not possible and the advance rate is correspondingly lower.

8.3.3 Special features

8.3.3.1 Shield skin and bentonite lubrication

The double shield machine, due to its machine concept with a long shield skin, is in danger of jamming if convergence of the rock mass occurs. This is not usually the clamping of the TBM along the entire length of the shield but more due to rock falls at the telescopic joint, which prevents the gripper shield being brought up. This possible danger to the double shield TBM is countered in the construction of the machine with appropriately stepped diameters and vertical offsets of the longitudinal axes of the cutter head, front and gripper shields. Bentonite lubrication of the shield skin reduces the friction on the shield during the boring stroke and as the gripper shield is brought up.

The high advance rate achievable with double shield TBMs can also be seen as advantageous with respect to convergence of the rock mass, since the machine constantly moves forward instead of standing still.

8.3.3.2 Telescopic shield

The layout of the telescopic shield demands particular attention because at this location sideways steering movements of the shield combine with the longitudinal movement during the boring stroke. Excavated spoil, which has been formed to a "hard cake" by the front edge of the gripper shield, gathers at this point and blocks the telescopic joint and the annular gap between shield skin and the sides of the tunnel. If the shield sections are angled to each other, this results in a one-sided opening of the telescopic joint. Based on the initial experience of tunnel driving with double shield machines, the sealing of the telescopic joint was ensured by a separate telescoping shield with its own cylinders (Figure 8-17) and material scrapers were provided [110], [112]. The sealing of the telescoping joint still represents a challenge in search of a solution for modern double shield TBMs, because the penetration of mud and water inside the shield has to be prevented when water inflows occur.

The inner telescoping shield in modern double shield TBMs can be extended along the tunnel axis by 600 to 800 mm and thus permits the opening of the TBM in exceptional cases and an exit for miners to undertake any support work required at the cutter head, as was necessary on the "Ginevra" TBM tunnel drive at the Evinos-Mornos Tunnel [113].

Provision for drilling ahead of the machine and drilling channels for the production of a pipe screen over the TBM are in line with this concept and are usually arranged in the shield tail area of double shields, so that the entire machine and especially the necessarily open telescoping area can be protected by the pipe screen.

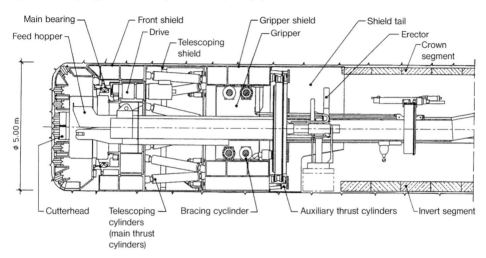

Figure 8-16 Double shield TBM ⌀ 5.0 m (Robbins) [280]

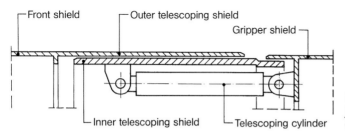

Figure 8-17 Telescoping joint of a double shield TBM (Herrenknecht) [122]

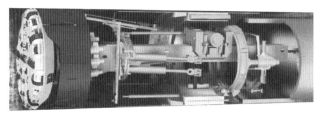

Figure 8-18 Model of the Robbins ACT TBM [135]

8.3.3.3 Examples

1) The ACT-TBM from Robbins [135]

All double shields have a few weak points, particularly in mixed ground conditions with hard rock and significant formation water quantities, or when unforeseen difficulties are encountered. Robbins has developed a new TBM concept, which permits more effective performance of the measures necessary to tunnel through unfavourable geological conditions. This new TBM concept is called the ACT TBM (All Conditions Tunneler – Universal TBM) (Figure 8-18). It combines the properties of a double shield machine and an open high-performance TBM, with particular attention being paid to ground conditioning measures directly behind the cutter head.

The ACT TBM is based on the idea that almost all geological conditions include unstable rock, and unstable conditions should be discovered and conditioned in advance. In order to implement this idea but still be capable of achieving high advance rates and the installation of concrete segments, the ACT TBM has the following properties:

– Retractable front shield to enable conversion to an open TBM in favourable geological conditions; in unstable conditions, the machine can be operated in double shield mode with limited possibilities for grouting,
– the working area directly behind the cutter head offers space for drilling and for the implementation of support measures, e.g. the installation of rock bolts, pre-excavation grouting, installation of steel beams or the application of shotcrete,
– floating gripper system to enable optimal advance rates even in the hardest rock,
– hydraulically expandable front shield, which supports the rock directly behind the cutter head and enables continuous steering and also the installation of temporary support,
– fully shielded TBM with segment erector in order to ensure a minimum open standup time between excavation and full lining with a fully sealed segment lining,
– fully electronic operating data recording system, which monitors all activities, records and simultaneously transmits the information to control stations above ground and the TBM control cabin in order to facilitate well-founded decisions between project management and operating crew.

Working area for ground treatment measures
Directly behind the cutter head is a working area for ground treatment measures, with high-performance percussion drills for probe drilling at an optimal distance behind the cutter head. These cover the entire perimeter (360°) and can also be used to drill holes for ground treatment if required by the geological conditions. Collaring holes in the cutter head permit advance fan drilling, holes parallel to the tunnel axis or drilling into the face.

(a)

(b)

Figure 8-19 Robbins ACT TBM
a) closed configuration (retractable shield is shut)
b) open configuration (retractable shield is open) [135]

Floating gripper system
The main beam and the gripper unit are mounted so that these assemblies can move independently of each other but still form a stiff construction in order to be able to react the full force of boring with 19" disc cutters. The gripper unit is constructed to enable free access to the working area for ground conditioning measures.

Fully shielded TBM
The ACT TBM is designed as a double shield TBM, permitting simultaneous boring and installation of segments inside the tail shield. The shields consist of a front shield around the cutter head bearing, a telescoping shield, which can be retracted, a gripper shield and a tail shield (Figure 8-19).

Maximum cutter head stabilisation and rock support
While boring in hard rock, it is very important to maintain a stable cutter head if high advance rates and reasonable disc costs are to be achieved. The ACT TBM is fitted with the classic Robbins Main Beam TBM stabilisation system, consisting of a solid bottom shoe, active side supports and an active roof shield. The active roof shield supports the tunnel crown in order to delay the loosening of faulted rock, overcut etc.

Performance forecasting when using the ACT TBM
It is assumed that the final lining should be complete as soon as possible after excavation. This also means that boring and lining should fundamentally be taking place at the same time. TBMs have to master difficult conditions but still maintain forecast advance rates in competent as well as poor rock. The ACT TBM is intended to achieve the following results:

- Consistently high advance rates in rock classes I and II (without hard rock or high temperatures). Average rates can be 25 to 40 m/d with peaks of up to 100 m/d while simultaneously installing the segmental lining.
- Predictable advance rates of 10 to 25 m/d with the simultaneous installation of the final lining in classes III and IV or in squeezing, rock bursting, blocky or jointed conditions. Similar advance rates should be achievable in class I and class II in extremely hard rock and high temperature conditions.

- In class V – fault zones, groundwater under pressure, unstable, running ground with high water inflows in combination with silt – measures are necessary to reduce the chance of the TBM jamming. Under such conditions, incremental advance of 2 to 5 m should be possible.
- In squeezing ground (class V), the TBM system should excavate an overcut and be able to install yielding support and yet still advance at 2 to 10 m/d.

2) Continuous advance at the Sophiaspoor Tunnel [157]

In order to implement continuous advance, various adaptations are required in the design of a shield machine (Figure 8-20):

- Elongated shield tail to provide space for two segment rings,
- provision of an articulated joint in the shield due to the resulting overall length of the shield,
- extension of the stroke of the thrust cylinders due to the increased distance to the location of the cylinders,
- design of the 28 thrust cylinders as single cylinders capable of being controlled separately,
- provision of a second erector to secure the invert segment against tipping when installing rings while boring continues,
- longitudinally mobile erector support cross coupled to the first backup in order to avoid hindrance to ring installation while boring,
- segment design: instead of the 7+1 system with small keystone used at the Botlek and Pannerdensch Kanaal Tunnels, a system called 7+0 is used, in which the keystone is not shortened.

In practice, only initial experience has been gained with this innovative technology. This has, however, made it clear that the implementation of new technology always poses challenges.

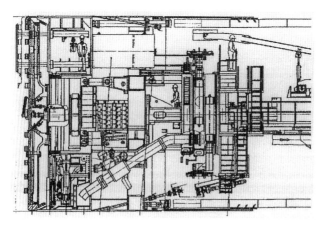

Figure 8-20 Longitudinal section through the shield machine for the Sophiaspoor Tunnel (Herrenknecht AG) [157]

9 Compressed air shields

The use of compressed air as a measure to hold back water has a long tradition. Its application in shield tunnelling was introduced in 1886 by Greathead, although Calladon had already suggested the use of compressed air to Brunel after the flooding of the Thames Tunnel in 1828 [268], [128]. Compressed air shields can be hand shields, shields with part face excavation or with full-face excavation with additional compressed air equipment and air locks to enable the use of the shield under open water or under the groundwater table.

The process is characterised by compressed air being fed into the tunnel in order to keep the working area free of water ingress. The compressed air reaches equilibrium against the hydrostatic pressure of the water. Resistance against ground pressure is not possible, and the ground has to be supported either by natural support or mechanically. One exception to this rule is the membrane shield. The areas of application for compressed air shields are therefore:

- soil types, in which the lowering of the groundwater table is not recommended for technical, economic or ecological reasons,
- where damaging settlement could result from groundwater lowering,
- tunnelling under open water.

The limits to the use of compressed air in tunnelling are essentially set by:

- the maximum permissible pressure of 3.6 bar according to the compressed air regulations; working under higher pressures requires an exception permit [25],
- the air permeability of the soil (upper threshold for the coefficient of permeability for water $k_w = 10^{-4}$ m/s) or the support,
- a minimum overburden depth above the tunnel crown (one-to two times tunnel diameter according to the soil type and bedding) to ensure safety against blowouts,
- shorter working times on site due to time spent in the air locks,
- reduced performance of the workers under compressed air (danger of compressed air sickness),
- the increased fire risk.

The disadvantages of compressed air tunnelling make it sensible to investigate whether holding back water with compressed air alone is the most suitable solution or if it would be possible to use a combination of partial lowering of the groundwater table and holding back the water with less air pressure, in order to mitigate at least some of the disadvantages of higher pressures.

The compressed air shield machine itself is an antiquated technology and has been superseded. Nonetheless, it is important to describe the technology in detail, as all the newer shield technologies like slurry or EPB shields still require compressed air support for access to the excavation chamber.

9.1 Functional principle

A traditional compressed air tunnel drive requires the enclosure of a large working area from the face back to the support provided by an (in the broadest sense) air- and watertight lining. This manner of holding back water demands uninterrupted provision of compressed air, since otherwise water could flow into the shield and the tunnel during a stoppage. The space filled with compressed air in a compressed air shield drive has to be closed off from atmospheric pressure. To this end, a pressure bulkhead in the tunnel or at the end of the tunnel closes it off from the outside air. In order to provide access to the working area under compressed air, personnel and material locks are provided in the pressure bulkhead, and the construction and operation of these has to comply with the compressed air regulations [25].

The compressed air is fed through the pressure bulkhead. The pressure has to be greater than or equal to the water pressure acting at the shield invert (Figure 2-5). As the water pressure, unlike the air pressure, has a pressure gradient, the crown will be under excess pressure, causing the air to penetrate into the soil forming a flow field with the associated air consumption (Figure 9-9). If the overburden is shallow, there is a danger that the soil particles lose equilibrium due to the air flow and the face collapses, which is termed a blowout. This process is described in detail in [128].

In a compressed air tunnel, the layout of the air locks is highly significant for construction operations. The layout is dependent on the geological and hydrological conditions, the structure (length, cross-section) and the later use, among other factors.

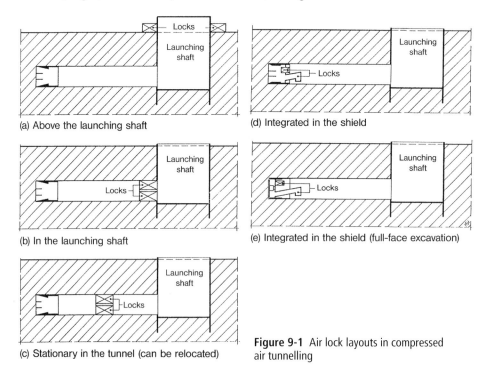

Figure 9-1 Air lock layouts in compressed air tunnelling

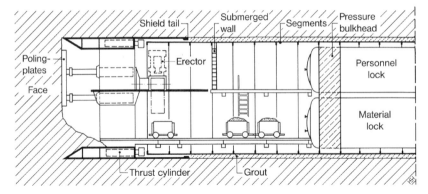

Figure 9-2 Personnel lock and material air lock for rail transport [293]

For short compressed air tunnels, it is possible to situate the air locks in the starting shaft (Figure 9-1 a, b). This enables non-waterproof tunnel support in the construction state. The layout is problematic for longer tunnels or larger diameters, as the air losses through the lining grow. It is also disadvantageous for construction operations that all work in the tunnel has to be performed under compressed air.

An improvement is the arrangement of stationary air lock facilities in the tunnel, which can be relocated (Figure 9-1 c). This means that only a certain forward part of the tunnel has to be pressurised, but requires a watertight lining behind the air lock.

The same applies to air locks integrated into the shield, resulting in only the tunnelling work in the pressure chamber in the shield being performed under compressed air while the installation of support can take place under atmospheric conditions (Figure 9-1 d).

The optimal solution regarding construction operations and occupational health and safety is to construct the shield with only the mechanical excavation equipment at the face being under compressed air (Figure 9-1 e). No personnel then have to work under compressed air under normal operating conditions. One precondition is, however, that the shield tail seal works reliably. In order to gain access to the excavation chamber at the front for repairs or maintenance purpose, a man lock is required in the pressure bulkhead.

Because of the disadvantages regarding health and safety and process technology, the use of compressed air shields is declining compared to more modern slurry or earth pressure balance shields. Special conditions on individual projects can mean that they are still used. For example, a 3.6 km length of underground railway tunnel in Mexico City was constructed using two compressed air shields (external diameter 6.24 m) between 1973 and 1986, but these were also replaced by a slurry shield in 1987 [256].

Compressed air tunnelling was also used under the Kiel Fjord in 1990 in connection with pipe jacking (Section 9.6.2).

9.2 Compressed air facilities

Compressed air facilities essentially consist of air locks and compressed air supply.

The purpose of an air lock is to separate the pressurised area from the area under normal atmospheric pressure. The purpose of the compressed air supply, with the compressor station as the central component of the compressed air facilities, is to ensure the supply of the tunnel with compressed air in sufficient quantities and pressure under all circumstances.

The compressed air regulations [25] include all the safety measures required for reliable and uninterrupted supply of compressed air and for the protection of the crew against damage to their health. The currently valid regulations were issued on 21 June 2005.

New regulations for working under compressed air are currently being prepared. The changes to be expected are essentially that the new regulations will be more stringent. The CEN standard "Compressed air locks; safety aspects" [72] will be revised by the working group CEN TC 151/WG 4 [45].

9.2.1 Air locks

Pressure bulkheads can be constructed of steel or reinforced concrete, according to the specification and the application. They must be designed to resist 1.5 times the operating pressure in the tunnel and permit all the supply pipes and cables (compressed air, electricity, telephone, control cable, hydraulic oil etc.) to pass through without loss of air. Air locks permit access through the pressure bulkhead.

The normal practice is to use separate personnel and material locks according to purpose, as these have to fulfil different requirements. The personnel or man lock is installed in the upper part of the pressure bulkhead and the material lock in the lower half. Figure 9-2 shows an example of separate air locks for personnel and material, in this case for rail trucks.

The compressed air regulations [25] include the requirement that air locks for personnel must not be used for material. The other way round, material locks must not be used for personnel to pass through. For combined air locks, the regulations for personnel locks apply. Another requirement is the provision of a treatment chamber from an excess pressure of more than 0.7 bar. These components are therefore now discussed separately. In order to protect the workers against any sudden inrush of water, a blanket (submerged wall) must be provided, which then acts as a diving bell (Figure 9-2).

Figure 9-3 Personnel lock with oxygen breathing apparatus [172]

Personnel locks

Personnel air locks, also called man locks, must have a minimum height of 1.60 m [25]. The personnel air lock must be sufficiently large to provide an air space of 0.75 m^3/person. Personnel locks intended for exiting with oxygen must be equipped with an oxygen breathing supply including oxygen masks. No oxygen should escape from the oxygen breathing apparatus into the air in the air lock. Further information about personnel locks can be found in [25], [27]. Figure 9-3 shows a personnel lock with miners breathing oxygen during decompression.

Material locks

According to the transport system, material locks may be constructed for use with muck cars, conveyor belt, pipeline transport etc. A requirement is that the doors of the material lock should be hung so that they are pressed into the sealing gaskets by air pressure.

In tunnels of larger diameter or to improve advance rates, a number of material locks may be provided. Both for moving material into and out of the pressurised area, the greatest possible acceleration of the locking process is important, and this purpose can be served by large-diameter valves unsuitable for personnel locks. Material locks are usually ventilated at high pressure.

Treatment chambers

The compressed air regulations require that all construction sites with working at overpressures of more than 0.7 bar must provide a decompression chamber for patients. Treatment chambers have to be at least 1.85 m high and consist of a chamber for patients and a separate chamber to act as inward and outward air lock (Figure 9-4).

One of the treatments for decompression sickness (also referred to as caisson sickness or the bends) is recompression at air pressures considerably higher than the working pressure in the tunnel. Compressed air sickrooms therefore have to be approved for an operating pressure of at least 3 bar for compressed air working at 1.8 bar, and above this for an operating pressure of 5.5 bar.

Figure 9-4 Treatment chamber [8]

In addition to the normal equipment of a personnel lock, the treatment chamber should also be provided with an intercom or equivalent telephone connection, internal and external operation of the air supply and extraction, a lock for medication, a couch, silencers for the air supply, toilet bucket with chemicals to prevent odour and equipment for breathing oxygen.

Fire protection measures

The increased content of oxygen in compressed air promotes the ignition of fires and accelerates their propagation. Rapid smoke propagation and the resulting problems for visibility and breathing also have to be expected.

Precautions must be taken to ensure that there is no unprotected readily flammable material in the air locks and that flammable materials in working areas are replaced by other materials as far as technically and economically possible.

For rapid reaction to fires, portable fire extinguishers suitable for operation in compressed air should be kept ready. For longer tunnels, it is also considered necessary to install a fixed extinguishing water pipeline with hoses ready for use, perhaps also water canons and rain curtains at certain intervals.

If welding or cutting work is carried out or any other work with naked flames or high temperatures, care should be taken that there is no flammable material in the vicinity and harmful gases should be extracted.

9.2.2 Compressed air supply

The compressor station is the central element of the compressed air facilities. It should be designed to ensure that the tunnel is supplied with compressed air in sufficient quantities under all circumstances.

The compressor station combines both high- and low-pressure compressors. This simplifies the operation and monitoring.

High-pressure compressors serve to supply air tools and also possibly the material air locks, which normally operate at a pressure of 7 bar [167]. Low-pressure compressors produce the compressed air to hold back the water from the face. The number and size of the installed low-pressure compressors is according to the air quantity required to hold back water in the tunnel, which is either calculated as described in Section 9.3 or estimated from experience.

The low-pressure compressors used to hold back water may be reciprocating piston, rotary piston or axial compressors, with the single-stage reciprocating piston compressor being most common. Figure 9-5 shows the stated compressor types.

Reciprocating piston compressors have a high degree of efficiency. Machines available on the market offer very varied performance in terms of the achievable pressure level (1 bar to many thousand bar) and the compressed air volume (a few litres to many hundreds of m^3/min).

Figure 9-6 shows a compressor installation consisting of water-cooled single-stage compressors. The installation was used for the construction of the underground railway in Frankfurt.

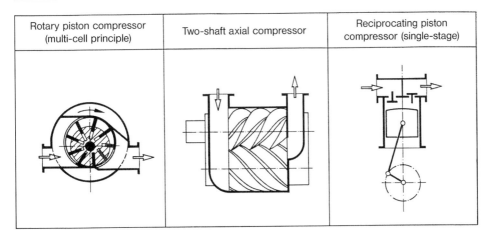

Figure 9-5 Compressor types

Figure 9-6 Water-cooled single-stage compressors of a compressor installation [102]

Compressors are generally differentiated into operating compressors, which produce the required quantity of air at the required pressure, and reserve compressors, which have to be available at any time in case of breakdown or a sudden unforeseen increase of the air requirement. For economic reasons, electric drive is mostly chosen for the compressors, in which case diesel-powered generators have to guarantee the compressed air supply to the tunnel in case of a power cut. For larger systems, it can be advantageous to provide the base load of compressed air with electrically driven compressors and provide the reserve compressors directly with diesel motors, which makes an emergency power supply unnecessary. The power consumption of a compressor station is sometimes so high that a dedicated connection to the mains is required. Figure 9-7 shows a scheme of a compressed air plant [247].

9.2.3 Compressed air regulations

The German regulations governing working under compressed air [25] (Compressed air regulations – DruckLV) came in to effect on the 4 October 1972. The valid issue is from the BG Bau, version June 2005. This also includes further details to the compressed air regulations, "Working in compressed air RAB 25", version from 12/11/2003 [238] and DIN EN 12 336 "Tunnelling machines" [73] and DIN EN 12110 "Tunnelling machines – Air locks" [72]. The following excerpts seem important:

Air with an overpressure of more than 0.1 bar is considered to be compressed air for the purposes of the regulations. According to § 9, the upper limit for its use is 3.6 bar. Workers under compressed air must not be younger than 18 or older than 50 years old.

According to § 12, section 1 sentence 2 of the compressed air regulations, a doctor appointed by the employer according to § 12 DruckLV and authorised by the responsible authority according to § 13 DruckLV must be present constantly at the place of work when work is carried out under a working pressure of more than 2.0 bar.

The following terms are defined:

- Working chambers are rooms, in which work under compressed air is performed,
- Personnel locks are accesses, which serve solely for the movement of personnel into and out of the working chamber,
- Material locks are accesses, which serve solely for the movement of materials into and out of the working chamber,
- Combined locks are accesses, which serve for the movement of personnel and materials into and out of the working chamber,
- Treatment chambers are rooms, which serve, independent of the operational pressure in the working chamber, for the treatment of patients suffering compression-related sickness and for trials of passing through air locks under medical supervision.

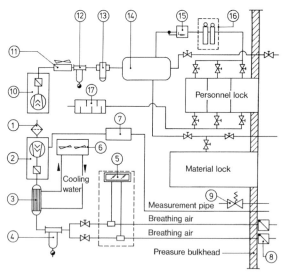

1) Intake filter
2) Low-pressure compressor
3) Heat exchanger
4) Condensate separator
5) Air quantity gauge
6) Water recooler
7) Control for the low-pressure compressor
8) Backflow flap valves
9) Pressure relief valve
10) High-pressure compressors
11) Compressed air cooler
12) Condensate separator
13) Air filter
14) Compressed air vessel
15) Pressure reducing valve
16) Breathing air filter
17) Exhaust silencer

Figure 9-7 Scheme of a compressed air plant [247]

The compressed air regulations consist of 26 individual clauses regulating administrative matters, requirements for operational facilities, safety measures for the reliable supply of fresh air to working rooms, medical tasks for the monitoring of the health of personnel working under compressed air and penalty provisions. They are supplemented by Appendices 1, 2 and 3, which are related to specific clauses. Appendix 1 is concerned with the configuration of working chambers and of compressed air sickness chambers and their operation and sanitary facilities. Appendix 2 includes the locking tables and Appendix 3 the instructions for air lock attendants.

Air locks and their electrical facilities may only be put into use after an officially recognised expert has issued a test certificate. There is no regulation concerning the supply of electric power, but protection measures and the provision of an emergency power supply, which starts automatically in case of a power cut, are required in every case. An emergency power supply system is not required if only air-cooled combustion motors and battery-operated emergency lighting are used. When the electricity supply is interrupted, the second electricity supply system must take over the supply automatically and immediately. If there is only one operational and one reserve compressor for a working chamber, each operational and each reserve compressor must be able to supply the required air quantity on its own.

With regard to the provision of fire extinguishers, it should be noted that they have to function under the highest permissible operating pressure. The same applies to warning and measurement equipment, which has to be installed when potentially dangerous gases are to be expected.

The detailed regulations for working under compressed air produced by the BG (Bau) in collaboration with the industry (RAB 25) [238] contains the following parts:

– Part 1: Recommendations for the approval of exceptions according to § 12 section 1 of the compressed air regulations,
– Part 2: Issue of a qualification certificate according to § 18 section 2 of the compressed air regulations,
– Part 3: Passing out through the air lock with oxygen breathing after working under compressed air in connection with § 21 section 1 and Appendix 2 of the compressed air regulations.

9.3 Air requirement

Due to the air escaping at the face, shield tail and through the tunnel lining, and also the air consumption from the operation of air locks, it is necessary to supply new air into the tunnel constantly in order to maintain the air overpressure and thus hold back water (Figs. 9-8 and 9-21).

9.3.1 Determination of air requirement

For a rough estimate of the probable air consumption Q_L in shield-driven tunnels under open water and thus the installed capacity (related to the intake air quantity) of the low-pressure compressors, the following rule of thumb formula was published as long ago as 1922 by Hewett/Johannesson [128]:

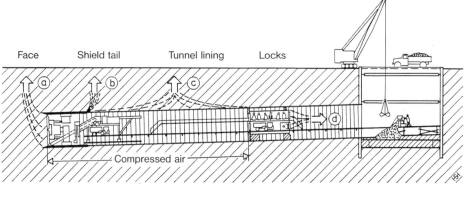

(a) Air loss at the face

(b) Air loss at the shield tail

(c) Air loss through leaks (joints) in the tunnel lining

(d) Air consumption through locking materials and personnel

Figure 9-8 Air loss during compressed air working [167]

$$Q_L = (3.66 \text{ to } 7.32) \cdot d^2$$

with:

Q_L [m³/min]	intake air quantity,
d [m]	shield diameter,
3.66	for soil of normal permeability (e.g. medium sand),
7.32	for very permeable soil (e.g. gravel or coarse sand).

This formula incorporates reference values for tunnel length and leakiness. It is related solely to the diameter of the tunnel. It can, however, be assumed that these reference values are included in the factor in the formula. The experiences of Hewett/Johannesson still produce reliable results today when the assumed preconditions apply to a shield tunnel drive. The preconditions will, however, not normally apply since the tunnel is not always under open water but may only be only under the groundwater table under a built-up area.

For a more precise determination of the air requirement (compressor capacity) for a compressed air shield tunnel, the work of Wagner/v. Schenck [290] is available (Table 9-1).

Although the precision of the theoretical relationships only apply to homogeneously permeable soils, these can also be applied to variable and stratified soils, if an average value of k_L applicable to the prevailing permeability can be inserted. The precondition is that an air flow field can become established, as in homogeneously permeable soil.

9.3 Air requirement

Table 9-1 general determination of the air requirement for compressed air shield tunnelling according to [290]

Under open water	In groundwater with an open surface below ground level
Air requirement in [m³/min] $Q_L = n \cdot c \cdot k_L \cdot A \cdot q_L + Q_S$	
$q_L = \dfrac{\alpha + \beta_i}{\beta_i}\left(\dfrac{P_T}{P_a} + 1\right)$	$q_L = \left(\dfrac{T}{\beta_i \cdot d} \cdot \dfrac{1-\alpha}{\beta_i}\right) \cdot \left(\dfrac{P_T}{P_a} + 1\right)$

n   component for the consideration of partial air losses at the face, shield tail and any leaks (Figure 9-9),
c   considers the influence of the three-dimensional flow field (for an overburden above the tunnel crown of one to two tunnel diameters: $c = 2$),
k_L [m/min]   Permeability of the soil to air,
A [m²]   area of the face,
P_T [bar]   air overpressure in the tunnel,
P_a [bar]   atmospheric pressure,
T, α, $β_i$, d   geometrical dimensions (Figure 9-10),
Q_S air consumption for air lock operation,
d   tunnel diameter.

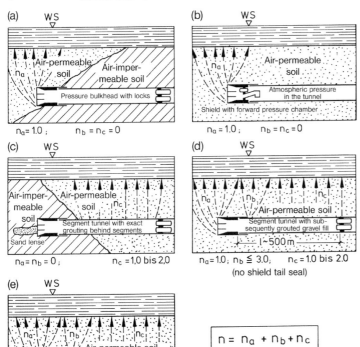

Figure 9-9 Factors for the consideration of partial air loss at the face, at the shield tail and through leaks in the lining [167]

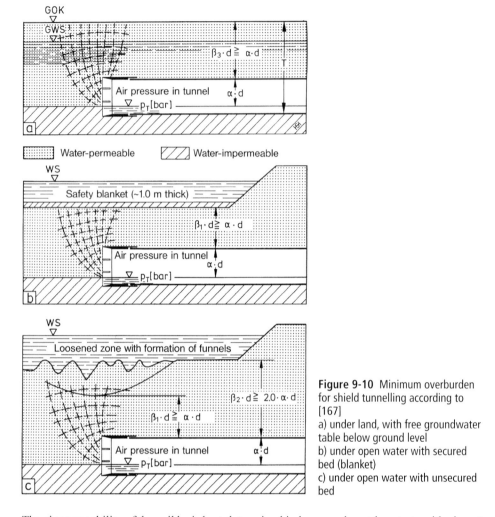

Figure 9-10 Minimum overburden for shield tunnelling according to [167]
a) under land, with free groundwater table below ground level
b) under open water with secured bed (blanket)
c) under open water with unsecured bed

The air permeability of the soil k_L is best determined in large-scale outdoor tests with air, not just through laboratory tests, The evaluation of completed compressed air tunnel drives in similar soil conditions is also very helpful. A rough estimate for approximate calculations is $k_L \approx 70 \cdot k_W$, where k_W is the water permeability of the soil according to Darcy.

The permeability k_W of the soil to water according to Darcy should be determined by groundwater lowering or pumping tests. It should not be forgotten that the water permeability may be different horizontally and vertically.

9.3.2 Verification of safety (blowout safety)

Avoidance of unintended air losses from the working chamber (blowouts) should already be considered at the planning stage. Considering the danger and damage, which blowouts can cause, attention to their avoidance or prevention should be paid in good time. Some simple preventative measures are:

- closing of any boreholes,
- design with adequate overburden.

Questions about adequate overburden depths are explained in [167], based on investigations by Wagner and Schenck [289], [290], [292] on the permeability of homogeneous soils. Simple equilibrium calculations show that with an effective density of the solid soil overburden of greater than 1.0 t/m³, a total overburden $\beta_1 \approx 1.0 \cdot \alpha$ offers sufficient safety against blowouts (Figure 9-10).

For compressed air tunnelling under open water, it has been observed in practice and also in model tests that an unsecured bed at the location of an air loss can lead to great alterations and loosening with the formation of a funnel. In order to ensure safety against blowouts in this case also, or to obtain a solid overburden $\beta_1 \approx 1.0 \cdot \alpha$, it is recommended [290] to select a total overburden of $\beta_2 \approx 2 \cdot \alpha$ (Figure 9-10 c).

Formation of a funnel is dependent on the soil properties, but also on the advance rate. It can be almost avoided if a well-graded blanket is applied, if necessary of material with a high specific density such as an ore. It is therefore recommended, if the overburden depth is relatively shallow, to provide a blanket as a precaution, to monitor it with soundings during construction and repair it if necessary. Blankets were constructed, for example, for the construction of the Rhine Tunnel in Düsseldorf [8], [126], the Rendsburg pedestrian tunnel [8] and the fourth bore of the Elbe Tunnel [29].

Outside Germany, impermeable layers on the river- or seabed are used. It may, however, be problematic to ensure the correct laying of the blanket [167].

There are no simple relationships for stratified soils. The necessary measures would have to be investigated and evaluated on site for the individual case.

For compressed air tunnelling under land (ground level above the groundwater table), the formation of funnels at the surface has not yet been observed. The overburden in this case should also be so deep that there is no danger of blowouts, especially as the soil above the groundwater table goes into the equilibrium calculation with a high density. For compressed air tunnelling in homogeneously permeable soil, a minimum overburden of $\beta_3 \approx 1.0 \cdot \alpha$ should be sufficient, even without a blanket [167].

9.3.3 Special processes

In order to be able to work at lower pressures or minimise air losses, other processes can be combined with compressed air working. The use of such processes is very dependent on the comparison of cost-effectiveness.

The simplest way of reducing the required pressure is to lower the groundwater table as much as possible with wells and only set the air pressure for the hydrostatic head of the lowered groundwater table [166]. This process can only be used where groundwater lowering is permissible and can be implemented with the necessary level of safety. This solution cannot be used under open water.

Air losses can also be reduced by altering the permeability of the soil through filling of sealing layers, blankets, soil exchange, grouting or freezing in part sections. These measures can also be used under open water.

Soil waterproofing through grouting is used to assist compressed air tunnelling in highly aquiferous ground or even to make it possible at all. The grout be used must have no side effects deleterious to health under compressed air, and must also comply with current soil and groundwater protection regulations. When loose ground is to be consolidated and at the same time the ingress of groundwater is to be reduced, or where grouting or chemical consolidation are not possible due to underground conditions or groundwater protection regulations, ground freezing would be more suitable.

9.4 Further developments

As mentioned at the start of the chapter, traditional compressed air tunnelling assumes the closing of a relatively large working space between the face and the installation of the final support with a waterproof lining. It also demands that nearly all, or just a part of the personnel working underground work constantly under compressed air, and personnel and material both have to be transported in and out of the working space through air locks. It is well known how much hindrance this entails for excavation and lining work. Compressed air shield tunnelling is also subject to limits imposed by the maximum permissible pressure required by the compressed air regulations, the minimum overburden above the tunnel crown and the air permeability of the soil. The first two of these criteria limit the vertical alignment of the tunnel downwards and upwards, while excessive air permeability of the soil (e.g in medium to coarse gravel) can put the feasibility of the process into question unless it proves possible to considerably reduce the k_L value through advance measures like sealing, grouting or ground freezing [172]. This results in considerable extra costs for the construction project.

Some further developments can be useful here. Worth mentioning are shields for full-face excavation, in which only the excavation chamber is pressurised, constructions with unpressurised working spaces and part face excavation, which requires remote control of the excavation machinery, and the membrane shield, with which the air permeability of the face is reduced by spraying it with bentonite suspension.

9.4.1 Compressed air shield with unpressurised working space and full-face excavation

Figure 9-11 shows a compressed air shield manufactured by Mitsubishi with 5.25 m external diameter closed cutting wheel. The material is pulled out of the pressurised excavation chamber through a ball valve air lock into the shield, which is under atmospheric pressure. The muck is then removed on a conveyor belt.

9.4.2 Compressed air shield with unpressurised working spaces and part face excavation

Figure 9-12 shows the principle of the Campenon-Bernard shield, which was used in 1963 in Paris for the Express Métro under the Seine connecting to the Seine crossing coming from Place de la Défense [172]. The structure for the crossing of the Métro with the Seine itself was constructed with a compressed air caisson.

9.4 Further developments 215

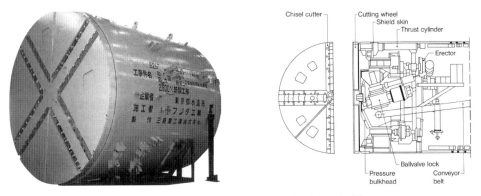

Figure 9-11 Compressed air shield with forward pressure chamber (Mitsubishi)

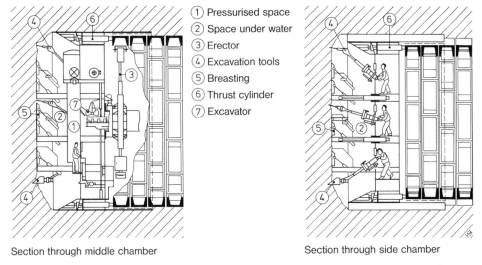

① Pressurised space
② Space under water
③ Erector
④ Excavation tools
⑤ Breasting
⑥ Thrust cylinder
⑦ Excavator

Section through middle chamber   Section through side chamber

Figure 9-12 Principle of a compressed air shield machine with unpressurised working space and part face excavation, Seine crossing for the Express Métro, 1963 [293]

The approximately 780 m long connecting section has an internal diameter of 8.70 m. Then soil was first excavated at the face pressurised to an overpressure of 2.25 bar with mechanical excavation machinery working parts of the face in three storeys, with the operating crew standing under atmospheric pressure about 2 m further back behind a pressure bulkhead. As the tunnel advanced, however, the method of working was changed to pure hand excavation.

For the construction of supply and drainage pipes with diameters in the range of 1.5 to 3.5 m, which are ideal for the pipe jacking process, Figure 9-13 shows an example of a compressed air shield with part face excavation and unpressurised working space.

The pipe acting as an air lock is divided vertically into personnel and material locks. The personnel lock also serves as the operator position and can be under atmospheric pressure most of the time. The excavation machinery is remotely controlled, and orientation is provided by a video camera in front of the pressure bulkhead in the excavation chamber.

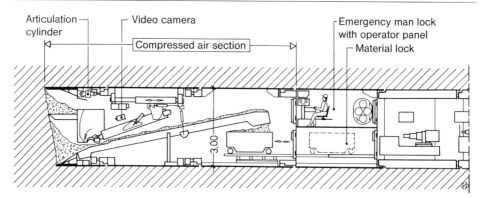

Figure 9-13 Compressed air shield with unpressurised working space and part face excavation for pipe jacking [11]

9.4.3 Membrane shield

The principle of the membrane shield process is the sealing of the face in the pressurised excavation chamber by spraying with bentonite suspension to form a membrane. As the ground is excavated by an excavator or roadheader, the locally broken membrane is continuously renewed. Figure 9-14 illustrates this process, a development of the company Züblin, which is however no longer in use today.

The suspension is configured so that after it has been sprayed onto the face, a thin layer sticks and closes the pore structure of the soil against the penetration of compressed air. The compressed air acts as a support force over the entire area of the face through this mostly impermeable membrane, i.e the ground pressure is resisted. Due to the meagre consumption of suspension for the membrane, no separation plant was required at that time to separate muck from the membrane suspension.

The good accessibility of the excavation chamber enabled relatively simple removal of obstructions compared to slurry shields. The membrane shield is therefore flexible and especially suitable for high-level tunnels, which naturally encounter more obstructions. The membrane shield has only been used until now in combination with pipe jacking (Chapter 13). There are, however, plans to use a membrane shield for the cross-sections of transport tunnels.

9.5 The use of compressed air with other types of shield

In contrast to the developments to slurry shields in other countries, the Hydroshield developed in Germany has two control circuits, which make the fluid pressure at the face independent of fluctuations of slurry flow and hold it practically constant with only ± 0.1 bar deviation. One control circuit adjusts the quantity of slurry supplied by the pumps while maintaining the set pressure level for the quantity of mix discharged. In case of volume changes in the excavation chamber with resulting slight fluctuations of the bentonite level, an air bubble acts as a second regulation circuit and pressure regulator (Chapter 10).

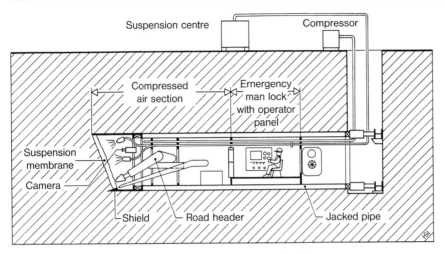

Figure 9-14 Pipe jacking with the membrane shield process [309]

Figure 9-15 Personnel lock of a Mixshield (Herrenknecht)

As it can never be ruled out that the excavation chamber of a slurry or EPB shield will need to be entered to remove obstructions or make repairs, these machines also require compressed air equipment, which has to comply with the requirements and limitations already discussed. Figure 9-15 shows a personnel lock integrated into the pressure bulkhead in the crown area of a Mixshield.

9.6 Examples

9.6.1 Old Elbe Tunnel next to the St. Pauli landing stage, 1907 to 1911

Project

In order to provide a fixed land link from the city to the free harbour area on the left-hand bank of the Elbe in Hamburg, the harbour administration decided in 1907 to construct a tunnel inspired by the construction of the Thames Tunnel in London using a compressed air shield. Alternative proposals for a fixed crossing of the Elbe, for example a high-level bridge, were rejected on grounds of danger to shipping, particularly in fog. Because of the topographical conditions, lifts were intended for access to the tunnel, installed in the starting and reception shafts for the shield drive. The length of the tunnel was 448.5 m. Two tunnel were driven simultaneously with the east bore about 100 m ahead [275].

Geological and hydrogeological conditions

The route of the compressed air shield drive runs through clay and sand formations with diluvial and alluvial deposition. The carriageway invert has a maximum depth of 21.3 m below the average high tide water level (Figure 9-17).

Tunnelling and machine concept

On account of the prevailing water pressure, an overpressure of 2.4 bar was planned for the compressed air shield drive. The shield diameter was 6.6 m. The face was divided into nine cells by two vertical and three horizontal walls, from which the miners worked. When working in sand, the top heading and bench areas were mechanically supported with hydraulic supports and the invert was naturally supported, while in clay the face remained unsupported. Some blasting was even necessary to excavate the clay.

One particular feature of the shield (Figure 9-16) was the hydraulically driven piles in the crown, which were driven ahead of the face support and served to secure the crown. This principle is strongly reminiscent of the current blade shield (Chapter 13). The thrust cylinders of the shield achieved a thrust force of 2,000 tonnes, and permitted a stroke length of 0.5 m.

The annular gap opened by the shield skin was grouted with cement mortar with the addition of slaked lime in two phases. This grouting, injected in the second phase through altogether three openings in the lining, mainly served to protect the wrought iron lining from rust.

The 2 m thick concrete pressure walls with personnel and material air locks were, except in the starting phase, positioned in the tunnel near the starting shaft. The lining of the tunnel consisted of wrought iron, riveted rolled beams, which could deal with changes of shape better than the cast iron used in England and America. The joints were sealed with lead. The tunnel ring was lined inside with concrete [275].

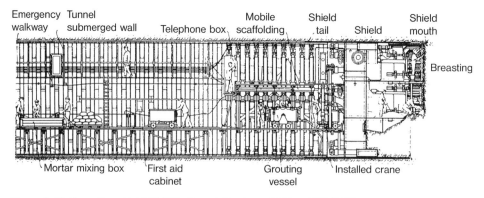

Figure 9-16 Tunnelling shield, Old Elbe Tunnel, Hamburg, 1907 to 1911 [275]

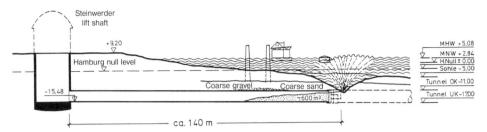

Figure 9-17 Blowout at the Old Elbe Tunnel during the shield drive

Progress of the project

In order to overcome the water pressure, it was necessary to produce an air pressure in the crown about 0.5 bar greater than the water pressure. This resulted in about 1 m^3/s of compressed air constantly loss through the face, leading to vigorous bubbling in the Elbe and obstructed shipping. Difficulties were also caused by the necessity of constantly adjusting the air pressure to the water level as it varied tidally.

An error in the pressure regulation in combination with the scouring of the riverbed resulting from the continuous leakage of compressed air led to a blowout after six months of tunnelling on 24/06/1909. The air suddenly tore through the face, carried the breasting with it, created a waterspout 6 to 8 m high in the Elbe and spread about 600 m^3 of soil on the riverbed. After the air had escaped for about 10 seconds, water and about 600 m^3 of soil penetrated into the shield [275].

Six weeks were lost until work could be resumed. Working very carefully, the tunnel was completed and opened for traffic in autumn 1911.

The average daily advance was 1.5 m/24 h, and he best day was 3.25 m/24 h [275].

9.6.2 Energy supply tunnel under the Kiel Fjord, 1989/90

Project

The city of Kiel is divided into two parts by the Kiel Fjord. The city utility company Stadtwerke Kiel decided in 1987 to construct an energy supply tunnel, mainly to supply district heating to the western parts of the city. The tunnel has a length of 1,368 m and an internal diameter of 4.10 m. The contract was awarded to the joint venture Fördetunnel Kiel.

Geological and hydrogeological conditions

The last Ice Age deposited thick layers of alluvial loam and granite blocks in Northern Germany. In the Kiel Fjord, loam with included boulders alternates with sandy-gravely soil. These are covered by deposits of ooze and mud (Figure 9-18).

The maximum water level in the fjord is almost 40 m above the tunnel invert.

220   9 Compressed air shields

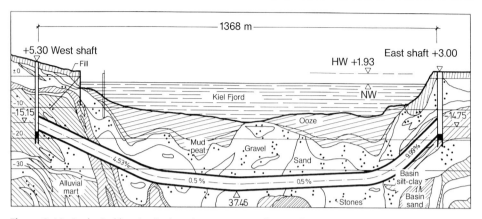

Figure 9-18 Geological longitudinal section, energy supply tunnel under the Kiel Fjord [133]

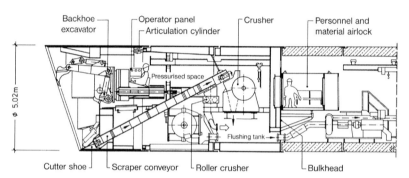

(a) Excavator mode

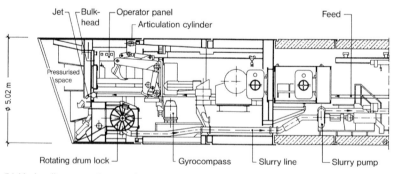

(b) Hydraulic excavation mode

Figure 9-19 Compressed air shield, energy supply tunnel under the Kiel Fjord (Westfalia Lünen, Howaldswerke Deutsche Werft)

Tunnelling and machine concept

The tunnel was constructed by pipe jacking (Chapter 13). In order to cope with the changeable soil conditions and water pressures, a tunnelling machine was used, which was capable of two different excavation processes (Figure 9-19). In the cohesive and solid soil formations, a powerful hinged backhoe arm with a tear-out force of 300 kN excavated the soil (Figure 9-20). The excavated material was then transported on an intermediate conveyor belt through a crusher and discharged into a flushing tank for hydraulic transport. The first 168 m and the last 235 m were driven in this "excavator" mode, with air pressures between 1.8 and 3.2 bar being necessary. The miners were able to work shifts of between 2½ and 6 hours in this mode to comply with the compressed air regulations [25].

In the sandy and gravely soils, which amounted to more than 70 % of the tunnel route, the excavator was withdrawn, the bulkhead closed and the machine was operated as a "flushing shield". The soil was then excavated hydraulically with water jets. The muck was sucked through a rotary hopper. This material lock enabled boulders with a side length of up to 600 mm to be taken out without having to enter the excavation chamber. During hydraulic excavation, the entire crew was working under atmospheric pressure, while the face was supported by air pressures of up to 3.9 bar [137].

The face support with compressed air at up to 3.9 bar was an extreme challenge for the design of pressure bulkhead, shield body and the personnel and material locks. All components had to be designed under the stringent requirements of the pressure vessel regulations and tested by the TÜV. The test pressure of 7 bar was a record for a tunnelling machine. The shield had an external diameter of 5 m and an overall length of 12.7 m. The jacking station in the jacking shaft could deliver 83,000 kN, and the presses in the necessary seven intermediate jacking stations 30,770 kN.

Figure 9-20 Compressed air shield for the energy supply tunnel under the Kiel Fjord, excavator mode (Westfalia Lünen, Howaldswerke Deutsche Werft)

Figure 9-21 Air losses under the Kiel Fjord during the tunnelling of the energy supply tunnel

Project progress

The conversion between the different excavation systems required a lot of work but proved to be successful due to the ability of adapting to changing soil conditions. The project, with a cost of about €19.7 m, could by completed in only eleven months. The average advance rate was between 6 and 8 m per working day in two-shift operation [133], [243].

Journals for the entire field of Structural and Civil Engineering

Volume 107, 2012
Impact factor 2010: 0,265

Volume 13, 2012
Journal for *fib* members –
International Federation for Structural Concrete

Volume 81, 2012
Impact factor 2010: 0,234

Design and Research
Volume 5, 2012
Journal for ECCS members –
European Convention for Constructional Steelwork

Zeitschrift für den gesamten Ingenieurbau
Volume 89, 2012
Impact factor 2010: 0,141

Fachzeitschrift für Führungskräfte der Bauwirtschaft
Volume 35, 2012

Volume 35, 2012
Official journal of the DGGT

Deutsche Gesellschaft für Geotechnik e.V.
German Geotechnical Society

Geomechanik und Tunnelbau
Volume 5, 2012
Journal for ÖGG members

Österreichische Gesellschaft für Geomechanik

Zeitschrift für Technik und Architektur
Volume 16, 2012

Wärme | Feuchte | Schall | Brand | Licht | Energie
Volume 34, 2012
Impact factor 2010: 0,173

Journal Online Subscription
All Ernst & Sohn specialist journals (from 2004 to the present) are available by online subscription.

WILEY ONLINE LIBRARY
www.wileyonlinelibrary.com

Order a free sample copy: www.ernst-und-sohn.de/journals

Ernst & Sohn
Verlag für Architektur und technische
Wissenschaften GmbH & Co. KG

Customer Service: Wiley-VCH
Boschstraße 12
D-69469 Weinheim

Tel. +49 (0)6201 606-400
Fax +49 (0)6201 606-184
service@wiley-vch.de

BOOK RECOMMENDATION

Recommendations on Excavations EAB

Deutsche Gesellschaft für Geotechnik e.V. (ed.)
Recommendations on Excavations – EAB
2nd revised edition
2008. 300 pg., 115 fig. 19 tab. Hardcover
EUR 69,-

ISBN: 978-3-433-01855-2

€ Prices are valid in Germany, exclusively, and subject to alterations. Prices incl. VAT. Books excl. shipping. Journals incl. shipping. 008358016_my

The aim of these recommendations is to harmonize and further develop the methods, according to which excavations are prepared, calculated and carried out.
Since 1968, these have been worked out by the TC "Excavations" at the German Geotechnical Society (DGGT) and published since 1980 in four German editions under the name EAB. The recommendations are similar to a set of standards.
They help to simplify analysis of excavation enclosures, to unify load approaches and analysis procedures, to guarantee the stability and serviceability of the excavation structure and its individual components, and to find out an economic design of the excavation structure.
For this new edition, all recommendations have been reworked in accordance with EN 1997-1 (Eurocode 7) and DIN 1054-1. In addition, new recommendations on the use of the modulus of subgrade reaction method and the finite element method (FEM), as well as a new chapter on excavations in soft soils, have been added.

www.ernst-und-sohn.de

Ernst & Sohn Verlag für Architektur und technische Wissenschaften GmbH & Co. KG
Für Bestellungen und Kundenservice: Verlag Wiley-VCH, Boschstraße 12, D-69469 Weinheim
Tel.: +49(0)6201 606-400, Fax: +49(0)6201 606-184, E-Mail: service@wiley-vch.de

10 Slurry shields

10.1 Development history

The historical origin of slurry shields can already be seen in the patents of Greathead (Figure 1-22) and Haag (Figure 1-23), which implemented the hermetic closing of the inside of the shield, thus creating an excavation chamber filled with fluid as a pressure chamber with naturally standing or sloping face [268].

Grauel was the first to equip such a shield with an excavator boom in 1912. In 1959/60, the first shield correctly described as a slurry shield was used in the United States for the construction of a 3.35 m diameter drainage tunnel. The machine, designed by E. C. Gardener and built by Eourtes Construction Methods & Equipment, featured hydraulic muck transport. The water used as a transport medium was not however pressurised, so no fluid support was provided to the face [268]. In 1960, Schneidereit introduced the idea of active face support using a bentonite suspension. This support method, as also the trend-setting patents of Lorenz, which described face support by suspension and mechanical excavation of the soil, had no immediate effect on tunnelling technology in Europe. Only the practical demonstration of supporting vertical soil faces with clay suspension in the rapidly developing technology of diaphragm walls created the necessary confidence to build an expensive tunnelling machine on this principle.

Three different lines of development of slurry shields can be differentiated: one Japanese, which led to the modern slurry shield; one English, which has now been abandoned; and one German, which led to the Hydroshields. The Japanese slurry shields also provided the starting point for the development of earth pressure balance shields (Chapter 11), while the German Hydroshield led to a range of construction variants like Mixshield, Thixshield and Hydrojetshield.

Japanese line of development

In Japan, Mitsubishi delivered the first trial shield with slurry support in 1967. After the first encouraging results with the 3.10 m diameter prototype, the first "large" 7.20 m diameter slurry shield was used in 1970 for the construction of the Keiyo railway line under a canal. Figure 10-1 shows one of the early slurry shields, built by Mitsubishi in 1973 with 5.05 m diameter and used by the Tekken Kensetu company.

Since than, this type of shield machine has been built by many manufacturers in Japan and has been used in remarkably high numbers from a European viewpoint. In Europe, the slurry shield was produced under Japanese license by Markham in Great Britain, Neyrpic Framatome Mécanique (NFM) in France and Fives-Cail Babcock (FCB) in France.

224 10 Slurry shields

Figure 10-1 Japanese slurry shield (Mitsubishi), 1973 [268]

English line of development

In England, the consultant Mott, Hay und Anderson was awarded a British patent in 1964 for a "Bentonite Tunnelling Process". The National Research Development Corporation and the London Transport Executive (LTE) purchased the right of use and supported the manufacture and testing of the first bentonite shield, which was built by the company Robert L. Priestley in 1971 and then used by the company E. Nuttall for a 144 m long test tunnel of 4.10 m diameter (Figure 10-2).

The system envisaged feeding the support fluid into the invert of the working chamber. Solids suspended in the support fluid were drawn off in the crown of the working chamber, and coarser material was lifted by the pockets of the cutting wheel and discharged into a central lock chamber, behind which the two material flows were combined again and pumped to above ground. The support pressure was intended to be kept constant by an over-pressure valve in the crown pipe.

Difficulties with the discharging of solids and pressure regulation led to much alteration and rebuilding, but satisfactory operation was still not achieved in two later applications in London and Mexico, although it has to be said these were under very difficult conditions. No further implementations are known.

German line of development

A third line of development of slurry shields started from 1972 in Germany. The company Wayss & Freytag developed the Hydroshield system and used the prototype in 1974 for the construction of the Hamburg-Wilhelmsburg main sewer under the harbour in Hamburg (Figure 10-3). The external diameter of the shield was 4.48 m. At the time of this contract, no reliable shield tail seal was available, so the entire tunnel was pressurised with compressed air. Because no experience was available for the operational behaviour of the new shield system, the design provided a number of fittings intended to enable the advance to continue if untested equipment broke down or failed, for example braking and steering shutters in the shield skin, an articulation joint in the shield and ten injection pipes for various purposes (including grouting, rolling control). After the essential teething problems had been remedied on the first drive, for example the active grid buckets were replaced by the screen in front of the transport system that is usual today for the removal of oversize blocks, the second drive was used intentionally for the development of a reliable shield tail seal. The intention of maintaining the inside of the tunnel at atmospheric pressure was achieved over the last 300 m using a new shield tail seal.

In Germany, the terms Hydroshield und Mixshield have become established for machines with full-face excavation and slurry support. Shields with fluid support and part face excavation (Thixshield, Hydrojetshield – Section 10.4.1.1) have not proved to be a successful process.

10.2 Functional principle

Tunnelling machines with slurry support use a pressurised fluid (suspension) to support the face. Bentonite suspensions have proved especially successful. The rheological properties of shear strength and viscosity, and the suspension density are adapted to suit the ground properties and have to be monitored continuously.

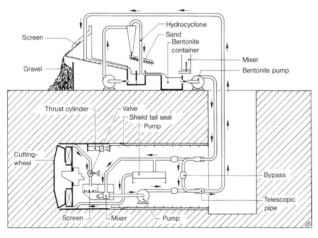

Figure 10-2 English slurry shield (R. L. Priestley), 1971 [268]

Figure 10-3 German Hydroshield, Hamburg-Wilhelmsburg main sewer, 1974 (Wayss & Freytag)

In order to provide face support in Hydroshields and Mixshields, the working chamber is closed from the tunnel by a pressure bulkhead. The required support pressure can be regulated very precisely with an air bubble behind the installed submerged wall and by setting the quantities pumped by the slurry and feed pumps. The required and maximum support pressures have to be calculated for the entire length of the tunnel before the start of boring (Section 2.1 – face support).

The soil is excavated over the full face by a cutting wheel equipped with tools and cleared hydraulically. The provision of a subsequent separation plant is essential. If it is necessary to enter the excavation chamber, for example to change tools, perform repairs or remove obstructions, the support fluid has to be replaced by compressed air. The support fluid then provides a relatively impermeable membrane to seal the face, termed a filter cake, whose lifetime is however limited by the danger of drying out (Figs. 4-1 b and c). This membrane permits the support of the face by compressed air and has to be renewed regularly if necessary (Figure 10-4).

The support medium can be lowered completely (full lowering) or partly (partial lowering) and replaced by compressed air. The maximum partial lowering is limited particularly by the need to provide a sufficiently large working space. This should be large enough to make safe working possible at all times and to provide a sufficiently large retreat space for the workers. Stones and rock banks can be broken down to an acceptable size either by disc cutters on the cutting wheel and/or by crushers in the working chamber.

In stable ground, the slurry shield can also be used in open mode without pressurisation with water being used as transport medium. Any additional mechanical support of the face by the cutting wheel or through poling plates should only be considered as an additional security, and their supporting effect should not be included in calculations to verify the support.

10.3 Scope of application

Slurry shields are mainly used in coarse- and mixed-grained soil types (Figure 10-5). The groundwater table should be above the tunnel crown with a sufficient margin of safety. Very permeable soils hamper the formation of a membrane and the transfer of support pressure. If the permeability is greater than $5 \cdot 10^{-3}$ m/s, there is a danger that the bentonite suspension flows uncontrolled into the ground. The scope of application can therefore be extended by

Figure 10-4 Slurry face with filter cake sealing

adding fine-grained material and filler or additives to improve the rheological properties. Alternatively, additional measures to reduce the permeability of the soil (like for example filling of the pore cavities) may be required. Stones and blocks, which cannot be pumped, are broken first by a crusher. High fines content can lead to difficulties with the separation plant. It should be borne in mind that the rheological properties of the support fluid are worsened by fine-grained material, as it is technically impossible to separate the clay fraction and bentonite.

10.4 Machine types

While the term slurry shield is normally used outside Germany for full-face machines with slurry support, the terms Hydroshield or often also Mixshield (Herrenknecht AG) are often used in Germany. The inventors of part face machines have also patented further descriptions for particular types. Even although part face machines with slurry support are no longer in use, a brief description of some significant inventions is included.

10.4.1 Full-face machines with fluid support

Slurry shield

The Japanese slurry shields were developed for use in the coastal cities of Japan. The covering strata consists of alluvial sands, mainly of weak, cohesive soils, whose thickness declines with distance from the coast. The sands below tend to be coarser, into the gravel range. The slurry shields are therefore primarily designed for use in sandy or silty soils. Clays with more solid consistency, which lead to blocking of the intake openings, and densely consolidated gravels, which lead to a large increase of the torque required to overcome the friction forces on the cutting wheel, lead especially for smaller shields to problems with the installation of suitable drive motors, and define the limits for the use of slurry shields (Figure 10-6) [170].

Particular characteristics of the slurry shields are the type of support fluid used – normally a clay suspension – the construction of the cutting wheel and the method of controlling and checking the support pressure.

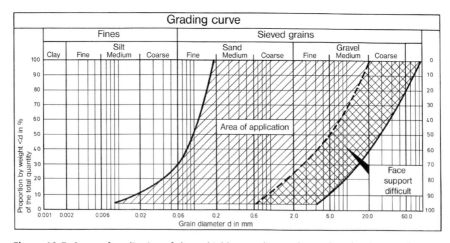

Figure 10-5 Scope of application of slurry shields according to the grading distribution of the soil [171]

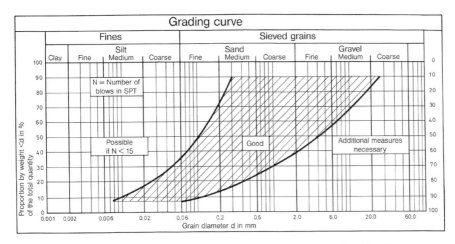

Figure 10-6 Scope of application of slurry shields according to the type of soil [307]

The cutting wheel of the slurry shields is flat and relatively closed. This provides a mechanical support effect during stoppages in addition to the slurry support. The only access to the face, for example to remove obstructions, is through a few windows, which are normally closed during operation. The excavation tools, generally scrapers or flat chisels, are arranged as rays in double rows, making excavation possible in each direction. The soil can pass through slots arranged parallel to the excavation tools, with the width of the slots being adapted to suit the expected maximum grain size. The slots also hold back any material too large for slurry transport (Figure 10-7).

Depending on the required cutting wheel torque, the cutting wheel can be driven centrally by a shaft (centre shaft type), around the perimeter (drum type) or through struts at about the quarter points of the cutting wheel (centre cone type) (Figure 11-15 a to c). The problem of cutting wheel bearings and their waterproofing is dealt with more fully in connection with EPB shields (Section 11.5.2).

The support fluid is fed into the top of the excavation chamber and the soil-suspension mixture is removed at the bottom near a mixing paddle intended to avoid settlement and achieve a homogeneous mixture for transport.

In slurry shield machines, the support pressure at the face is directly controlled by the regulation of the feed and return pumping of the support fluid in the excavation chamber (Figure 10-8). The support pressure is measured by an electrical pressure cell in the excavation chamber and in the feed and slurry pipes and compared by a computer with the theoretical calculated support pressure. The pumps and valves in the suspension circulation are controlled correspondingly.

As the face cannot be viewed, its stability can only be checked by comparing the mass of the theoretical and actual excavated volumes. The actual excavated volume is determined by measuring the density of the support suspension, and the theoretical excavated volume is calculated from the unit weight, the consolidation and the void ratio. These have to be estimated from by probe drilling carried out in advance. [169].

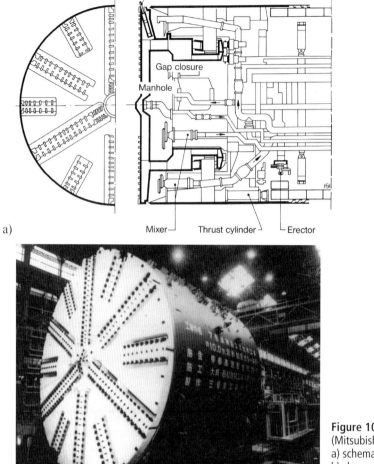

Figure 10-7 Slurry shield (Mitsubishi)
a) schematic diagram
b) slurry shield, external diameter 10.0 m

Hydroshield

All the Hydroshields on the market today are derived from the development work of Wayss & Freytag. The most prominent features of the conventional Hydroshield are the star-shaped cutting wheel and the division of the excavation chamber into two parts by a submerged wall (Figure 10-9).

The soil is excavated from the full face by the cutting wheel (1) rotating in bentonite and is mixed into the bentonite. The area of the shield, in which the cutting wheel rotates, is called the excavation chamber (2) and is separated from the part of the shield under atmospheric pressure by the pressure bulkhead (3).

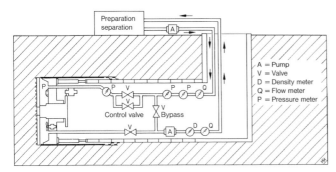

Figure 10-8 Support pressure control in a slurry shield [169]

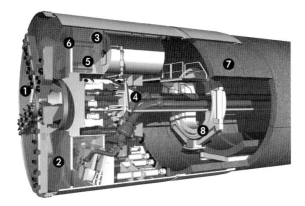

Figure 10-9 Functional principle of a slurry shield as Hydro shield (in this example a Mixshield from Herrenknecht AG)

The bentonite suspension supplied through the feed pipe (4) is pressurised in the excavation chamber by an air bubble (5) with a pressure appropriate for the prevailing ground and water pressures. This prevents uncontrolled entry of soil or the loss of stability at the face. The support pressure in the excavation chamber is not controlled directly through the pressure of the suspension, but through a compressible air bubble. For this reason, the excavation chamber is divided from the pressure bulkhead behind the cutting wheel by the submerged wall (6). The area between the submerged wall and the pressure bulkhead is described as the pressure chamber or working chamber.

The essential difference of this process from the Japanese slurry shield is the active control of the suspension pressure through the air bubble. The decoupling of the control of support pressure from the quantity of suspension circulating in the slurry circuit and the advantages of air as a compressible medium in case of volume fluctuations have proved successful.

Sudden losses of support medium, for example when entering fault zones, can be absorbed without the collapse of support pressure to the face over a relatively wide tolerance range, which approximately corresponds to the volume of suspension in the rear chamber. It is also possible to increase or alter the volume of suspension in circulation without a direct effect on the support pressure if this is necessary to improve the removal of muck. The air bubble and submerged wall enable access to the working chamber through an air lock located in the upper part of the shield (Figure 9-15), which makes the clearance of obstructions easier in comparison to a Japanese slurry shield. In order to carry out repair or maintenance work to the cutting wheel, the bentonite suspension is run out of the excavation chamber and replaced by compressed air. The filter cake temporarily seals the face and permits it to be supported by compressed air alone. The filter cake shrinks in contact with air and has to be periodically refreshed in order to limit air losses (for example by spraying or flooding the chamber).

The open form of the cutting wheel as an open star with freestanding spokes enables the muck to flow immediately behind the cutting wheel into the excavation chamber; this also decouples the support and excavation functions. Because of the open cutting wheel, a screen is necessary in front of the intake to the suction pipe in order to keep out material with a grain size too large for pumping.

Mixshield

The concept of the Hydroshield was further developed by the company Wayss & Freytag in collaboration with Herrenknecht to produce the Mixshield, a shield machine with convertible face support. Due to the ring gear drive with floating bearing of the cutting wheel and the modular construction of the machine, the operating mode can be changed according to the soil conditions encountered in the tunnel. All combinations of slurry, hydro-, earth pressure balance, compressed air or open shield are generally possible for convertible Mixshields (Figure 12-3).

The various operating modes enable a scope of application in a wide geological spectrum. Most of the tunnelling machines described as Mixshields are however not rebuilt during the boring of a tunnel but are used exclusively as slurry shields. Figure 10-10 shows the first Herrenknecht Mixshield.

Current Mixshields are fitted with a hydraulically driven jaw crusher in front of the screen to reduce larger stones to a size capable of being pumped. The building up of material in front of the screen is prevented by directed flushing jets of suspension. Various types of cutting tools (Chapter 4) can be integrated into the cutting wheel according to the prevailing soil type.

Figure 10-10 The first Herrenknecht Mixshield, ⌀ 5.95 m, Hamburg Hera, 1984

Further details of Mixshields, in particular the conversion of operating mode during a tunnel drive, are explained in Chapter 12.

10.4.2 Part face machines with slurry support

The slurry shield machine with part face excavation has not proved a successful process. The following disadvantages can be stated:

- In contrast to an open part face machine, the excavation process cannot be followed visually.
- The formation of a filter cake and transfer of support pressure cannot be sufficiently controlled and monitored.
- Muck transport is problematic, particularly when obstructions are encountered.

The Thixshield (Figure 10-11), developed by the company Holzmann at the end of the 1970s, was used on two projects (1978 in the Hamburg harbour, Transportsiel Winterhude, diameter 4.20 m, and in 1980 for the construction of a contract of the Stadtbahn Gelsenkirchen, diameter 7.29 m).

The Hydrojetshield, a development patented by Wayss & Freytag AG in 1979 (Figure 10-12), has also not become established. Instead of mechanical excavation by a cutting wheel, the ground is excavated by directed jets of fluid in the excavation chamber.

The excavation jets are arranged at right angles to the shield centreline inside the blade, so that the excavation jets act immediately in an area limited to the space inside the shield skin. The diameter of the swan-neck shaped swivelling jets is about 10 to 24 mm, with a loading pressure of about 10 bar. The jets have a range of up to 1.5 m.

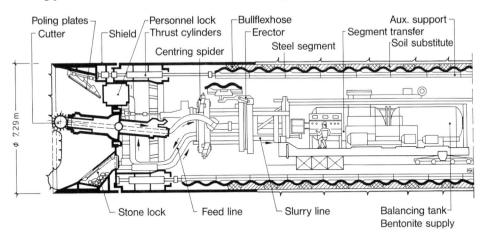

Figure 10-11 Thixshield, Stadtbahn Gelsenkirchen, Trinenkamp contract, 1980 (Ph. Holzmann/Wirth) [134]

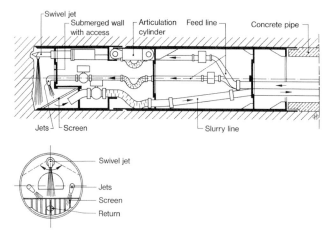

Figure 10-12 Principle of the Hydrojetshield (Wayss & Freytag)

10.5 Machine and process technology

The following details refer solely to the Hydroshield and Mixshield.

10.5.1 Soil excavation

The open star (Figure 10-3) rotating within the excavation chamber is the classic type of Hydroshield cutting wheel. The high degree of openings was found necessary at the time in order to transfer the support pressure through the bentonite suspension in sandy soil. The cutting wheel arms (spokes) were traditionally fitted with drag picks, and overcutting was only possible within limits.

On particular disadvantage was the lack of support to the face during compressed air interventions. For this reason, some tunnelling machines between 1990 and 2007 were also fitted with hydraulically extended poling plates and a perimeter rim. The use of poling plates however often turned out to be problematic and had more disadvantages then advantages (Figure 10-13).

Figure 10-13 Rimmed cutting wheel of a Mixshield, ⌀ 11.34 m, Westerschelde Tunnel, 1998

Figure 10-14 Closed cutting wheel, ⌀ 7.7 m, East Side CSO, Portland, 2006

The first generation of Hydroshields often experienced problems with sticking at the centre in fine-grained soils. The independently rotating centre cutter was developed to combat this problem and was first used with success on contract 34 in Essen in 1990.

But the centre cutter (Figure 10-13, right) does represent an additional risk, particularly if obstructions, blocks and stones are encountered. For this reason, closed cutting wheels rotating in front of the shield blade (Figure 10-14), which hardly differ from the cutting wheels of EPB shields (Chapter 11), are often used in heterogeneous ground with blocks or rock inclusions. The perimeter rim improves the supporting effect, particularly during compressed air interventions. The muck openings should be designed to improve flow as much as possible. The size of the openings in the cutting wheel and the degree of openings should be defined as the "best compromise" considering the stone size and the inclination of the soil to stick. The soil is excavated by scrapers, disc cutters and buckets.

10.5.2 Muck transport

The flow of material through the cutting wheel and inside the excavation chamber and the hydraulic transport are often hindered in fine-grained soils by material sticking to the excavation and transport equipment. Sticking leads to a reduction of advance rate and the degree of utilisation, as the cutting wheel and excavation chamber may have to be laboriously cleaned by hand under compressed air. Sticking combined with a high content of abrasive minerals can also lead to heavy wear to the cutting wheel and in the excavation chamber.

The sticking potential of a soil is evaluated from the liquidity index I_c and the plasticity index I_p [%]. In addition to the tendency of the soil to adhesion or cohesion, its dispersal behaviour is also affected by further process-related factors and these should also be considered. Adhesion describes the capability of the soil to stick to the metal surfaces of the excavation and transport equipment, which hinders material flow around the tools and through the openings in the cutting wheel. Cohesion together with the consistency defines the capability of sticking and tendency of the pieces of soil to combine to form clumps. When muck accumulates in front of the intake in the shield invert, there is a danger that the hydraulic transport flow is interrupted. Dispersal behaviour describes the property of the soil to dissolve into the support and transport medium and thus alter the rheological properties of the suspension.

Stiff and semi-solid clays have a particularly marked sticking potential [281]. No standardised test methods are yet available for the measurement of sticking potential. In addition to the mineralogical composition of the soil and the Atterberg limits, process-related factors have a significant effect on the risk of sticking. The following are worth mentioning:

– The size, configuration and coating of the openings in the cutting wheel,
– the suction pressure and the geometrical layout of the intake area,
– the feed scheme, the suspension circulation quantity or the flow speed in the endangered zones,
– the type and geometrical details of the excavation tools,
– penetration and advance rate.

The risk of sticking can therefore be reduced significantly by designing the excavation and transport equipment to optimise flow conditions, optimising the hydraulic feed and transport scheme and through the correct selection of excavation tools.

For a machine working in clay, the measures listed in Figure 10-16 are particularly suitable:

– Suspension feed through the cutting wheel, the central cone and through the pressure bulkhead,
– suspension feed through an independently rotating centre cutter, tangential feed in the invert of the excavation chamber,
– provision of agitators in the area of the intake and rotors on the back of the cutting wheel,
– isolation of the intake area from the remaining working chamber (isolated invert segment).

An increase of the quantity of suspension circulating reduces the density of the transport flow and therefore also the risk of sticking. The intention of the measures listed above is also to increase the flow speed in critical areas of the excavation chamber and avoid accumulations of material, particularly in the invert. In the intake area, where there is a

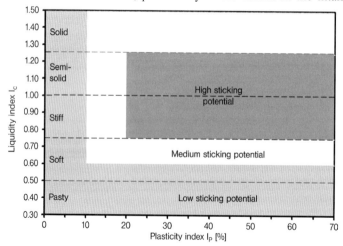

Figure 10-15 Evaluation of the sticking potential according to Thewes [281]

special danger of material sticking, the formation of a compartment with closing plates leads to a considerable improvement of the suction effect. Pressure communication with the face is maintained by active balancing pipes. This has the effect of reducing material accumulations in the invert and cleaning work (compressed air interventions) [244].

10.6 Examples

10.6.1 Westerschelde

Project

One of the most important underground structures in the Netherlands with a cost of about 767 m € is the "Westerschelde Tunnel", which provides a fixed road link between the two parts of the Dutch province of Zeeland, Zuid-Beveland and Zeeuwsch-Vlaanderen. The two parallel bores, each 6.6 km long, were driven with a slight time lapse by two Mixshields from Herrenknecht. The completed Westerschelde Tunnel has an internal di-

Equipment elements of a slurry shield related to the sticking potential of the ground					
No.	Equipment element	Implementation	Sticking potential		
			low	medium	high
1	Cutting wheel	closed	o	o	−
		open	+	++	++
2	Cutting arm cross-section	box-shaped	o	o	−
		flow-optimised	+	+	++
3	Distance cutting wheel-submerged wall	short	o	−	−
		long	+	++	++
4	Tools	drag picks	o	−	−
		Y chisels	o	o	−
		scrapers	+	+	++
5	Active centre cutter	not installed	o	−	−
		installed	+	++	++
6	Submerged wall opening	large, > 4 %	o	o	−
		small, < 4 %	+	+	++
7	Junction shield-submerged wall	sharp angle	o	−	−
		rounded with tapered metal	+	++	++
8	Intake area	screen without jaw crusher	o	−	−
		screen with jaw crusher	o	o	−
		agitator + roller crusher	+	+	++
		isolated invert segment	o	+	+
9	Fresh suspension feed	mainly in intake area	o	−	−
		intake area + excavation chamber	+	+	o
		mainly in excavation chamber	+	++	++
10	Arrangement of bentonite jets	intake area	++	++	++
		invert of excavation chamber	+	++	++
		tangential in the excavation chamber	o	+	+
		axial in the excavation chamber	o	+	+
		cutting wheel centre, radial	+	++	++
		between face and cutting wheel centre	+	++	++
11	Design of slurry circuit for increased suspension flow		+	++	++
12	Increase of the suspension flow in the shield through recirculation		o	+	++
13	Raised pressure at flushing points through additional pumps in the shield		o	o	+
Legend:			Symbol	Detail	
Remark: not all elements can be combined with each other without limitation. Small diameters < 3.0 m limit the availability of some components			++	very favourable	
			+	sensible	
			0	possible	
			−	unfavourable	

Figure 10-16 Measures to counter sticking in Hydroshields [281]

ameter of 10.10 m. The 7 + 1 segments of the single-layer lining are only 45 cm thick. The construction of the works was awarded to a consortium of the companies BAM Infrabouw BV, Heijmans NV, Voormolen Bouw BV (all Netherlands), Franki BV (Belgium), Philipp Holzmann AG and Wayss & Freytag AG (Germany).

Geological and hydrological conditions

The geology in the construction area under the River Schelde is dominated by sand and clay formations. The upper 20 to 30 m mainly consist of medium and fine Quaternary sands. On the north side, there are Tertiary sands containing glauconite. Under these sand beds are the "Boomse Klei" (Boom clay) strata. This material is a particularly stiff clay, which had already caused the engineers to rack their brains during the design phase because of its pronounced stickiness. The 6.6 km long drive of the two tunnelling machines started along the specified alignment from Zeeuwsch-Vlaanderen on the south bank of the Schelde, initially with a gradient of 4.5 % until below the Pas van Terneuzen. This shipping channel has a bed depth of about 35 m and the tunnel at this location has its crown about 50 m below sea level in sand containing glauconite. Depending on the state of the tide, support pressures of up to 6.5 bar were necessary in the crown. Then the machines crossed under the Middelplaat, an extended sandbank. The excavation cross-section here lay for a long stretch completely in the Boom clay, or else a mixed face of sand and clay had to be excavated. The tunnel drives then passed below another shipping channel, the Pas van Everingen, at a depth of 40 m below sea level followed by the final uphill stretch to Zuid-Beveland through various sand formations.

Tunnelling and machine concept

The high water pressure in the very permeable sand formations and the stickiness of the clay forced the engineers to subject all the components of the Mixshields to a fundamental aptitude test. Investigations into the adhesion behaviour of the Boom clay finally led to a special construction of the cutting wheel and excavation chamber of the tunnelling machine and the flushing concept. The cutting wheel was designed as an open star with a rim, and the shape of the cutting arms was designed to optimise flow as far as structural requirements permitted. On the back of the cutting wheel in the outer area were two rotors similar to ploughs, which were intended to carry the muck in the invert to the intake area. A centre cutter with a diameter of 2.5 m was mounted on the centreline of the machine, equipped with its own suspension feed, a transport pipe and a high-pressure cleaning system in order to counteract any sticking. The excavation chamber was rounded throughout to avoid sharp corners where soil could stick. In the intake area of the excavation chamber, a roller crusher and two agitators were fitted.

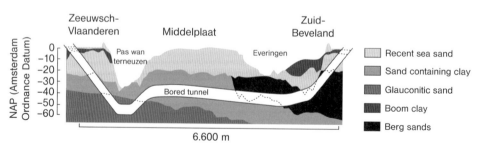

Figure 10-17 Geological longitudinal section of the Westerschelde Tunnel

Although no stones were to be expected from the geological forecast for the tunnel alignment, it was decided to install a roller crusher because this is also capable of breaking down larger clumps of the sometimes semi-stiff clay to a suitable size for pumping. The rollers in front of the intake also assist the suction effect.

Another special feature was the flushing concept for the excavation chamber and intake area at the Westerschelde. In contrast to conventional Hydroshield tunnelling machines, it was possible to aim the bentonite solution being used as a transport medium at individual points: the centre of the main cutting wheel, the centre cutter, the perimeter of the excavation chamber, the middle of the excavation chamber, the opening in the submerged wall and the intake area. A large part of the fresh bentonite supply of max. 2,000 m^3/h could be supplied to the shield and controlled by volume. In addition, a recirculation pump was installed in the shield, which could increase the circulation capacity in the excavation and working chambers by about a third. The concept used required numerous ball valves and pumps to control the flow of suspension. Various scenarios were therefore configured in the control system, each of which activated a grouped switching of pumps and ball valves, and could be selected by the operator of the machine. This enabled a choice between feeding the entire feed quantity in front of or behind the submerged wall and various intermediate stages.

Table 10-1 Essential tunnel and machine specifications

Tunnel	Westerschelde
Length	2 x 6,600 m
Segmental lining	
Internal diameter:	10.10 m
Thickness:	45 cm
Ring length:	2 m
Shield	
Diameter:	11.34 m without hard facing
Maximum thrust force:	120,687 kN
Nominal torque of cutting wheel drive:	12,900 kNm
Breakout moment of cutting wheel drive:	15,050 kNm
Slurry circuit quantity	1,800 to 2,000 m^3/h
Length of the backup:	approx. 182 m
Cutting wheel	
Scrapers:	64
Buckets:	2 × 12 (left + right)
Centre cutter	
Scrapers:	10
Buckets:	2 × 4 (left + right)

Project progress and experience

All parties to the project were clear that the crossing of the Westerschelde would be a technical innovation. As part of the planning of risk management, all conceivable mishaps were analysed before the start of the tunnel drives and solution plans were produced. Personnel, material and machines were prepared. Nonetheless, it did not prove possible to avoid a few critical problems; successful implementation hung by a thin thread. The design and construction of the Westerschelde Tunnel is extensively documented in [120].

The first tunnelling machine (east bore) started in July 1999 from the southern starting shaft and reached the north bank in Zuid-Beveland in February 2002. The second machine (west bore) followed three months behind and reached the south bank shortly before the east machine. The first difficulties already appeared while breaking through the sealing block. The block, constructed in the starting excavation out of a compacted sand-cement mix, proved to be much harder than expected. The tools could not break off large pieces, suffered heavy wear due to the grinding loading and had to be extensively modified.

After boring through the sealing block, the machine dived into the natural soil. After 12 rings, the vertical deviation had reached 100 mm, but it was driven back to the original alignment. In the first kilometre of boring through sandy-silty soil, the advance rate varied between 0 and 90 m/week due to the usual modifications and difficulties in the learning period.

In May 2000, both tunnelling machines had to stop temporarily as the stock of segments was insufficient. While standing still, deformation appeared in the shield tail. The machines at this time were at the deepest point of the vertical alignment, approx. 65 m below sea level. After restarting, the thrust forces increased over-proportionately. In order to reduce the friction, it was decided to fit larger buckets in order to enlarge the overcut. This was carried out by divers working under saturation in the suspension with shuttle transport and breathing trimix [68]. After the completion of assembly work, the west machine was carefully restarted. In order to resume the east drive, additional pressure pads were inserted between the completed ring and the shield tail. The cause of the deformation was the great depth of the tunnel and the glauconitic sands, which were in an overconsolidated stress and density condition due to heavy Ice Age preloading.

Shortly after the west machine had restarted, a leak occurred at ring 547 caused by the breaking of a 2" ball valve, which was sheared off as the thrust cylinders were being withdrawn for ring installation due to a drainage pump that had not been removed. This resulted in 100 m³ per hour of sand-water-bentonite-grout mixture pouring into the machine, reaching a level of 2 m in the shield. Luckily, this could initially be pumped away as fast as it entered. Since the entire machine subsided, the annular gap and therefore the valve junction clogged up almost independently. After further great efforts, the crew at last managed to close the junction. The penetration of water and soil had displaced the machine by 110 mm, which proved to be favourable by bending back the deformation of the shield tail and temporarily destressing the shield, but was critical for the functioning of the brush seal. The affected area of the brushes was temporarily sealed by grouting and the drive was resumed with an additional temporary sealing chamber, in order to achieve a suitable starting position for the repair of the first two rings of the brush seal consisting of altogether four rows.

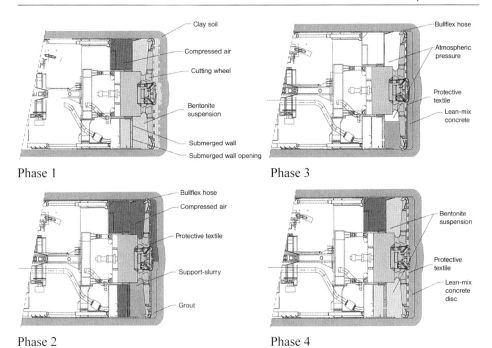

Figure 10-18 Working phases to form a sealing wall in front of the cutting wheel [38a]

It had already been discovered in March 2000 that the sealing system of the cutting wheel drive was leaking. In order to continue driving the tunnel, the leakage chamber and gear chamber were pressurised and the repair of the seals was prepared, and this was carried out at ring 1922 in the protection of the Boom clay with the aid of a so-called sealing wall in front of the cutting wheel (Figure 10-18). First, the cutting wheel was advanced to its full forward travel in order to make room for the sealing wall (Phase 1). Then the cutting wheel was withdrawn, the submerged wall gate was closed and compressed air at 3.5 bar was applied for support in the excavation chamber. The front of the cutting wheel was covered with a fibre-reinforced protective foil with an integrated radial Bullflex hose and a lean mix concrete wall was concreted in front of the cutting wheel (Phase 2). The protection provided by these measures enabled the forward seals of the cutting wheel drive to be inspected and exchanged under atmospheric pressure (Phase 3). Finally, the protective foil was removed, the excavation chamber filled with suspension and the sealing wall was bored through (Phase 4).

The further progress of the tunnel through the upper layers of the nearly impermeable Boom clay continued successfully. For safety reasons, a decision was made to use sufficient support pressure all the way. The newly developed measures to counter sticking (see also Section 10.5.2) proved successful. The advance rate was limited to about 21 mm/min to avoid accumulation of muck. Fine material from the hydrocyclones could be discharged directly through a pipeline into the Schelde, which was advantageous for the overall performance.

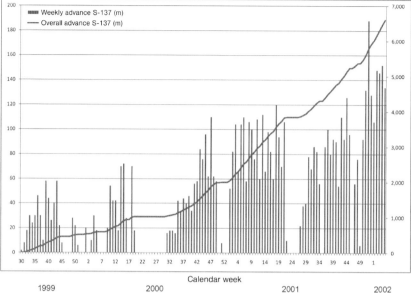
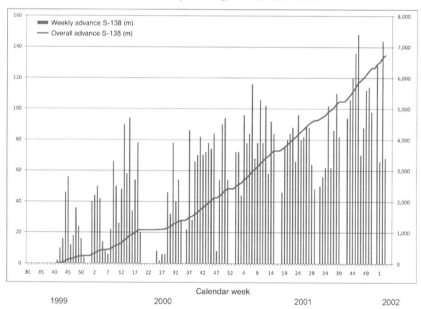

Figure 10-19 Advance rates at the Westerschelde Tunnel; top east bore, bottom west bore

Despite the success of the double crossing of the Westerschelde, the conditions on the project should be regarded with great respect as being on the limits for the use of tunnelling machines in loose ground. The experience during the breakdowns shows how close success and catastrophe can be under pressures of 6 to 7 bar.

10.6.2 Lower Inn Valley railway, Münster/Wiesing Tunnel, main contract H3-4; Jenbach/Wiesing Tunnel, main contract H8, 2007 to 2009

Project

The construction of the first section of new two-track railway line in the Lower Inn Valley is part of the northern approach route for the Brenner railway axis with a total length of 40 km. The main construction contract H3-4, Münster/Wiesing Tunnel with a length of about 5.8 km and one section of main contract H8, Jenbach/Wiesing Tunnel with a length of about 3.5 km were driven by shield machines with slurry support to the face [20], [21].

The route of both contracts crossed beneath the Inn Valley autobahn A12 and the lines of Austrian Railways ÖBB many times. The tunnels also passed under existing bridges and the buildings of the Zillertalbahn line. Sealing blocks 500 m apart constructed using diaphragm walling, bored piles and jet grouting, which served as starting shafts for a pipe jacking machine with slurry-supported face (4.85 m external diameter), had to be bored through. The overburden was between 6 and 45 m. The tunnels were lined during tunnelling work with reinforced concrete segments 50 cm thick.

The works were carried out by a consortium consisting of Porr/Bögl (H3-4) and STRABAG/Züblin/Hochtief (H8).

Geological and hydrological conditions

The alignment of both tunnels mostly ran through heterogeneously bedded loose ground. At the end of the contracts near the Wiesinger zoo, a transition to solid rock had to be overcome. The sections through loose ground were dominated by fluvial sediments of the bed of the Inn valley, the alluvial fans of the Kasbach streams and the Bradler alluvial fan from the Rofangebirge mountains. The soils were fluvial sediments (gravels) and fine-clastic, lacustrine sediments (sands or silts) and brown or grey sands. In the transitions from Inn sediments to alluvial fan sediments and within the alluvial fan facies, permeabilities of up to $5 \cdot 10^{-3}$ m/s were forecast. Under the Wiesinger zoo are the Wetterstein limestones and dolomites [108].

Figure 10-20 Mixshields from Herrenknecht for the Lower Inn Valley railway, contract H3-4 (left), H8 (right), 2007 to 2009 [20], [21]

The groundwater table is only a few metres below ground level and fluctuates with the level of the River Inn. Maximum groundwater pressures of 3.6 bar for H3-4 and 3.0 bar for contract H8 had to be expected for the selected vertical alignment [244].

Tunnelling and machine concept

Almost identical Mixshields (SM-V4) from Herrenknecht were used for both contracts [244] (Figure 10-20).

The essential machine specifications are summarised in Table 10-2.

Table 10-2 Essential tunnel data and machine specifications in the Lower Inn Valley

Tunnel	Lower Inn valley contract H3-4/H8
Length	5,840/3,470 m
Segmental lining	
Internal diameter:	11.63 m
Thickness:	50 cm
Ring length:	2 m
Shield	
Diameter:	13 m
Maximum thrust force:	102,446/93,100 kN
Nominal torque of cutting wheel drive:	15,490/24,442 kNm
Breakout moment of cutting wheel drive:	24,780/31,572 kNm
Slurry circuit quantity	2,800 m³/h
Length of backup:	approx. 100 m
Cutting wheel	
Disc cutters/rippers:	70
Scrapers:	268
Buckets:	2 × 8 (left and right)

Contract H3-4

The Herrenknecht Mixshield for contract H3-4 was built with the classic two-chamber system with the face supported through an air bubble in the working chamber. To cope with maximum groundwater head heights of up to 36 m, the machine was designed for an operating pressure of 5 bar. Two double-chamber personnel locks and one material lock were installed to permit entry to the excavation/working chambers under compressed air.

The shield had a bored diameter of 13.03 m and a length of about 10 m. With a total excavation area of about 133 m², the relatively closed rimmed cutting wheel had a degree of openings of about 30 %. Overcutting was made possible by the installation of an articulated joint. The cutter head with a total weight of 265 tonnes was fitted with single- and double-row disc cutters, also scrapers and buckets. Some of the disc cutters were replaced by rippers. The cutting wheel was driven by a continuously adjustable electric drive of 3,200 kW. If stones or blocks were encountered, a jaw crusher was fitted capable of coping with lengths of up to 1,200 mm. In order to reduce sticking, separate central flushing was provided.

The altogether 56 hydraulic cylinders were capable of a maximum thrust force of about 103.000 kN. The lining of the tunnel was 2 m wide and 50 cm thick reinforced concrete rings consisting of seven segments and a tapered keystone.

Eight grout lines were fitted for the grouting of the annular gap between the segmental lining and the surrounding ground. These were fed by four double-piston pumps. In order to prevent penetration of water, soil or grout at the shield tail, a three-row brush seal was fitted.

The approximately 100 m long backup of the tunnelling machine provided space for necessary equipment like controls, segment handling, electrical cabinets, slurry circuit, hydraulics, grouting, cable extension, ventilation, water, HV cable, fire protection rescue chamber, probe and grout drilling equipment [244]. A 300 m^3 bentonite storage tank was also installed on the backup, which enabled large quantities of fresh bentonite to be supplied quickly in case of losses. The excavated muck was pumped to the separating plant through a 500 mm nominal diameter pipeline with relay pumps. The separation plant was designed for a maximum circulation quantity of 2,800 m^3/h (4 × 700 m^3/h).

Contract H8

The Herrenknecht Hydroshield machine for contract H8 was identical to that for contract H3-4 except for the drive and the intake area. The machine had electro-hydraulic drive units with an installed power of 4,000 kW. In comparison to electric drive, the efficiency is lower and the required drive power is slightly higher. The intake area in the excavation chamber could be closed from the remainder of the working chamber with shutters, and pressure communication with the face support was ensured by active balancing pipes. This was intended to minimise material accumulation and sticking with the resulting cleaning work in the invert (compressed air interventions). In order to balance high loss of suspension in the expected round gravel layers, a separate 200 mm nominal diameter pipeline was installed directly from the separation plant instead of a storage tank.

Project progress

From the relevant starting shafts of contracts H3-4 and H8, the tunnels run towards the Wiesinger zoo. Both machines were dismantled underground in the rock section of the conventionally mined Tiergarten tunnel (contract H3-6).

The tunnel drives of 5.8 km (contract H3-4) and 3.5 km (contract H8) were bored by the two tunnelling machines with the performance data given in Table 10-3.

Table 10-3 Advance rates

		H3-4, L = 5.8 km	H8, L = 3.5 km
Average daily advance	[m/day]	9.8	6.6
Maximum daily advance	[m/day]	30	24
Average monthly advance	[m/month]	274	171
Maximum monthly advance	[m/month]	534	352

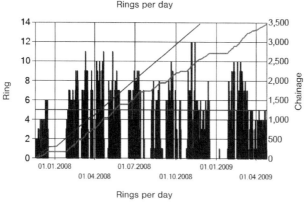

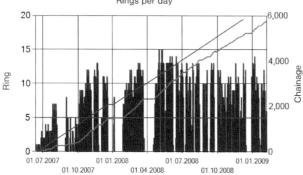

Figure 10-21 Advance rates; top: contract H8, bottom: contract H3-4

The crossing beneath the River Inn, the TIGAS pipelines, the Inn Valley autobahn, the railway line and the Wiesinger bridge were achieved without problems on contract H3-4. The maximum settlement due to the tunnel passing beneath the Inn Valley motorway A12 was a maximum of 18 mm, without additional ground improvement or safety measures.

Contract H8 passed under the Lower Inn Valley motorway A12 at a skewed angle along a length of 500 m with a minimum cover of only a half diameter, and this also proceeded without problems. A jet-grouted canopy was used to keep the settlement down to 3 mm. The following section very near to the Inn led to high bentonite losses but it proved possible to compensate for this without requiring any additional bentonite supply. The crossing of the secured TIWAG tailrace channel, the second crossing of the Inn Valley motorway A12, the railway tracks near the station at Jenbach, the secured main road L7 and the mixed-face section all went without great difficulties.

The surface settlements on both project contracts were in the range 5 to 20 mm and thus significantly less than the forecast settlements of 25 to 50 mm.

The advance rates and costs were determined to a significant extent by tool wear and soil separation. Compressed air interventions were required at an average of every 70 to 90 m at up to 3.1 bar to check the tools, and tools had to be changed at an average of every 120 m. On contract H3-4, the change from disc cutters to drag picks led to a reduction of tool wear.

Only localised losses of bentonite occurred despite to the high permeability of the Inn gravels. The grain fractions were well mixed in the excavation chamber, and the fines content created by the grinding action of the disc cutters presumably had a favourable effect on the sealing of the face.

In the end, all types of soil and tunnelling scenarios were successfully overcome. After the two tunnelling machines had reached their targets at the eastern and western ends of contract H3-6, the dismantling began. The shield skin of the H3-4 and H8 tunnelling machines stayed in the tunnel permanently.

10.6.3 Fourth bore of the Elbe Tunnel

In order to extend the capacity of the existing three bores of the tunnel under the River Elbe, a 2,560 m long tunnel was bored by a 14.20 m diameter slurry shield machine. The horizontal and vertical alignments of the new bore were mostly determined by the existing tunnel bores and the requirement for a maximum gradient of 3.7 %.

Geological and hydrogeological conditions

In addition to the extreme diameter, the main challenges were the high water pressure of 4.2 bar in the middle of the river and the shallow overburden of only 6 m under the riverbanks. The whole variety of Ice Age deposits had to be bored through; sand, glacial drift, basin silt and Lauenburger clay and in addition stones ranging up to boulders, pebble fields and Tertiary mica clay.

Tunnelling and machinery concept

In the planning of the machine and tunnelling concept, a Mixshield from Herrenknecht (Figure 3-1) was chosen, the main advantages being seen as the sensitive control of pressure in the excavation chamber leading to low settlements under built-up areas on the banks of the Elbe, the integration of a rock crusher and the secure mastering of water pressure.

Considering the diameter of 14.20 m and the special requirements of the project, the following innovations were developed for the machine:

– Cutting wheel design: Atmospheric tool changing from the accessible main arm of the cutting wheel (Figure 10-23) without having to lower the suspension level; combined tool equipment with scrapers and double-ring disc cutters. Due to the vertical alignment and the associated low cover at high water pressures, entry to the excavation chamber was only possible for divers in some parts of the Elbe. The system of changing tools from the accessible cutting wheel was a great help here.
– Active center cutter ($\varnothing$ 3 m) with separate flushing circulation and integrated cone breaker. The centre cutter could be extended forward by 600 mm. The muck flow in the centre was optimised to avoid accumulation of material as could be expected in the sections in glacial drift. With hindsight, the lack of access for tool changing from behind can be seen critically.
– Seismic advance investigation, which used reflections from acoustic signals to detect obstructions in front of the cutting wheel.
– Support pressure regulation independent of the water level in the Elbe, the suspension level behind the submerged wall and the pressure conditions in the excavation chamber.
– Ring former: Integration of a ring former consisting of invert shoe and bracing arches, which was intended to stabilise the ring in its position until securely bedded by the grouting of the annular gap. This was considered necessary with the self-weight of the segments being up to 20 tonnes. The quality of the segmental lining later in the tunnel

drive showed, however, that there were no quality defects in the lining where the ring former had not been used. The ring former was therefore not used for most of the tunnel, resulting in more working space for ring erection.

- Backup loads: a walking mechanism was developed to distribute the loading evenly over many segment rings in all operational circumstances. Due to the size of the machine and the starting situation (shaft length 40 m), it was not possible to construct a long backup. The installed machinery and the reduction of the length of the entire system to 60 m led to a high loading per metre, which led to the use of a walking mechanism for the first backup to even out the loading in the front part of the backup. The walking mechanism proved successful during the driving of the tunnel; any tendency of the backup to roll could be countered through an appropriate walking movement.

An analysis of possible incidents had already been undertaken during the tendering phase. Risks had been evaluated and their distribution clarified. These were integrated as appropriate solutions in the machine and tunnelling concept into the specification of the tunnelling machine and the tunnel handbook, which stated detailed procedural instructions for the tunnelling crew.

For example, the tunnelling concept provided measures for the creation of a sealing wall in front of the cutting wheel and the production of a waterproofing canopy from the machine in order to overcome unusual or unforeseen obstructions at the face.

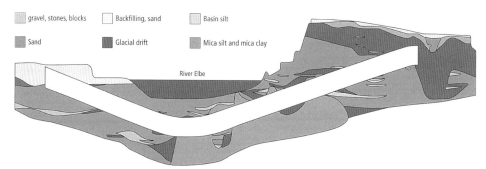

Figure 10-22 Tunnel vertical alignment and geology, fourth bore of the Elbe Tunnel

Figure 10-23 Accessible main arm of the cutting wheel for atmospheric changing of disc cutters and scrapers (design and cutting wheel left); Internal view of changing tools in the main arm (right)

Further points in the tunnelling concept [9] were:

- ground improvement in the tunnel route area of the River Elbe through vibro-compaction with the addition of steel slag,
- filling of the Elbe riverbed with copper slag to prevent the scouring of the riverbed and to ensure operational safety in the construction state,
- construction of two maintenance stations of overlapping bored pile walls in unreinforced concrete B5, in front of the south bank and directly on the beach of the north bank,
- test fields for the observation of surface movements to collect data for the boring of the tunnel beneath the built-up area on the Elbe bank (observed settlements of the order of 20 to max. 29 mm with about 10 m of sand cover),
- securing of the existing building substance with multi-phase grouting intended to produce heave where the tunnel passed under the northern Elbe bank.

Project progress

After the tunnelling machine had set off from the southern starting shaft in the middle of October 1997 and started to bore under the Elbe and reached about TM 700, a slab-shaped collapse of the face occurred shortly before the completion of repair work to the cutting wheel in June 1998. The compressed air escaped uncontrolled upwards and formed a blowout. Despite the quickly rising bentonite level, the men working in the excavation chamber managed to retreat into the personnel lock before the face collapsed completely and the cutting wheel was blocked with a mixture of soil and slag fed by the chimney-shaped funnel reaching up to the Elbe.

In order to remedy the collapse, grouting was first carried out from the shield skin into and around the funnel. After the determination of the soil compaction values with penetration tests, steel slag was used to backfill the crater resulting from the collapse and a covering of large-area sand packets wrapped in geotextile was laid before the cutting wheel and excavation chamber could be flushed out.

A second blowout further along the tunnel was recognised in time from the increased compressed air losses and avoided through countermeasures (restarting, compressed air reduction and grouting).

Figure 10-24 Shield at the breakthrough into the flooded reception shaft

There were problems in the further course of the tunnel with sticking in the clays and marls, which built up in the working chamber and between the spokes. Very variable advance rates were achieved depending on the geology. Whereas 25 mm/min could be achieved in sand with new tools, the advance rate in clay with partially worn tools fell to 5 mm/min. Increased wear to the gauge tools in marl with sand and gravel inclusions led to more frequent changing of the tools, which could be reached through the main arms under atmospheric pressure although this turned out to be very time-consuming. After 1,050 m of advance, the wall thicknesses of the cutting wheel steelwork were measured, which showed that these were also affected by wear; the original metal thickness of 80 mm had been reduced to 16 mm. This made strengthening of the steelwork by welding on plates from the inside necessary.

After 28.5 months of tunnelling, the breakthrough into the northern reception shaft was made at the end of February 1999 (Figure 10-24). Due to the difficulties already described and the resulting delays, the actual average advance rate was 5.9 m/d.

The experience from the fourth bore of the Elbe Tunnel fully confirmed the measures taken – accessible cutting wheel with tool changing under atmospheric conditions and the planned integration of diving work into the machine concept, but problems of sticking and wear lead to additional measures and delays. It should finally be pointed out that only the basic motivation and will to succeed of all parties to the project in design, tendering and construction enabled the implementation of this large project and the overcoming of all difficulties.

10.6.4 Chongming

In order to provide a crossing under the River Yangtze (Chang Jiang) within the urban area of Shanghai, a twin-bore, 7.16 km long road tunnel was built with three lanes in each bore. The new tunnel together with a bridge connects the islands of Changxing and Chongming to the motorway network and the urban area. Another level is integrated below the level of the carriageways, which is used for service and safety equipment and also provides room for a future metro line. The tunnel has an external diameter of 15 m. The rings consist of 9 + 1 segments with a block length of 2 m. The segments have a wall thickness of 64 cm with individual weights of up to 16.7 t.

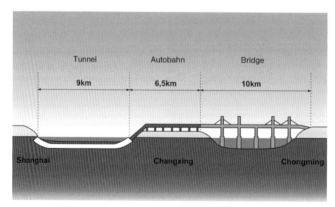

Figure 10-25 Overview of the Chongming project

Geological and hydrogeological conditions

The geology of the tunnel was, due to the location in the river delta, dominated by soft, silty clay deposits with thin layers of sand.

Tunnelling and machine concept

The two slurry shields from Herrenknecht for this project were based on the experience with the Mixshields used for the fourth bore of the Elbe Tunnel and the subsequent further developments of large shields for work under high water pressures. With a shield diameter of 15.43 m, the two machines are currently the largest ever built (Figure 10-26).

The Mixshields had the following mechanical features:

- The steelwork of the shield was designed for an expected operating pressure of 6 bar at the center line. Due to the intended use for an underwater crossing and the elongated alignment of the future motorway tunnel (R_{min} = 4,000 m), no articulation joint was provided in the shield.
- The invert area of the shield was fitted with two agitators (∅ 1,900 mm) intended to assist the flow of muck to the screen and the intake with nominal diameter 500 mm. The Mixshields were also fitted with submerged wall gate, bentonite jets, flushing of the cone and stator, all adapted for the conditions in soft sediments and the possible sticking potential.
- The double-walled shield tail was fitted with a three-row brush seal with emergency seal and spring steel packet. Ground freezing ducts were also integrated into the shield tail. These were intended to enable the freezing of the surrounding ground in case of a problem in order to minimise the risk of water ingress during brush changing or repair work.
- The cutting wheel was constructed as flat wheel with six main arms accessible under atmospheric pressure (Figure 4-25). In order to reduce the number of compressed air interventions, a complete set of tools acting on the entire face could by changed from inside the cutting wheel under atmospheric pressure. In line with the expected geology, the cutting wheel was completely fitted with scrapers. Two hydraulically extendable copy cutters created a 40 mm overcut to the radius. In consideration of the distance to be tunnelled of more than 7,000 m, the front and external areas of the cutting wheel and the back as well were designed to be particularly durable and wear-resistant. As an additional level of safety, both Mixshields were fitted with all necessary equipment for entering the chamber under saturation diving conditions such as air locks and installations.
- The installed cutting wheel drive power was 3,750 kW and the bearing diameter 7.6 m. The resulting torque of the variable, frequency-controlled electric cutting wheel drive was 174,057 kN. Both shields were designed for a maximum advance rate of 45 mm/min.
- The three-part backup amounted to a total length of 118 m and was divided into a primary backup, an intermediate bridge and a pipe layer. The first three-storey backup housed all the machinery to supply the shield (hydraulics, electrics, pumps, grouting) and the control cabin. In order to distribute the wheel loads, the backup had an integrated wheel and walking system. The central backup had a 67 m long bridge, under which a prefabricated invert slab element could be placed. The necessary bridge construction served as a gantry for the supply crane (segments, grout skip transfer and other materials). The third section of the backup, or second backup, provided all extension functions (water supply, air, bentonite, HV cable, ventilation) and also housed workshop places and further secondary equipment.

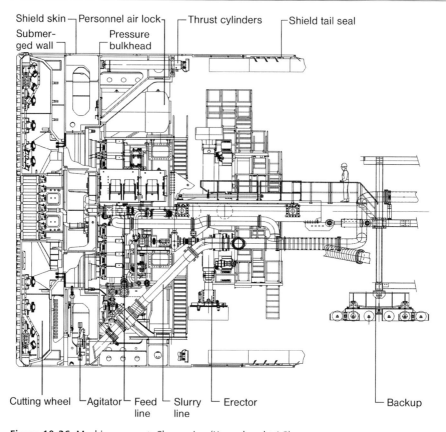

Figure 10-26 Machine concept, Chongming (Herrenknecht AG)

- The segments and grout were transported to the machine on wheeled trucks, which carried either segments or grout skips in convoys. The segments were reloaded by the segment crane and segment feeder. The grout was loaded from the transport skip into the first backup.

The shield steelwork and the assembly of the 132 m length of tunnelling machinery with a total weight of 2,300 t were carried out in Shanghai. The cutting wheel and further main components were manufactured in Germany and shipped to China. After the workshop acceptance, the tunnelling machinery was dismantled and transported about 6 km to the starting shaft.

Project progress

Tunnelling of the first bore started in September 2006 and the second bore in January 2007. The constant weekly advance rates ranged between 90 und 120 m/week after the initial learning curve. The first machine broke through in May 2008, and the second arrived in September 2008. The performance of both machines through the ooze deposits of the Yangtze Delta turned out to be remarkably good, and neither machine required changing of excavation tools despite the tunnel length of 7 km. The two large tunnels were structurally complete ten and twelve months ahead of the planned schedule (Figure 10-27).

10.6 Examples 253

Figure 10-27 Breakthrough of the first Mixshield, Chongming, May 2008

Whatever your challenges are

In the construction of new space underground, MEYCO® provides more than purely equipment and chemicals for sprayed concrete. Its new solutions range from the field of TBMs and injection to waterproofing and fire protection, all supported by the expert engineering knowledge of our global team.

www.meyco.basf.com/tbm

□·**BASF**
The Chemical Company

Expanding Horizons

Underground

MEYCO

Fundamental Ideas.
Visionary construction.

Tunnelling . Soil Injections . Boring . Civil Engineering
www.tunnelvortrieb.de

WÜWA Bau GmbH & Co. KG
Am Kiefernschlag 30
91126 Schwabach, Germany
Phone +49 9122 9973-11900
Fax   +49 9122 9973-11908
info@tunnelvortrieb.de

BOOK RECOMMENDATION

Hardrock Tunnel Boring Machines

This book covers the fundamentals of tunneling machine technology: drilling, tunneling, waste removal and securing. It treats methods of rock classification for the machinery concerned as well as legal issues, using numerous example projects to reflect the state of technology, as well as problematic cases and solutions. The work is structured such that readers are led from the basics via the main functional elements of tunneling machinery to the different types of machine, together with their areas of application and equipment. The result is an overview of current developments.

Close cooperation among the authors involved has created a book of equal interest to experienced tunnelers and newcomers.

Maidl, B. et al.
Hardrock Tunnel Boring Machines
2008. 343 pages with
256 figures, 37 Tab. Hardcover.
€ 99,-
ISBN: 978-3-433-01676-3

Ernst & Sohn
Verlag für Architektur und
technische Wissenschaften GmbH & Co. KG

www.ernst-und-sohn.de

For order and customer service:

Verlag Wiley-VCH
Boschstraße 12
69469 Weinheim
Deutschland

Telefon: +49(0) 6201 / 606-400
Telefax: +49(0) 6201 / 606-184
E-Mail: service@wiley-vch.de

11 Earth pressure balance shields

11.1 Development history

The development of earth balance shields started in the early 1970s in Japan and one of the first applications took place in 1974 in Tokyo. In the following years, the process was known under various descriptions such as Earth Pressure Shield, Pressure Holding Shield, Slime Shield, Soil Pressure Shield, Earth Pressure Balanced Shield, Confined Soil Shield, Mud Pressurized Shield or Muddy Soil Shield. All these terms essentially refer to the same process and the term "Earth Pressure Balance Shield (EPBS)", or "EPB-Shield", has become established internationally.

The development was derived from the blind shields used in cohesive soils with very high plasticity. The name blind shield denotes that the undisturbed soil of the face cannot be seen (Figure 11-1). The soil is not excavated mechanically but the viscosity of the material is exploited and the soil is forced through an opening under the pressure applied by the thrust cylinders, the rate being regulated by a gate valve in the pressure bulkhead of the shield.

Control of low-settlement tunnelling is achieved by comparing the mass of the soil removed against the theoretical excavation volume determined from the advance rate (Chapter 11.2). The volume of soil removed is regulated by the gate valve, which has to be designed so that it can be opened and the soil removed manually if the plasticity is insufficient.

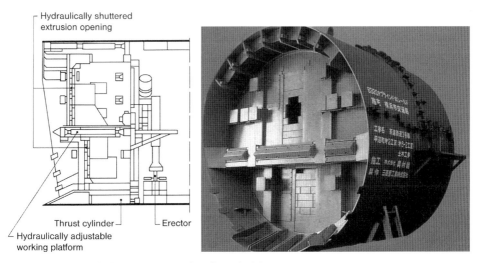

Figure 11-1 Blind shield with two chambers (Mitsubishi)

11.2 Functional principle

In tunnelling machines with earth pressure support (Figure 11-2), the face is supported by slurry formed by remoulding the excavated earth. The excavation chamber (2) of the shield is closed off from the tunnel by a pressure bulkhead (3). A more or less closed cutting wheel (1) fitted with tools excavates the soil. Mixing buckets on the back of the cutting wheel (rotors) and on the pressure bulkhead (stators) help to remould the soil into a suitable consistency. The pressure is measured by pressure cells distributed over the front of the pressure bulkhead. A pressure-tight screw conveyor (5) removes the soil from the excavation chamber. The support pressure is controlled through the revolution speed of the screw conveyor and the injection of a suitable conditioning agent, controlled according to pressure and volume. The pressure gradient between excavation chamber and tunnel results from friction in the screw conveyor. This means that either the soil in the screw conveyor has to ensure the sealing of the muck removal system or else alternative mechanical measures have to be taken. Complete support to the face, particularly the upper part, only succeeds if the soil can be remoulded into a soft to semi-stiff plastic mass to act as a support medium. This is significantly influenced by the content of fines (grain size < 0.06 mm). Soil conditioning, for example with bentonite, polymers or foam, can be used to extend the scope of application of an earth pressure balance shield, in which case attention needs to be paid to the environmental acceptability of tipping the material.

Figure 11-2 shows the principle of an EPB shield with its most important components.

11.2.1 Support pressure measurement and control

The pasty, often even stiff consistency of the materials places more stringent requirements on the measurement and control of the support pressure than the regulation in a Hydro- or Mixshield through the air bubble or the suspension pressure. With increasing density and increasing friction angle of the support medium, the pressure distribution of the earth slurry at the pressure bulkhead and at the face is subject to considerable fluctuations. Earth pressure balance shields do however have the advantage over slurry shields that no sudden, large-scale collapse of the face can occur into the excavation chamber filled with pasty and condensed material.

The support pressure is measured with earth pressure cells installed flush in the pressure bulkhead and in the cutting wheel.

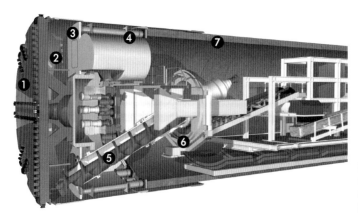

Figure 11-2 Principle of the construction of an earth pressure balance shield (Herrenknecht AG)

Figure 11-3 shows an earth pressure cell of a type often used in modern German earth pressure balance shields. A pressure cushion connected to an electric sensor contains a closed system containing a fluid. When the cushion is loaded, the resulting hydraulic pressure on the membrane is transferred to the electric sensor and converted to a voltage proportional to the loading.

The measurement of earth pressure is part of a complex data recording system, which serves to document the entire progress of the advance and to enable the control of certain parameters in the control circuit. Based on these measurements, the support pressure is controlled/regulated through the thrust cylinder extension rate, the screw conveyor revolution speed or the injection of conditioning agent according to pressure and volume.

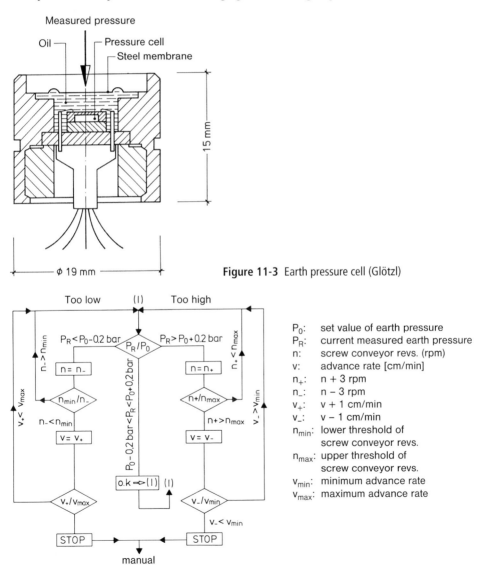

Figure 11-3 Earth pressure cell (Glötzl)

P_0: set value of earth pressure
P_R: current measured earth pressure
n: screw conveyor revs. (rpm)
v: advance rate [cm/min]
n_+: $n + 3$ rpm
n_-: $n - 3$ rpm
v_+: $v + 1$ cm/min
v_-: $v - 1$ cm/min
n_{min}: lower threshold of screw conveyor revs.
n_{max}: upper threshold of screw conveyor revs.
v_{min}: minimum advance rate
v_{max}: maximum advance rate

Figure 11-4 Regulation flow diagram of an earth pressure balance shield

Figure 11-4 shows a flow diagram of the regulation scheme for an EPB shield without consideration of the soil conditioning. The intention is to maintain the regulated parameter P_R (momentarily measured earth pressure) at the set value P_0. An increase of the thrust cylinder extension rate or a reduction of the revolution speed of the screw conveyor will have the effect of increasing the earth pressure; a reduction of the thrust cylinder extension rate or an increase of the revolution speed of the screw conveyor will have the effect of reducing the earth pressure. If the first set value (n = screw conveyor revs.) is outside a defined working range ($n_{min} < n < n_{max}$), then the second set value (v = thrust cylinder extension rate) is altered. If this value is also outside a defined range ($v_{min} < v < v_{max}$), then the advance is stopped and the screw conveyor gate valve at the end of the screw conveyor is closed.

It should be mentioned here that the cause-effect relationship in the excavation chamber does not fit with any stringent laws of soil mechanics and that the experience of the operator is still of essential importance. The use of a control system enables the simple control actions of the TBM operator to be reduced to a minimum in normal operation, but continual visual monitoring of the most important instruments should never be neglected.

The support pressure can nowadays be controlled actively through foam conditioning with the same precision as with a slurry shield. The set and the actual values of all foam parameters are monitored by process control systems (Figure 11-5).

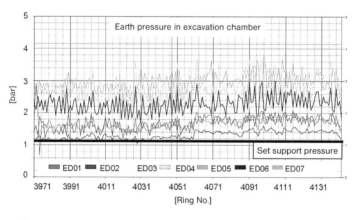

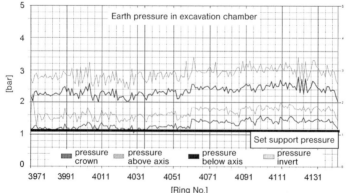

Figure 11-5 Comparison of set and actual values of support pressure with the software Maidl-PROCON

11.2.2 Soil conditioning

In order to use the soil that was removed from the face as a support medium, it should possess the following properties:

- good plastic ductility,
- pasty to soft consistency,
- low internal friction,
- low water permeability.

Few soils have these properties in their natural state and most therefore have to be conditioned, i.e. processed. The good ductility and pasty to soft consistency of the material ensures that the support pressure acts on the face as uniformly as possible and that the flow of material into the intake of the screw conveyor is continuous. This avoids any blocking of the cutting wheel or blockages of the excavation chamber in areas of low pressure gradient (Figure 11-6). In addition, the drive torque of the cutting wheel and the screw conveyor remain within economic limits. The soil consistency should not however be too soft, in order that the muck can be transported along the tunnel on conveyor belts or in muck cars (unless high density slurry transport is being used).

Figure 11-6 shows the flow of soil through the excavation chamber and through the screw conveyor in idealised form. The schematic illustration assumes a constant support pressure. The rotation of the earth plug, the opening ratio of the cutting wheel and gravity are not considered.

The earth slurry is removed from the pressurised excavation chamber by the screw conveyor and discharged into the tunnel under atmospheric pressure. It should be noted that the pressure measured at the pressure bulkhead is not exactly the same as the pressure on the face (Figure 11-6). Particularly around the screw conveyor, a lower pressure is measured at the bulkhead due to the potential gradient in the excavation chamber. In order to transfer the material from the screw conveyor to a conveyor belt without an air lock, the soil must have an appropriately low permeability and sufficiently stiff consistency. Uncontrolled flow through the screw conveyor has to be avoided.

The conditioning process has to be appropriate for the prevailing type of soil and is thus influenced by the soil parameters grading curve, water content (w), flow limit (w_L), plasticity index (I_p) and liquidity index (I_c). These parameters can be affected by:

- the addition of water,
- the addition of bentonite, clay or polymer suspensions,
- the addition of foam.

In the planning and selection of a process, the probable volume of conditioning agent should be determined. The consistency of the soil should not be to liquid to ensure onward transport and tipping of the material without additional measures if at all possible. The conditioning of the soil should take place directly during excavation at the face in front of the cutting wheel, in order to prevent the material sticking to closed cutting wheels with small openings.

11.2.3 Mass-volume control

In earth pressure balance shield machines of the first generation, the so-called mass control was often used to check the stability of the face. In this process, the excavated volume or weight is measured and compared with the theoretical volume or weight determined from the advance rate and excavation diameter [49], [171]. The control parameters were the advance rate of the shield and the revolution speed of the screw conveyor. This process has the following disadvantages:

- It delivers no information about the actual degree of filling or compaction of the soil in the excavation chamber, only a quantitative statement without factor of safety
- The determination of the mass of excavated material is difficult, inexact and hinders the clearing of muck from the tunnel.
- A qualitative statement is only possible after a delay; the momentary state of the face remains unknown.
- When conditioning agent is added at the same time, no precise retracing of the origin of the mass is possible.
- The bulking factor and the swelling effect of the transported soil are additional unknowns.

Figure 11-7 shows the mass control process. In practice, this makes it possible to analyse and interpret extra excavation quantity and overcutting, but the process is not suitable for the control of support pressure and the intention to minimise settlement. In addition to the classic belt scales method shown at (a), newer technologies have been developed in recent years.

In the ultrasound method shown at (b), ultrasound sender, receiver and processor are used to measure the distance and calculate the cross-section of the material on the conveyor belt. If the speed v of the conveyor belt is also measured, the volume flow of the transported material can be calculated and compared with the theoretical excavated volume. The alteration of the speed of the ultrasound waves can be compensated by measuring the temperature over the processor.

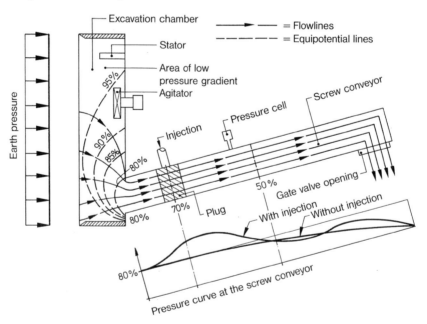

Figure 11-6 Illustration of the flow of the soil in the excavation chamber [205]

In the laser method shown at (c), the volume is calculated using a camera coupled to a processor. The optical visualisation of the surface is enabled by a compartmented laser beam, which is projected onto the surface by a reflector. The latter two methods demand relatively flat surfaces, so their practical use is problematic.

The greatest uncertainty in all the methods is the determination of the ideal value for the theoretical transport volumes/masses. The decisive soil parameters like dry density, water content, pore volume and assumptions related to the process regarding the displacement of the water in the pores, the ingress of formation water, the overcut and the determination of the conditioning agent quantities are all subject to greater variation than the measurement precision of the systems.

As part of the process control, it is also possible to record the entire mass balance including the grouting and conditioning materials. Figure 11-8 shows as an example the mass control for the EPB shield at the Katzenberg Tunnel.

(a)
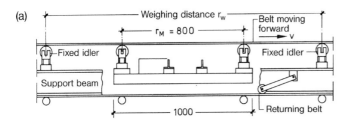

(b)
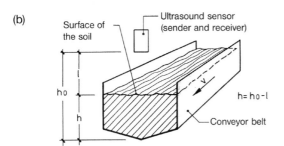

(c)
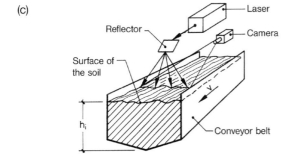

Figure 11-7 Mass control processes
a) Conveyor belt scales method
b) Ultrasound method (Mitsubishi)
c) Laser method (Mitsubishi)

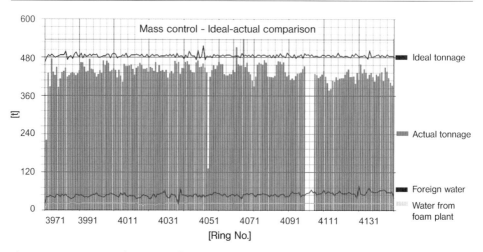

Figure 11-8 Mass control using Maidl-PROCON at the Katzenberg Tunnel

11.3 Areas of application

Machine types with earth pressure support are particularly suitable in soils with fine-grained (< 0.06 mm) content of over 30 %. In coarse- and mixed-grained soils and rock, the contact force and the cutting wheel torque increase over-proportionately with increasing support pressure. The flowing properties of the excavated soil can be improved by the addition of suitable conditioners like bentonite, polymers or foam. For active support pressure control and to ensure low-settlement tunnelling, soil conditioning with foam is recommended outside the predestined area of application.

Optimal preconditions for the use of an earth pressure balance shield are offered by clay-silt and silt-sand soils with pasty to soft consistency. Depending on the condition of the soil encountered, no addition or only small quantities of water is necessary. The mechanical effect of agitators and kneading tools in the excavation chamber can turn even highly cohesive soil into plastic slurry. Figure 11-9 shows a silty clay (w = 32 %; w_L = 38 %; wp = 23 %), which has been conditioned mechanically without the addition of any conditioning agent.

With increasing coarse-grained content, the addition of water alone does not work. There is no reduction of the angle of internal friction but there is a risk of separation of the earth slurry. The increased permeability makes sealing of the screw conveyor problematic. The fines content has to be supplemented by adding clay or bentonite suspension. Pore water can be bound by the suspension, which is capable of swelling, and the excavated muck turns into plastic slurry with good flowing properties and reduced permeability. Good results can also be obtained by injecting a strongly water-adsorbing polymer suspension [13].

The entire spectrum of application of earth pressure balance shields is shown in Figure 11-10. Above a widely graded threshold line (grading curve 1) with a minimum fines content of 30 %, there are practically no limits to the grading distribution of the soil. These soils are predominantly impermeable to water with a consistency determined by their water content.

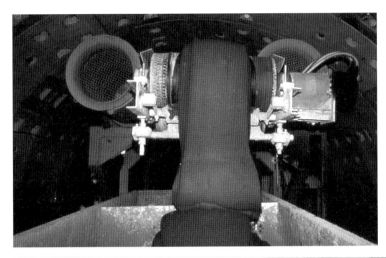

Figure 11-9 Muck produced by an earth pressure balance shield, Taipei Metro, 1992 (Herrenknecht)

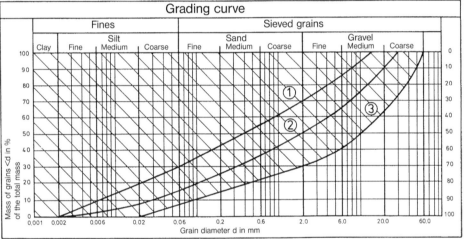

Zone	Preconditions	Conditioning agent
1	I_c support medium = 0.4 – 0.75	water clay and polymer suspensions tenside foams
2	$k < E^{-5}$ m/s water pressure < 2 bar	clay and polymer suspensions polymer foams
3	$k < E^{-4}$ m/s no groundwater pressure	high-density slurries high-molecular polymer suspensions polymer foams

Figure 11-10 Area of application of earth pressure balance shields depending on the grading distribution of the soil [205]

When the soil has stiff consistency ($I_c > 1$), high cohesion and low water permeability, it is normally possible to work without support pressure. If however face support is required, then the consistency of the soil should be pasty to soft ($I_c = 0.4 – 0.75$). Depending on the mineralogical composition, water and low-viscosity suspensions (bentonite, polymer) can be used as a conditioning agent, but also foam.

Under the threshold line, the water permeability and the internal friction of the material increase sharply. The limits of application are determined by the coefficient of water permeability k and the prevailing groundwater pressure. For practical application, the coefficient of water permeability should not exceed a value of 10^{-5} m/s and the pressure should not exceed max. 2 bar. [171]. If the simplified relationship $k \cong d_{10}^2$ [d in cm] is used for the coefficient of water permeability, then at least 10 % of the soil must have grain diameter less than about 0.03 mm. In Figure 11-10, this zone is above grading curve 2. In the area between grading curves 2 and 3, earth pressure shields should no longer be used in groundwater pressure. Under grading curve 3, the permeability is too high and the use of conditioning agents has no effect because they drain unhindered in front of the face and it is not possible to build up any support pressure. The diameter and proportion of stones should also be limited; in contrast to slurry shields, no rock crushers can be fitted inside the excavation chamber, with the result that the screw conveyor can easily be damaged.

The only suitable conditioning agents for soils below grading curve 2 are high-viscosity clay suspensions (high-density slurry) or polymer foams. If the liquid proportion of conditioning agent exceeds a value of 40 to 45 % of the excavated volume, then the consistency is normally fluid and muck transport on a conveyor belt will have to be replaced by hydraulic transport.

The range of application of EPB shields can therefore be extended far into the area of application of slurry shields by the use of soil conditioning measures. Non-cohesive soils should however not be regarded as a predestined area of application for an earth pressure balance shield. In non-cohesive soils with low fines content, the slurry shield is mostly more advantageous than the EPB shield and should be preferred. Further information about areas of application can be found in Chapter 19 [54].

11.4   Operating modes and muck transport

The special advantage of an earth pressure balance shield is its flexibility. Figure 11-11 shows five different operating modes, which differ according to face support and muck transport. There are no internationally or nationally harmonised tunnelling classifications or definitions of the operating modes.

There is, however, agreement that no measurable and actively controllable support pressure acts in the excavation chamber in open mode. It is an open system under atmospheric conditions. Even if the excavation chamber is partially or completely filled, no supporting effect from the earth slurry can be assumed in verification calculations. The verification of closed mode is a measurable support pressure at the crown due to a complete filled excavation chamber.

11.4.1   Open mode (screw conveyor – conveyor belt)

In rock or stable loose ground, earth pressure shields often work in open mode, i.e. without support against ground or water pressure. The cutting wheel respectively the cutter head can be designed and fitted with disc cutters alone or combined with scrapers. In aquiferous zones, it is possible to apply compressed air in the excavation chamber during stoppages

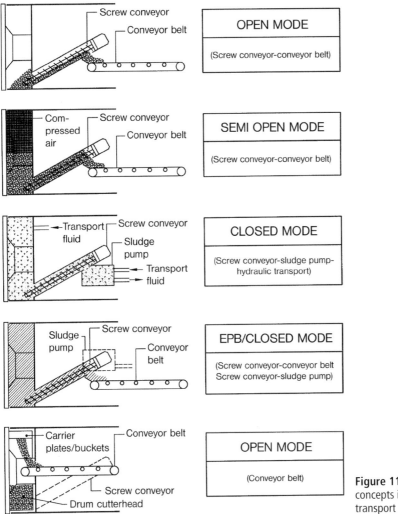

Figure 11-11 Machine concepts including muck transport

after closing the gate valve of the screw conveyor. The removal of soil from the invert requires special measures to protect the screw conveyor, cutting wheel and excavation tools against wear.

11.4.2 Semi open mode (screw conveyor – conveyor belt)

In temporarily stable ground with formation water ingress, the machine type described above can be operated without modification in closed mode with partial lowering. The upper part of the excavation chamber is filled with compressed air so that water in joints and pores is displaced and the ingress of water can be reduced. This process can result in problems with uncontrolled blowouts through the screw conveyor. Sufficient sealing to ensure a closed system can only be assumed in clay or heavily weathered rock. Without foam conditioning, there is high risk of sticking. Depending on the rock and ground parameters, heavy wear has to be expected.

11.4.3 Closed mode (hydraulic mucking circuit)

The danger of blowouts at the screw conveyor discharge can be avoided by connecting a closed transport system. Sealing and air lock systems for dry material handling in the form of transfer boxes, cycled locks, rotary feeder and even double screw conveyors have proved to be very unreliable in practice. Hydraulic transport systems have, however, been used successfully with piston pumps as well as slurry pumps. The fluid transport medium can be either fed directly into the pressurised excavation chamber or into a flushing box situated immediately after the screw conveyor. Depending on the rock strength, abrasiveness and cuttability of the rock mass, the cutting wheel and the screw conveyor have to be matched to each other. Roller crushers should also be provided at the interface of the screw conveyor to transport pump.

11.4.4 EPB mode (screw conveyor – conveyor belt or screw conveyor – piston pump)

Many shields with muck removal by screw conveyor, which are used in rock (e.g. Channel Tunnel, Athens metro etc.) are described as earth pressure balance shields but are not really operating in EPB mode. Operation of a shield with muck removal by screw conveyor in EPB mode in rock requires extremely elaborate conditioning measures because of the insufficiently ductile properties. For this reason, soil transport and soil disposal should be considered in the decision process or the selection of the best tunnelling process in addition to face support.

11.4.5 Open mode (conveyor belt)

Systems being developed by various manufacturers offer the option of operating the shield in closed mode with screw conveyor or in open mode with direct muck removal out of the centre on a conveyor belt. Operation in open mode is possible when the ground has sufficient stand-up time and no excessive water ingress is to be expected during the advance. The transport belt can normally be retracted very quickly during stoppages and the centre closed pressure-tight. Relief valves and pump systems then make enable resumption of the advance in open mode.

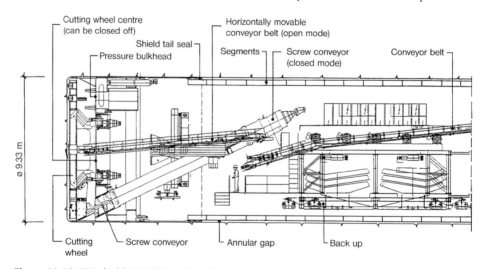

Figure 11-12 EPB shield S-165 (Herrenknecht) with open mode, Madrid Metro, ∅ 9.33 m, 1997 [122]

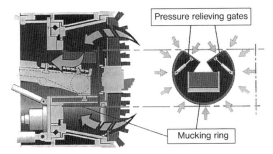

Figure 11-13 EPB mucking ring-conveyor belt system (Lovat) [179]

The cutter head is often constructed as a rotating drum and fitted with steel plates or shovels to pick up the excavated material. The excavated rock then falls under gravity through a hopper in the centre and onto the conveyor belt. However, the construction of the cutter head as a drum has disadvantages in EPB mode, as the conditioning of the ground is significantly more difficult.

Many manufacturers currently offer combinations of the excavation and transport systems described above. Figure 11-12 shows the EPB shield from Herrenknecht AG for a Metro project in Madrid. The machine can operate in open mode or in closed mode.

The Lovat system with the mucking ring and conveyor belt system typical for the manufacturer has proved extremely robust and successful in practice (Figure 11-13). Additional possibilities for closing and material dosing are offered by the patented pressure relieving gates, which enable almost closed operation. Conversion to screw conveyor transport out of the excavation chamber is also possible although time-consuming. For precisely regulated earth pressure operation, however, a precondition for this system is very good ductility of the soil, which is often not found.

The planning of a project using a shield with screw conveyor mucking requires detailed analysis of the ground conditions. In particular, operation in closed mode in rock is often problematic, as the possibilities for successful conditioning are extremely limited.

In open mode, practical experience of screw conveyor mucking is entirely positive. The problem of wear in abrasive ground should, however, countered by appropriate measures.

11.5 Components

11.5.1 Cutting wheel

The cutting wheel has to excavate the soil from the face cleanly and also knead the soil already in the excavation chamber as it rotates to convert it into the required plastic slurry. As the soil should have already been conditioned during excavation from the face, injection nozzles for the addition of suitable conditioning agents should be provided on the front of the cutting wheel.

The size of the openings in the cutting wheel as a percentage of its area can be varied practically at will, as long as the structural load-bearing capacity is guaranteed. An open cutting wheel (open star) has the advantage over a closed one that the risk of sticking is considerably less, the distribution of the support pressure to the face is more even and the flow pattern to the screw conveyor is better. On the other hand, according to the type of soil, the disadvantages are the increased risk when working in the excavation chamber under compressed air due to the lack of mechanical support, an increased risk of settlement where the cover is shallow and that the cutting wheel is less stable and stiff. If the open star is surrounded by a rim, this can improve the stability of the cutting wheel. The remaining free areas between the cutting wheel arms can be closed by hydraulically powered shutters to secure the face [237]. Practical experience shows, however, that complicated hydraulic constructions are exposed to much higher attack from the more abrasive support medium in closed mode operation than with other processes, so this construction is extremely prone to give trouble and is often not fit for use for its intended purpose. Streamlined construction of all rotating and fixed components should also be a design aim for working with high support pressures, an intention which is also opposed by the stated construction.

Since the removal of stones wedged in the screw conveyor is extremely difficult, all manufacturers today use closed cutting wheels, with only a few exceptions. The width of openings through the cutting wheel should be limited to the maximum stone diameter, which can be transported out of the excavation chamber by the screw conveyor. Larger stones are broken down to the slot diameter in front of the cutting wheel, which reduces the danger of the screw conveyor being damaged or blocked. The design of the openings demands special care in planning and design. A problem, which occurs constantly in practice, is the cutting wheel sticking up because the opening widths are too small or the slots are too deep without tapered enlargement to the inside. The excavated material cannot fall into the excavation chamber filled with fluid as with slurry shields, but rather has to be pressurised into the screw conveyor intake following the flow lines (Figure 11-6). Closed cutting wheels or drums only offer limited protection for the personnel who enter into the excavation chamber to undertake necessary work. Particularly in soils, which can flow out, working under compressed air is associated with considerable risks, as shield there is no sealing of the face by a filter cake with an earth pressure, in contrast to a slurry shield. Closed cutting wheels or closed drums are used in soils with an increased tendency to face instability and prevent large pieces being loosened from the soil structure.

Figure 11-14 Open and closed cutting wheels of earth pressure balance shields
a) Japanese type (Mitsubishi)
b) Madrid M-30, D = 15.2 m (Herrenknecht AG)

A disadvantage of the closed cutting wheel is the much less uniform distribution of the support pressure over the area of the face. The earth pressure values measured at the pressure bulkhead do not always agree with the actual earth pressure at the face and are also dependent on the position of the cutting wheel and the direction of rotation of the cutting wheel.

The stated problems with closed cutting wheels can now be overcome with foam conditioning.

The cutting wheel can, as described in Chapter 4, be fitted with drag picks, scrapers, round-shafted chisels or disc cutters according to the geological conditions. It should be possible to remove the tools to the side or backwards in order to enable checking, changing and combining. Depending on the alignment and the shield construction, the cutting wheel should be fitted with the appropriate copy cutters to produce the required overcut. Attention should be paid to adequate wear protection, particularly in abrasive ground.

The necessary cutting wheel torque increases exponentially with increasing diameter. On method of reducing the resultant torque is the use of two cutting wheels rotating in opposing directions. Figure 11-14 b) shows the cutting wheel of the EPB shield for the Madrid M-30 project.

11.5.2 Bearing and drive construction

The same bearing and drive constructions are used as known from hard rock and slurry shield machines. The demands placed on the technology used in EPB machines are, however, considerably different. The accessibility and sealing systems of the bearing construction are critical, as they are subjected to extreme mechanical attack in abrasive soils, in which the earth slurry has the characteristics of grinding paste. Major damage to the seals is known to have occurred [193].

The earth pressure balance shields built until now had one of the three basic layouts of drive and bearings shown in Figure 11-15.

The advantages of a closed or open cutting wheel with central drive (Figure 11-15 a) are simple construction and the relative lack of obstruction to the flow and conditioning processes by mechanical parts in the excavation chamber. Another advantage is that all seals are accessible and can be repaired from the shield space. This has to be balanced against the disadvantages that a central cutting wheel drive mechanism occupies the centre of the shield and the screw conveyor can only have a slight gradient, which results in the shield having to be longer. The routing of pipes into the rotating cutting wheel is also technically laborious. When the diameter is larger (d > 4.0 m), the drive shaft diameter increases for structural and constructional reasons. This makes mucking openings in the centre of the machine impossible, so the muck has to be forced outwards in front of the cutting wheel in order to reach the excavation chamber.

In the drum type (Figure 11-15 b), the cutting wheel and excavation chamber form a unit and rotate together. The bearings and the drive are at the perimeter of the drum. The following advantages of this system can be stated:

– Stable guidance of the cutting wheel even with variable contact pressure resulting in one-sided loading when the geological conditions of the face are inhomogeneous,
– the possibility of using heavy cutter heads for hard rock.

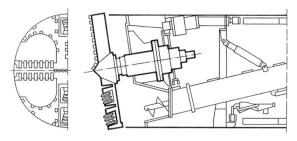

(a) Center shaft type

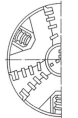 

(b) Drum type

Closed cutting wheel   Open cutting wheel

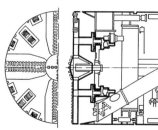 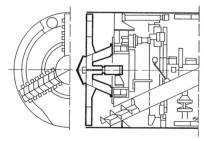

(c) Center cone type

Figure 11-15 Types of cutting wheel drive and bearings (Mitsubishi)

On the other hand, the disadvantages are:

- the danger of the muck sticking and rotating with the drum are greater due to the completely rotating excavation chamber, which worsens the conditioning of the material and thus the flow to the screw conveyor,
- in larger diameter machines, the high perimeter length of the seals that have to be sealed by grease lubrication systems,
- the outer sealing systems between the rotating drum and the non-rotating parts of the shield skin are not accessible for repair or changing.

The closed or open cutting wheel with centre-free drive (Figure 11-15 c) combines the advantages of centre and perimeter cutting wheel drives. The drive bearing diameter is about half of the shield diameter. The advantages are stable guidance, lower risk of sticking (the material cannot rotate with the fixed excavation chamber) and assistance of the conditioning process by the kneading effect of the support beams. Additional struts further increase the support and kneading effect.

11.5 Components 271

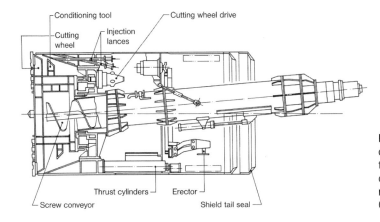

Figure 11-16 Mechanical conditioning tools in the excavation chamber, Toshori-Tsumori Drainage Tunnel, Osaka (Mitsubishi)

11.5.3 Excavation chamber

The excavation chamber, which can also be called the plenum chamber, describes the part of the machine between the shield blade and the pressure bulkhead. In the excavation chamber, the material excavated from the face has to be remoulded into a plastic earth slurry with good ductile properties, unless this has already been done at the face, in order to keep the support pressure to the face as constant as possible.

The design of the excavation chamber is essentially determined by the location of the screw conveyor, the man lock that may be provided and the type of cutting wheel drive. The excavation chamber should be designed to be as favourable to flow as possible in order to avoid the formation of clumps and ensure continuous flow of material to the screw conveyor.

The length of the excavation chamber determines the length of time the muck stays depending on the advance rate. The excavation chamber can be fixed in the shield or in the cutting wheel. As already described in Section 11.5.2, the danger of blockage is considerably worse if the excavation chamber rotates with the cutting wheel (drum type), since the mixing and kneading effect decreases with increasing viscosity of the material.

The quality of the conditioning effect results from good mechanical stirring of the earth slurry. In case the mixing effect of the cutting wheel and its constructional elements (Figure 11-16) are not sufficient, the following measures can lead to an improvement:

− Provision of rotors on the back of the cutting wheel,
− Provision of stators on the pressure bulkhead,
− Provision of mixing paddles/agitators on the pressure bulkhead.

In larger diameter shields, there is an increased danger of sticking particularly in the centre of the excavation chamber, as the peripheral speed of the rotating kneading tools decreases towards the centre. Rotors and stators have no effect here at the usual cutting wheel rotation speeds of 1 to 2 rpm.

11.5.4 Screw conveyor

The screw conveyor has to undertake the following tasks in an earth pressure balance machine:

- Transport the muck out of the pressurised excavation chamber into the tunnel under atmospheric pressure,
- seal against pressurised groundwater in permeable soils,
- control the support pressure in the excavation chamber through controlled material removal.

Screw conveyors are continuous conveyors (DIN 15 262) [76], which consist of a screw-shaped flight rotating inside a fixed casing. The screw is normally fixed to a rotating shaft but can also be constructed without central shaft so that larger stones can be transported through the centre (Figure 5-10).

The drive can be either central or peripheral. With a central drive, the muck is transported through a lower gate shutter, whereas a peripheral drive enables muck discharge directly from the end of the casing.

The screw conveyor can normally be extended into the excavation chamber with displacement cylinders, which improves the muck feed. When the screw conveyor is in the retracted position, the earth slurry has to be moved to the intake by pressure. The material taken into the sleeve pipe is prevented from rotating by friction with the casing and is thus moved along inside the conveyor.

Low-lying screw conveyors have proved successful, as the earth slurry is helped towards the intake by its self-weight. When a screw conveyor is located in the centre, the material collected at the bottom has to be picked up against gravity. With low-level screw conveyors, the emptying of the excavation chamber under compressed air is much less problematic and it is possible to lower the earth slurry to a lower level. For this reason, compressed air locks should be located as high as possible.

Sealing against groundwater under pressure is provided by various closing or sealing systems. Figure 11-17 shows various systems for centrally and peripherally driven screw conveyors. In silty and sandy soils, the technically simple gate valve construction has

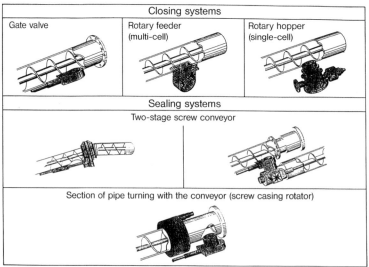

Figure 11-17 Closing and sealing systems of screw conveyors (Mitsubishi)

proved successful, but with increasing grain size and increasing water permeability, this system no longer offers a satisfactory solution. The main problem with the rotary feeder is the increased risk of sticking and blockages.

The sealing system works on the principle of mechanical throttling at a uniform rotation speed of the screw conveyor and compaction of the muck in the casing as a result of continuity. With increasing density, a plug of low permeability forms, and this prevents the entry of water from the excavation chamber.

The formation of a sealing plug should take place at a defined location, and this should not be right next to the material discharge opening. Two independently controlled screw conveyors can create a material jam in the transfer zone if the second conveyor is turning more slowly or even in the opposite direction. The two screw conveyors can be arranged in the tunnel one behind another or next to each other.

In the screw casing rotator, a section of the casing rotates with the screw conveyor leading to a reduction of the quantity transported and thus to a compaction of the muck in the previous section of the casing due to the lack of skin friction.

Pressure cells enable the TBM operator to get a subjective impression of the degree of compaction of the muck in the screw conveyor (Figure 11-6). Injection of water or bentonite through the openings provided enables the formation of a sealing plug and reduces the required drive torque.

11.5.5 Foam conditioning

Without foam conditioning, no significant extension of the scope of application would have been possible. The first experience of its use was in the 1980s in Japan. Today, foam is produced and injected according to precise recipes monitored by control systems. The foam is mixed with the excavated muck and takes up a permanent place in the grain skeleton.

Figure 11-18 shows the principle of the construction of a foam plant installed on the backup of a tunnelling machine. The foam solution is mixed with compressed air in a foam generator (Figure 11-19) and injected through openings in front of the cutting wheel or into the excavation chamber. The foam solution consists of water with the addition of a surfactant substance acting as foaming agent (tenside) and a stabiliser (polymer).

The air bubbles in the grain skeleton result in a lower bulk density of the earth slurry and reduce the grain friction. The stiffness modulus of the soil mixture is reduced to such an extent that the soil has ideal plastic properties over a wide range of deformation, which enables much better control of the support pressure at the face. When the pressure in the excavation chamber falls, the gaseous phase in the grain skeleton expands and the soil expands like a spring; when the pressure increases, the air voids initially contract and any sudden rise of the drive torque is effectively damped. In effect, the air bubble used in Hydro-/Mixshields to control the support pressure is contained inside the grain skeleton of the soil.

The use of anion-active, highly water-absorbent polymers surrounds the solid particles with a slip layer, and a continuous three-phase system is formed. Figure 11-20 shows a sand stuck together by the injection of a polymer foam, which forms a durable water-impermeable but still plastically deformable matrix as a continuous plug when subjected to water pressure from below.

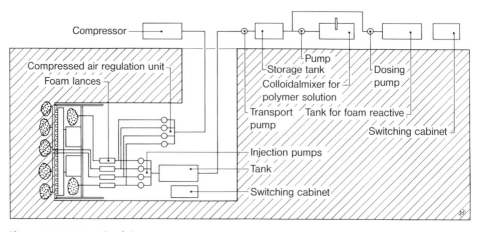

Figure 11-18 Principle of the construction of a foam plant installed on the backup

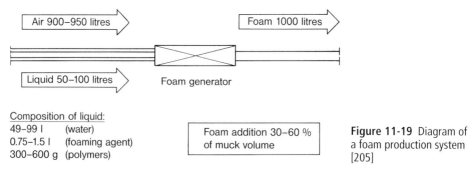

Composition of liquid:
49–99 l   (water)
0.75–1.5 l (foaming agent)
300–600 g   (polymers)

Foam addition 30–60 % of muck volume

Figure 11-19 Diagram of a foam production system [205]

Figure 11-20 Sand treated with foam under water pressure

Free pore water is displaced from the pore system by the injection of foam in front of the cutting wheel or adsorbed by the polymers; the water permeability of the newly created three-phase system is considerably less than that of the natural undisturbed soil.

Since the internal friction of the soil matrix is reduced, all drive measures (cutting wheel, thrust cylinders, screw conveyor) can be designed much more economically. If the recipe of the foam is correct, the soil still has solid coherence through chemical and physical bonding forces, similar to the apparent cohesion of sands, so it can be transported on conveyor belts without problems.

The fact that the foam often consists of over 90 % air, which escapes from the muck after a few days and returns it to its original consistency, offers decisive advantages for tipping or recycling of the excavated soil. No laborious and expensive separation is required.

In order to determine the required quantity to add Q, the following formula is given by the Japanese company Obayashi:

$$Q(\%) = \frac{\alpha}{2} \cdot \left[\left(60 - 4 \cdot X^{0,8}\right) + \left(80 - 3,3 \cdot Y^{0,8}\right) + \left(90 - 2,7 \cdot Z^{0,8}\right) \right]$$

where:

X passing sieve size 0.075 mm (proportion < fine sand)
Y passing sieve size 0.420 mm (proportion < coarse sand)
Z passing sieve size 2.0 mm (proportion < fine gravel)
α correction factor

For an approximate calculation, the values for X, Y and Z can be taken from the grading limits according to DIN 4022 (X < 0.06 mm, Y < 0.6 mm, Z < 2.0 mm). The correction factor α takes into account the curve and should be assumed as follows depending on the degree of uniformity $U = d_{60}/d_{10}$ as follows:

U < 4 $\alpha = 1.6$
$4 \leq U \leq 15$ $\alpha = 1.2$
U > 15 $\alpha = 1.0$

This formula does not take into account the parameters pore volume, water content, water permeability, consistency or the level of the required support pressure in the excavation chamber.

In Figure 11-21, the grading curve (1), for which no addition of foam is required (Q = 0), is displayed according to the formula stated above. Also shown are the grading curves for the gravel and sand formations (the area between grading curves 2 and 3) from the Passante Ferroviario project in Milan (Italy), where foam was used for soil conditioning. The quantity required according to the formula for the middle grading curve 4 is about 30 %. The tunnel was bored by an EPB shield of 8.03 m external diameter and runs above the groundwater level for its entire length.

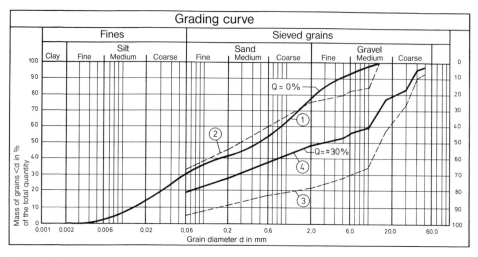

Figure 11-21 Determination of the required quantity of added foam, in this example from the Passante Ferroviario project in Milan

11.6 Examples

11.6.1 Katzenberg Tunnel on the new railway line Karlsruhe – Basel, 2005 to 2007

Project

The Katzenberg Tunnel is the largest single structure on the new railway line from Karlsruhe to Basel and the longest rail tunnel in Germany bored by a tunnelling machine. Of the total of about 9.4 km, two bores each 8,984 m long were driven by two earth pressure balance machines from Herrenknecht. The overburden was between 25 and 100 m. The section with shallow overburden passed below the village of Bad Bellingen and was monitored with an extensive measuring scheme. The muck excavated from the tunnel (2.5 mio. m^3) was transported along housed conveyor belts directly to a quarry 2.5 km away. The bidding consortium of Ed. Züblin AG, Wayss & Freytag Ingenieurbau AG, Marti Tunnelbau AG and Jäger Bau GmbH were awarded the contract with the shield variant.

Geological and hydrological conditions

In addition to Quaternary overlay in the area of the tunnel entrance, the site investigation mainly found Tertiary sedimentary rocks like claystone, siltstone and isolated sandstone in various stages of weathering. The rock strength was mostly low. In the southern part of the tunnel, dense Jurassic limestone was forecast. Isolated karst structures at the transitions to the Tertiary had to be expected – and thus great variation of rock strength over the face.

The water ingress expected in the tunnel was forecast as low, although the entire route of the tunnel lay below the groundwater table. The natural water table was to be re-established after completion, so the tunnel lining had to be designed to resist a water pressure of up to 9 bar. 60 cm thick segments were used for the tunnel lining with a segment width of 2.0 m. The rings were assembled from 7 (6 + 1) segments.

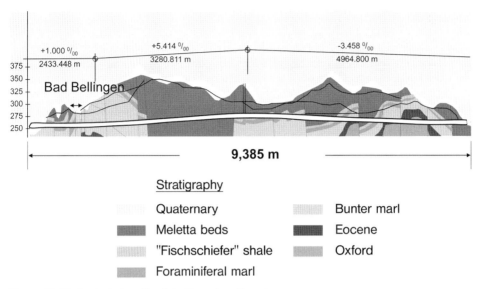

Figure 11-22 Geological profile of the Katzenberg Tunnel

Tunnelling and machine concept

The chosen machine concept was based on an earth pressure balance shield (D = 11.16 m) capable of active face support and screw conveyor mucking. The soil conditioning used water, foam or bentonite suspension. Eight foam injection points in the cutting wheel and four injection points in the excavation chamber were used for the addition of foam.

The material transport in the tunnel was along a conveyor belt to an intermediate tip on the site and then further to landfill in the quarry at Kapf. In order to improve the consistency, lime was added before tipping in the quarry.

Table 11-1 shows the essential machine specifications.

Figure 11-23 Tunnelling machine and transport system to the intermediate tip (Herrenknecht AG)

The tender documents described two tunnelling classes along the route. For predominantly stable rock, i.e. in the dry and hard limestone of the White Jurassic, open mode was intended. Closed mode was to be used where open mode tunnelling proved impossible due to significant rock falls causing sustained hindrance to excavation and mucking. Closed mode was specified compulsorily to ensure low settlements under the village of Bad Bellingen. The entire quantity of water from the face was to be transported out of the tunnel with the muck and disposed of when working in closed mode. In open mode, the intention was to collect all water inflow at the face or in the shield and pump it out of the tunnel in adequately large pipes. Soil conditioning could be with water, bentonite or polymer suspension.

Table 11-1 Project data, Katzenberg Tunnel

Tunnel	Katzenberg
Length	2 × 8,984 m
Segmental lining	
Internal diameter:	9.60 m
Thickness:	60 cm
Ring length:	2.00 m
Shield	
Diameter:	11.16 m
Maximum thrust force:	88,660 kN
Nominal torque of cutting wheel drive:	19,350 kNm
Starting moment of cutting wheel drive:	23,650 kNm
Screw conveyor diameter:	1.25 m
Conveyor belt width:	1.20 m
Length of backup:	approx. 220 m
Cutting wheel	
Disc cutters:	64
Scrapers:	164
Buckets:	2 × 8 (left + right)

Project progress

The first machine started work in May 2005. After a month-long learning phase, the advance rate was continuously improved at the start of July 2005. The second identical tunnelling machine started work in September 2005.

Tunnelling in open mode with water ingress proved problematic. Sticking and fluidisation of the muck led to delays to the advance. In closed mode, full filling with insufficient conditioning led to high contact pressure, high torque and poor advance rates. It was, however, demonstrated that all soil and rock types encountered could be conditioned given the appropriate conditioning and that working in closed mode was even possible at support pressures of over 2.5 bar. Due to the unforeseen presence of core stones, high-strength limestone blocks, the penetration in the soil-type Meletta beds had to be restricted to below 20 mm/rotation to protect the excavation tools.

11.6 Examples   279

While passing under the village of Bad Bellingen in the northern section, the support pressure was increased due to a sliding slope. The provision of an "analysis section" enabled the conditioning to be optimised in advance and the section under Bad Bellingen was then successfully bored in closed mode in conglomerate strata similar to rock.

The average advance rates were about 65 minutes for a 2 m stroke in open mode without obstructions and between 100 and 170 minutes in closed mode or in difficult sections. The best daily performance was 17 rings (= 34 m) for both drives.

The most significant performance data is shown in Table 11-2.

Table 11-2 Advance rates at the Katzenberg Tunnel

Drive	Average advance rate	Advance time	Ring building time	Down time
East	24.8 mm/min	81 min/ring	45 min/ring	135 min/ring
West	24.5 mm/min	82 min/ring	50 min/ring	107 min/ring

The average advance rates were (including all stoppages) 5.5 rings per calendar day on the east drive and 6 on the west drive.

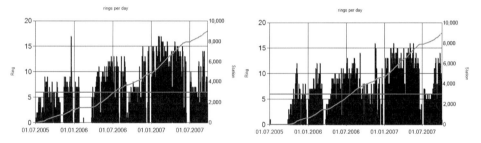

Figure 11-24 Advance rates, east (left) and west (right)

Figure 11-25 Breakthrough of the tunnelling machines in autumn 2007

The contractual schedule of 20 months was exceeded by 8 months and the breakthrough occurred in autumn 2007 (Figure 11-25).

In order to clarify the spheres of risk, the client DB Projektbau GmbH south-western region used the process controlling system Maidl-PROCON. The purpose was to record the complex interactions between ground and process technology in real time. Critical operating conditions could be recognised early and countermeasures introduced in good time. The complete documentation of the progress of the tunnel drive and real-time analysis provided important information for the clarification of the sphere of risks and were used for the collaborative technical evaluation of variations.

11.6.2 Madrid M-30 (Bypass Sur Tunnel Nord)

Project

The new urban motorway M-30 is intended to relieve the traffic in the urban area of Madrid from 2007. The southern bypass consists of two three-lane road tunnels 8,344 m long, of which 7,212 m were bored by a 15.16 m diameter tunnelling machine and 632 m constructed by cut-and-cover. The shield drive passed under densely built-up areas and some railway and underground stations had to be passed within only 6.5 m without any "side-effects". The two 3.65 km long single-layer segment tunnels have an internal diameter of 13.37 m. The 9 + 1 segments are 60 cm thick and 2.00 m long. The minimum curve radius is only 500 m. The contract for the north tunnel with a volume of 340 m € was awarded to a consortium of the companies Necso and Ferrovial Agroman. The works were carried out using a Herrenknecht machine. The south tunnel was awarded to a consortium of FCC and ACS Dragados for 429 m € who used a Mitsubishi machine.

Geological and hydrological conditions

The ground in Madrid had already been shown to be predestined for EPB shield machines during the construction of the underground. The tunnel passes through the Peñuela, Tosco formations and sandy-clay sediments. Peñuela describes a consolidated soil or rock formation of shale, which is created by the over-consolidation or carbonised cementation of highly lithified clay. A blue-green colour with brown sprinkling is characteristic. There are extremely hard layers over 2 m thick. The rock structure is full of splits and gouges, which give the Peñuelas their typical soapy appearance. These are easily formable loams containing clay. The plasticity index is over 40 and the percentage of very fine material is over 80 %.

Figure 11-26 Road tunnel for the urban motorway M-30 in Madrid, earth pressure balance shield (Herrenknecht AG)

The Tosco consists of ochre- or brown-coloured loamy sediments containing clay and sandy beds. In the upper parts, fine gravels can also be present. There can also be sporadic rather harder layers and sections containing lime. These are clays and silts with a proportion of over 60 % of fractions smaller than 0.074 mm, containing more or less sand. The plasticity index varies between 20 and 25. The sandy Tosco consists of sands with a high content of clay and feldspar and clays with high sand content. The fractions above 0.074 and the fraction below this size are very similar. The very fine fraction mostly consists of bentonite and illite in very similar proportions and a small amount of kaolinite. The very fine fraction amounts to 40 to 60 %. The Tosco sand mainly consists of feldspar (20 to 60 %) and the remainder is quartz, plagioclase and heavy minerals. The very fine fraction varies between 25 and 40 %. The "Arena de miga" are sandy agglomerations containing quartz and feldspar with isolated edges as well as platelets with rounded edges streaked with quartzite, gneiss and granite, a sandy and to a slighter degree sandy-clay formation, generally with lower plasticity and high compaction. The groundwater is between 5 and 20 m above the tunnel crown. There are confined groundwater layers.

Tunnelling and machine concept

Two of the largest EPB shields in the world were used to bore the two 3.65 km long tunnel bores. The manufacturer Mitsubishi decided on a conventional concept consisting of an earth pressure balance shield with a main cutting wheel and a screw conveyor in the invert.

As there was still absolutely no experience with diameters of over 12 m and support pressures of over 3 bar at the time the order was placed, Herrenknecht decided on a mechanically complicated and innovative machine concept. New was the use of inner and outer cutting wheels to reduce the effective torque. The 430 t cutting wheel was therefore fitted with an outer and an inner cutting wheel (diameter 6,780 mm) working concentrically. The inner cutting wheel was designed as a flat, closed cutting wheel installed in the centre of the outer cutting wheel. It was mounted flush with the steelwork of the outer wheel, but could be extended forward independently by up to 400 mm to improve the conditioning of the excavated soil. The hydraulic cutting wheel drives of each wheel were completely independent of each other and thus permitted different revolution speeds and rotation directions. The drive power of the outer cutting wheel of 12,000 kW was provided by an arrangement of 50 cutting wheel drive motors in two rows of 29 and 21 units.

The advantages of the two-part cutting wheel were in the excavation process and also the soil conditioning by the turning cutting wheel, even if the installed drive power of the outer wheel was only partly used in order to reduce the electricity consumption. The construction with two cutting wheels achieved better adaptation of the revolution speed regarding the penetration depth of the excavation tools and also had a positive effect on tool wear.

Altogether 32 foam generators with a maximum capacity of 28,260 Nl (standard litres) of foam per minute were installed to reduce the torque. This enabled the soil to be conditioned into a plastic, flowing slurry at an advance rate of over 70 mm/min. Eight independent foam injection nozzles were arranged in the inner cutting wheel, and 14 foam lances supplied the outer cutting wheel. This meant that the inner foam nozzles covered 36 % of the volume. The rest was supplied through the outer cutting wheel or through nozzles along the pressure bulkhead. Adapted to suit the local geology, the cutting wheels were fitted with a total of 380 narrow scrapers, each with favourable weight for tool changing, and 57 17" two-ring disc cutters. Two hydraulically extendable copy cutters could produce an overcut of 20 mm in the radius.

Table 11-3 Essential machine specifications

Tunnel	Madrid M-30 centre/outer
Length	3,539 m
Segmental lining	
Internal diameter:	13.45 m
Thickness:	60 cm
Ring length:	2.00 m
Shield	
Diameter:	15.16 m
Maximum thrust force:	315,880 kN
Nominal torque of the cutting wheel drive:	8,450/96,000 kNm
Starting torque of the cutting wheel drive:	10,890/125,268 kNm
Screw diameter:	0.60 m/2 × 1.25 m
Conveyor belt width:	2 × 1.60 m
Length of the backup:	approx. 114 m
Inner and outer cutting wheel	
Disc cutters:	16/50
Scrapers:	88/292
Buckets:	2 × 8/2 × 12 (left + right)

Altogether three screw conveyors transported the conditioned soil out of the excavation chamber. In order to estimate the transport volumes, the material flow through the excavation chamber to the screw conveyors had already been analysed with a FE simulation. Two large screw conveyors with diameters of 1,200 mm were installed in the invert. The third conveyor located in the centre had a nominal diameter of 600 mm. Material transport continued along a conveyor belt with 1,600 mm belt width. The installed thrust force was 276,390 kN. The TBM was designed for a maximum advance rate of 65 mm/min. The 57 thrust cylinders could be adjusted by about 200 mm to compensate any rolling of the shield with a countermoment.

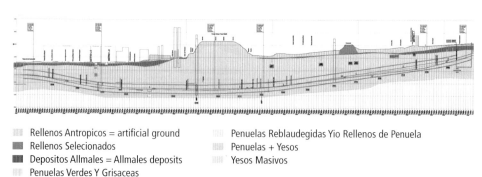

  ▧ Rellenos Antropicos = artificial ground ▨ Penuelas Reblaudegidas Yio Rellenos de Penuela
  ■ Rellenos Selecionados   ▨ Penuelas + Yesos
  ▤ Depositos Allmales = Allmales deposits   ▨ Yesos Masivos
  ▨ Penuelas Verdes Y Grisaceas

Figure 11-27 Geological longitudinal section

The twin-walled shield tail was fitted with brush seals in four rows. To cope with the specified curve radius of only 350 m, a joint was provided in the shield with articulation cylinders (stroke: 300 mm). The annular gap was grouted through 12 injection points (+ 12 reserve); at the end of the shield tail, an external spring steel plate was mounted around the upper 270° as a grout stop. The four-part backup extended to a total length of 91 m divided into two supply backups, an intermediate backup for belt transfer and an extension backup. Mounted on the first three-storey backup was all the machinery for the supply of the shield (drive hydraulics, electricity, mucking pumps, grouting) and the control cabin. In order to distribute the load uniformly, the backup had an integrated walking support system. The second backup housed further hydraulics (tank, main hydraulics, auxiliary plant) and the compressors with a capacity of 4×15 m^3/h. This backup was already running on radially arranged carriages with Vulkollan wheels. On the third backup was the material transfer from the machine conveyor via a transfer belt to the tunnel conveyor. The last backup looked after all extension functions (water supply, air, belt extension, ventilation) and also housed further secondary equipment like crew rooms and transfer pumps for grout handling. The machine concept and the soil conditioning concept are shown in Figure 11-28.

The tunnelling work was monitored based on knowledge produced by the real-time data analysis system Maidl-PROCON. All machine data was transmitted over a VPN link and the interaction of the tunnelling machine with the ground was analysed by a team of experts. Support pressure was controlled actively through the foam injection (Figure 11-29).

Project progress and experience

The superb preparation of the project, but also the particular suitability of the ground under Madrid for the use of an earth pressure balance shield, set new standards and led to speed records for shield tunnelling. After the works acceptance in May 2005, the machine was dismantled, shipped and assembled on site in three months. Some maintenance work was also undertaken, which could not be undertaken during normal operation. The scheduled construction time of twelve months was shortened by a third with completion in only eight months. The Mitsubishi was indeed assembled two months later, but profited from the experience with the Herrenknecht machine. The tunnel drive could be completed in less than 35 weeks.

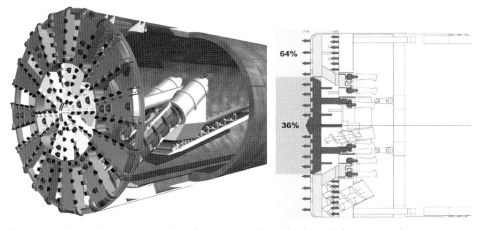

Figure 11-28 Machine concept M-30 with two-part cutting wheel, muck clearance on three screw conveyors (left), soil conditioning concept (right)

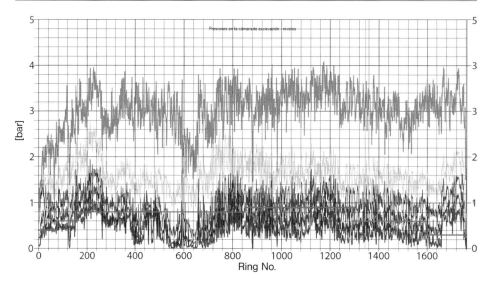

Figure 11-29 Support pressure analysis system Maidl-PROCON

The best daily advance of the Herrenknecht machine was 34 m (Figure 11-30) and advance rates were often over 70 mm/min. The soil conditioning concept turned out to be excellently suited to the ground conditions in Madrid. The effective torque was less than expected. In contrast to the original planning, it proved possible to drive the tunnel with considerably lower support pressure values (Figure 11-29). This not only significantly reduced the drive forces but also tool wear.

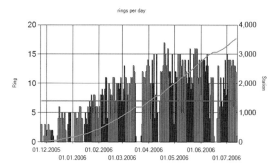

Figure 11-30 Advance rates M-30 south: Herrenknecht EPB, D = 15.16 m

11.6.3 Heathrow

In order to improve the accessibility of the aircraft parked in outlying areas and to connect the future Terminal 5 at London Heathrow Airport, two 1.3 km long road tunnels (segment internal diameter 8.1 m) were needed, the Airside Road Tunnel. The main challenges of the project were the requirement to avoid disturbance to airport operations and the previous negative experience of the airport operator following the collapse of a tunnel excavated using the shotcrete method.

Geological and hydrogeological conditions

Minimal settlements and no impairment of airport operations were the main factors determining the vertical alignment of the tunnel. The tunnel was aligned to bore through London clay, which is generally considered stable. The chosen vertical alignment provided a cover of between 0 and 11 metres below the overlying aquiferous gravels, which were regarded as critical. In the central section, the vertical alignment was raised to safely pass over the existing Heathrow Express Tunnel, leaving 3 m between the crossing tunnels.

Tunnelling and machine concept

Because of the particular demands of the project, a machine concept was developed in collaboration with all parties to bore the tunnel with an earth pressure balance shield (diameter 9,150 mm) in closed mode, or with mechanical face support and compressed air. Preliminary full-scale trials showed that the London clay could be processed into a pasty mass with foam conditioning, but this was difficult and associated with considerable energy consumption. Additional tests with compressed air application also showed that the unconditioned clay would not form the required sealing plug in the screw conveyor.

The solution to the requirements of the project proved to be a sludge pump after the screw conveyor (analogous to Figure 5-15) and a special design of cutting wheel. The clay at the face could be supported mechanically and by compressed air:

- The following double-piston pump was designed for an advance rate of 50 mm/min. The pumping capacity was therefore 400 t/h with a piston diameter of 750 mm. For the application described here, the use of this pump is the largest of its type.
- The cutting wheel (Figure 11-31) was constructed as an internal, closed cutting wheel (opening ratio 30 %) fitted with scrapers and different copy cutters. With the cutting wheel extended forward, this could produce a radial overcut of 63 mm (drag picks) or 10 mm (disc cutters). The cutting wheel also had four pressure cells to measure support pressure at the face.
- The construction of the upper, rearward part of the shield as a separate ring space (safety chamber) enabled the operationally safe pressurisation of the excavation chamber with compressed air.
- The minimised annular gap could also be supported with bentonite suspension through the tapered shield skin using two rows of 10 bentonite nozzles arranged around the perimeter and fed from a bentonite tank by a feed pump.
- The annular gap was filled through six grouting lines. The grouting pressure in the annular gap was measured with special pressure sensors.

Project progress

Tunnelling work was started in June 2002. The face was pressurised with compressed air at pressures between 0.5 and 2 bar according to the overburden. The first bore was completed successfully after only six months in the middle of December 2002. The advance rates in the first bore with completely installed backup were 12 m/d with peak days of 20 m/d. After the tunnelling machinery had been relocated, the project was successfully completed with the breakthrough of the second bore in June 2003.

Figure 11-31 Retractable flat cutting wheel of EPB shield, Heathrow

The specification of minimised settlements of at the most 35 mm or 0.8 % volume loss was considerably bettered by the actual values of maximum settlement/heave of 15 mm or 0.35 %, with an average volume loss of 0.2 %.

11.6.4 DTSS Singapore

A tunnel system 50 km long was driven for the first phase of a new drainage system, the Deep Tunnel Sewer System in Singapore. The approximately 50 km of deep tunnels formed the basis for a later unpressurised sewer. The contract T-05, the 8.08 km and 4.54 km long Kranji Tunnel (bored diameter 4.88 m), intended a segmental lining with external diameter of 4.1 m and an internal diameter of 4.1 m. The 4.02 and 5.58 km long Queensway Tunnel (segmental lining with external diameter 4.20 m and internal 3.75 m) was driven by an EPB shield with a bored diameter of 4.45 m.

Geological and hydrogeological conditions

For contracts T-05 and T-06, tunnelling under mixed-face conditions had to be expected. The tunnel passed constantly between sections in granite and in loose ground, the soft clays and sands of the Kallang formation (Figure 11-32), under water pressures of up to 4 bar. The loose ground was also in a highly cemented form (Old Alluvium) in places, which can cause severe wear. The granite, in addition to the described fresh state, also showed various degrees of weathering and inclusions of claystone, siltstone and sandstone, which have low strengths but the abrasiveness is just as bad.

The adjacent tunnel in contract T-01 (Chanji Tunnel $\varnothing$ 7.2 m; L = 5.7 km) lies in the geologically more favourable Old Alluvium sections and the Kallang formation and is only mentioned at this point for the sake of completeness.

Tunnelling and machine concept

The design of the four EPB shields for contracts T-05 and T-06 or T-01 intended the removal of muck solely in a screw conveyor in addition to a closed cutting wheel fitted with disc cutters. The machines had the following mechanical features:

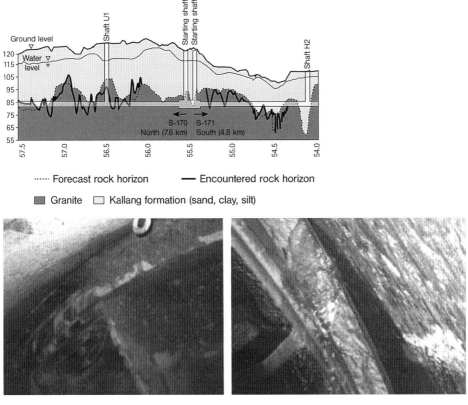

Figure 11-32 Geology of DTSS Singapore, contract T-05 Kranji Tunnel (section); location of the vertical alignment in the transition between granite and loose ground (top); face: "mixed face conditions" – loose ground intruding in the crown (left), water ingress along the shield skin in granite (right)

All the cutting wheels (Figure 11-33) were similar to hard rock cutter heads with almost closed fronts (degree of openings ≈ 10 %) fitted with disc cutters, wear plate facing in the gauge area and buckets arranged for mucking in one direction. The maximum loading per cutter roller was 25 kN. The cutting wheel drives of all machines were designed for working in open mode in hard rock at a revolution speed of up to ≈ 10 rpm.

All shields were built in two parts. A steering joint with 2 × 4 steering cylinders separated the front shield with cutting wheel, drive, pressure bulkhead and stabilisers from the center shield with advance probe drilling, thrust cylinders, bulkhead, erector and the articulated shield tail. In all machines, the area between the pressure bulkhead and the rear bulkhead served as lock chamber.

All shields featured a cylinder adjustment of 200 mm stroke, in order to compensate the rolling resulting from the limitation to only one mucking direction.

The machines for contract T-05 were initially fitted with a relatively short screw conveyor (L = 7 m; ∅ 700 mm); the shields for contract T-06 had a comparatively longer screw conveyor (L = 13 m; ∅ 700 mm).

Figure 11-33 EPB shields DTSS Singapore, contract T-05 Kranji Tunnel ⌀ 4.76 m (left), contract T-06 Queensway Tunnel ⌀ 4.5 m (right)

Project progress

Work on the tunnel sections of contract T-06 were started in April 2001. The machines for contract T-05 started some time later in September and December 2001.

As the tunnelling work on contract T-05 progressed, the layout of the vertical alignment in the wave-shaped transition between the peaks of the very hard fresh granite with uniaxial compression strengths of 200 to 300 MPa and the weathered valleys with the full head of hydrostatic groundwater pressure of up to 52 m turned out to cause severe wear on the machines. Because the tunnel was designed to carry water with an unpressurised open fall, it was not possible to change the vertical alignment to run through more uniform geological conditions. In order to avoid overbreak in the loose ground, which was capable of flowing, the TBM operator was forced to make immediate changes from open mode to closed mode with the full face in rock as soon as the first signs of loose material appeared in the crown.

Encounters with granite banks in loose material caused impact wear on the disc cutters; sticking of loose material followed by chips accumulating in the dics cutter housings led to the rollers jamming and discs being ground flat on one side.

Due to the geological conditions, the lifetime of the disc cutters had already been forecast at less than 100 m³ of excavation per disc cutter. Experience in the tunnel, however, showed that the forecast lifetime had to be reduced even further in the actually encountered "mixed face conditions". All parties therefore agreed new guidelines for the TBM operator based on operational experience, disc cutter consumption and the wear already experienced. These guidelines were intended to optimise advance and wear rates by operating the EPB shield with very sensitive control of the main parameters of support pressure and contact force at the cutting wheel.

Entries to check and exchange tools were necessary at least every fourth ring under the conditions described. Repair and maintenance was also demanded by the excessive wear on the screw conveyor and the steel structure of the cutting wheel and shield. Most of the interventions on contract T-05 had to be carried out under compressed air at up to 3.4 bar with the resulting locking times. In order to limit the pressure levels, the groundwater was lowered as much as possible. On one occasion, the ground even had to be frozen in order to achieve stable face conditions for an intervention.

While the machine coming from the south on contract T-05 bored its 4.8 km long tunnel section with scarcely any technical alterations, considerable modifications were undertaken to the TBM for the 7.6 km long northern section. For example, in September 2003 rebuilding work was undertaken in an intermediate shaft with the fitting of a longer screw conveyor and a new cutting wheel with larger mucking openings, which took two months.

The south machine on contract T-05 successfully completed its drive with breakthrough in March 2004. The north machine had to overcome further abrasive sections with mixed-face conditions. The screw conveyor had to be completely renewed due to wear. The tunnel section was finally completed at the end of January 2005 after 7,466 m with an average advance rate of 77 m/week (after rebuilding) and peak rates of 210 m/week.

The earth pressure balance shields used in section T-06 achieved the best monthly performance of 625 m in the northern section; average weekly rates were achieved of 59 m and 55 m respectively. The geological conditions in this section were also dominated by full-face granite with sections through loose material, but the percentage of transitions was considerably less. The two drives on contract T-06 were successfully completed in September 2003 and February 2004.

12 Convertible shields or multi mode machines

To be able to tunnel through any formation using one and the same machine has long been a wish of tunnel builders. Machine manufacturers and contractors have created numerous terms for this idea in recent years, each representing the product of the company. This led to terms such as "Polyshield", "Mixshield" or Convertible shield machines (KSM according to DAUB, Chapter 1.2) or hybrid and multi mode machines.

The types of shield machine described in the former chapters have been able to considerably widen their fields of application due to the rapid progress in all areas (machinery, process technology, electronics, logistics, surveying instruments) and not least through practical experience, including the mishaps. Ecological and economic aspects, however, set additional secondary limits to the practical application of all these processes, for example the separation of the suspension/soil mixture produced by hydraulic transport or the disposal of chemically conditioned soil from an earth pressure balance shield.

A convertible shield is designed so that mechanical modifications can be undertaken to adapt to different geological and hydrogeological conditions. This will normally be a change of the method of supporting the face, but consideration also needs to be paid to the required excavation tools, mucking system and all other equipment.

In addition to the implementation of combination 1 (compressed air shield – open shield), only combinations 2 (slurry shield – open shield) and 3 (EPB shield – open shield) have actually been implemented. The first spectacular example of the combination of a slurry shield with an open shield is the Grauholz Tunnel, which is described in detail in Section 12.3.1. The best example of the combination of an EPB shield with an open shield is the machines used under the sea on the French side of the Channel Tunnel. Combination 4 (slurry shield – EPB shield) was to be used for the first time in 1994 for the Duisburg-Meiderich light railway. The machine was designed to enable the conversion, but coped with the ground conditions encountered so well in slurry shield mode that the conversion to EPB

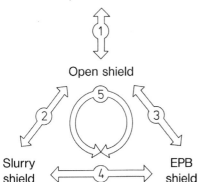

Figure 12-1 Possible combinations of shield machine types

mode was never carried out (Figure 12-4). Conversion from slurry to earth pressure support was carried out on the Socatop Project, which is described in Section 12.3.3.

The combination 5 (open – slurry –EPB shield) would ultimately represent a universal shield tunnelling machine. It should be mentioned in this connection that slurry shields and earth pressure balance shields are equipped for compressed air support. This is necessary for entry to the excavation chamber, for example to repair the cutting wheel or remove obstructions, which often only requires a partial lowering of the support medium.

The consideration of a machine concept for a convertible shield poses conflicting aims for client, contractor and machine manufacturer alike, because the practicality of converting the machine has to be considered while time and cost factors remain as ever the essential criteria for judging the success of a construction project.

This conflict of aims was already shown by the example of the Galleria Aurelia (Rome) in 1983. After starting with very good advance rates in stable clays, the open shield with an external diameter of 10.64 m was converted underground into a hydroshield, as the geological forecast showed the tunnel to be nearing a transition to highly permeable sand soils. The conversion took three months and unluckily, it turned out against expectations that the geological transition was only actually encountered after a further 100 m. Another conversion had to be ruled out due to the length of time required so that the section in clay had to be driven with the elaborate hydraulic transport installation and separation plant (e.g. an upstream separation stage installed on the backup and further expensive separation stages above ground).

The example shows that the length of time required for the conversion of operating mode is of great importance in addition to the selection of a suitable base machine and the suitable operating mode in the tunnel. This time is determined by safety requirements in addition to construction practicalities and may only be a few hours in extreme cases. Auxiliary measures (e.g. grouting) and equipment facilitate boring through geological changes and transition zones as well as the conversion of the machine and thus extend the practical scope of application. Such measures can be used to prepare the prevailing ground conditions so that the existing machine can overcome them.

New developments in bearing and drive systems have given a great boost to the development of convertible shield machines in recent years. Many shield machines are now fitted with peripheral drive, which leaves the centre of the machine free and thus accessible for different installation options. The final implementation of a universal shield, which ultimately demands the combination of the "slurry shield" and "earth pressure balance shield" operating modes, depends not least on the readiness of the client and the contractor to bear the additional expense, and also the risks.

In Japan, clients, contractors and machine manufacturers often work together to further development, which has considerably assisted innovations. As the involvement of clients in the risk of such new developments seems impossible not only in Germany but also all over Europe, contractors and machine manufacturers are justifiably very cautious regarding the provision of universal machines for all possible geological formations in order to cover the entire geological risk.

12.1 Development strategies

The construction of a universal shield machine according to combination 5 in Figure 12-1 forces, in addition to the capability of operating as an open shield, a combination of the slurry shield with the earth pressure balance shield. There are currently two contrasting development strategies:

- The integration of all the components required for both operating modes in the construction of a shield machine,
- the construction of a shield as a building block system, in which the components for each operating mode can be exchanged.

12.1.1 Convertible shield with integrated components for multiple operating modes

Consideration of the slurry and earth pressure operating modes (Chapters 10 and 11) shows that the individual assemblies for the two processes are different. Comparison of the equipment for each operating mode for the example of a circular tunnel with full-face excavation and segmental lining shows, however, that the fundamental differences only affect a few assemblies.

Cutting wheel

It is possible to design a cutting wheel for both slurry and EPB operating modes. Exchanging the cutting wheel is also possible, but this requires a shaft (station or similar).

Material transport

The method of clearing the excavated material out of the excavation chamber is generally different for the two processes, slurry support and earth pressure support (screw conveyor – slurry circuit), so a convertible machine will have to possess the necessary components for each process. However, the provision of both mucking systems with the optimal size and capacity is scarcely practical for tunnels of less than about 8.5 m diameter for simple geometrical reasons. It is therefore better to decide a preferred operating mode depending on the geological formations to be driven through.

For the construction of the Botlek Tunnel in 1999, an EPB shield was tendered after a slurry shield had been tendered, but a hydraulic muck transport system was developed instead of the usual conveyor belt. The consistency of the excavated soil certainly influenced this decision. Figure 12-2 shoes an illustration of the principle and further details can be found in Chapters 10 and 11 [157].

It is not practical for a stone crusher to be mounted or to remain in the excavation chamber of a machine working in earth pressure balance mode. This means that appropriate equipment has to be provided at a later stage of material flow in slurry shield mode, or else lumps too large for hydraulic transport have to be removed manually. It is, however, possible to install a roller crusher after the screw conveyor of an EPB shield in order to enable hydraulic transport of coarse-grained soil or rock.

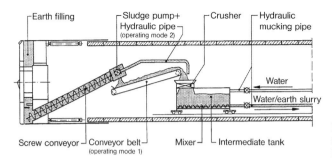

Figure 12-2 Material transport scheme of the EPB shield for the Botlek Tunnel [157]

Face support

The different methods of mucking and face support lead to different requirements for the shield structure. The layout of submerged wall and pressure bulkhead in a slurry shield is not possible in an EPB shield; the submerged wall becomes a pressure bulkhead and has to be designed and constructed to resist the resulting high loading.

The opening below the submerged wall is closed in slurry shield mode. If a telescopically extendable screw conveyor (retracted in slurry shield mode) is used with a so-called submerged wall gate, this process can be partially mechanised, although a small amount of manual work (release and final locking in position) remains unavoidable.

The necessary connections in the pressure bulkhead for both operating modes have to be provided in the base machine, although the influence of the preferred operating mode also has to be considered here.

Installation of support

The remaining components of the shield machine and the segmental lining are suitable for both modes of operation. It is only necessary to provide sufficient space in the erector zone for the simultaneous presence of two mucking systems.

Backup system

In the backup area, two parallel material transport systems would have to be provided (hydraulic transport pipework and usually conveyor belts) and also the installations necessary for each operating mode (pumps, pipe layers, conveyor belt, tanks and pumps for conditioning agents etc.). This additional equipment naturally reduces the space remaining for the equipment required in all operating modes (machinery to supply the machine, segment handling) or leads to more elaborate constructional solutions. This also makes the determination of a preferred operating mode unavoidable; just one method of muck transport should be used with the appropriate preparation of the soil.

The above considerations make clear that the conversion of a shield in the tunnel between slurry and different EPB operating modes is fundamentally possible, given the appropriate design of the machine and the provision of all necessary components. A realistic estimate of the time required for the conversion of a tunnelling machine designed on this basis would be about eight hours, according to the technical elaboration included in the base machine for the purpose. The changing of operating mode "at the press of a button" has to be regarded as not practical at the moment.

12.1.2 Building block systems

In parallel to the development work aimed at providing universal machine technology, modular construction of tunnelling machines has become more established in recent years. This trend is not only driven by the intended capability of boring through different ground formations by modifying the machine, but also by the increasing demand for manufacturers to buy back machines after completion of the tunnel. This has led to continuing efforts by machine manufacturers to design and develop assemblies independent of machine type. A palette of individual assemblies is available, which are only bound to a particular diameter to a limited extent and whose use and reuse offers the most economic solution in an individual case because the cost of providing them is much less than for new parts.

If all the appropriate components for the specified operating modes are available as modules in a building block system, then it is no longer necessary to keep additional assembles available that are not required for the currently operated mode. For example, it is not necessary to install a screw conveyor and a hydraulic mucking system in tandem. It is a significant advantage of the building block system that scarcely any limitation of the individual operating modes is required.

Because of the restricted space available in the shield and in the tunnel, the building block system does make it necessary to undertake the rebuilding work in a shaft along the route of the tunnel. Particularly for the construction of inner-city underground railways, however, contract sections tend to be divided into sections of 0.5 to 2 km length between underground stations. If certain compromises are made in the design, this makes it possible to convert the base machine at suitable locations in the tunnel, i.e. in the intermediate shafts required in any case such as for stations. If the base machine is designed for modular rebuilding, the conversion process can normally be undertaken in about one to three weeks during the normal stopping time in the shaft.

A solution like this is the optimal economic solution in many cases. The availability of different operating modes, like slurry and earth pressure support, during the tunnelling time means not only the parallel installation of the machine assemblies but also the necessary parallel installations in the tunnel and above ground. Compromises with the individual operating modes during tunnelling are not necessary due to the possibility of complete rebuilding. According to the equipment of the machinery, the price of the "conversion set" can be estimated at about 15 to 25 % of the basic price of the machine.

12.2 Machine concepts

The development of convertible shield machines has been and is still being driven forward by German machine manufacturers, predominantly due to the prevailing geological conditions in Europe. Even though a considerable part of the development and process technology in shield tunnelling comes from Japan, there are no known concepts or practical applications of convertible shield machines from Japanese manufacturers. This may be due to the homogeneous nature of the geological formations for applications in Japan, but the sales strategy also seems to concentrate on the perfection of existing process technologies with the intention of being able to tunnel different soil types with one and the same machine without conversion of the operating mode.

12.2.1 Mixshield

In Germany, the firms Wayss & Freytag and Herrenknecht patented a convertible shield machine under the name "Mixshield" in 1985. The tunnelling machine shown in Figure 12-3 is based on the Hydroshield also patented by company Wayss & Freytag (Chapter 10) with the well-know arrangement of submerged wall and pressure bulkhead. Conversion is possible to an EPB shield and also to a compressed air shield with rotary feeder.

The essential construction feature of the Mixshields that enables conversion to different operating modes is the cutting wheel drive with free centre and the characteristic Mixshield articulated support of the cutting wheel in a large spherical bearing (Figure 4-29), with the position of the cutting wheel being secured by synchronously controlled displacement cylinders.

The patented version of the Mixshield has been used until now almost exclusively with slurry-supported face. The conversion of a Mixshield from slurry support to open shield has been undertaken for the first time during the construction of the Grauholz Tunnel (Section 12.3.1).

The conversion of the operating mode of a Mixshield from slurry to earth pressure support on the same construction project was first planned in 1994. For construction sections TA7 (crossing under the River Ruhr) and TA8a (Meiderich Süd) of the Duisburg light railway (Figure 12-4), the necessary components were to be installed in the machine in open excavations. The shield was prepared for the required conversion in its basic design. It was also the intention to use foam for soil conditioning in the earth pressure balance version for the first time in Germany on this project. The conversion during the construction phase was not actually carried out because it proved possible to successfully overcome the soil conditions encountered with slurry support.

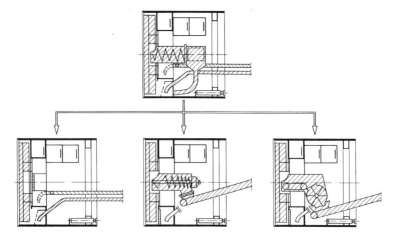

Figure 12-3 Patent of the Mixshield (Herrenknecht/Wayss & Freytag), 1985

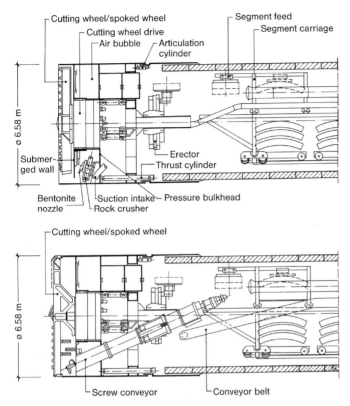

Figure 12-4 Convertible shield machine for operation with slurry and earth pressure support, Duisburg light railway, 1994 (Herrenknecht)

12.2.2 Polyshield

The basic version of the Polyshield (Voest Alpine) is similar to the Mixshield in that the hydraulic material transport circuit can be dismantled to convert it into an earth pressure balance shield with screw conveyor or into an open shield with conveyor belt projecting into the excavation chamber.

Since this type of shield has only been used a few times, the production has been abandoned. Further information can be found in the first edition of this book [196] and [241].

12.3 Examples

12.3.1 Grauholz Tunnel, 1990 to 1993

Part of the Swiss government project "Eisenbahn 2000", the 6.3 km long Grauholz Tunnel was intended to replace the existing railway line north-east of Bern through Zollikofen and enable rail traffic with speeds of up to 200 km/h between the Löchligut and Mattstetten junctions. The client was Swiss Federal Rail (SBB).

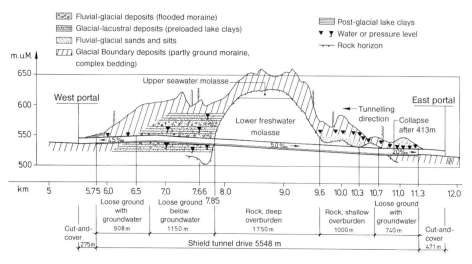

Figure 12-5 Geological longitudinal section, Grauholz Tunnel, 1990 to 1993

Geological and hydrogeological conditions

Figure 12-5 shows the geological longitudinal section of the selected alignment between the west portal near Löchligut and the east portal near Mattstetten. After a cut-and-cover section 471 m long at the east portal, the mechanised tunnel drive started in a 740 m long loose ground formation formed in the Ice Age consisting of over-consolidated clay/silt combinations with interbedded moraine material. Then followed a 1 km long section in the Tertiary fresh water molasse consisting of sand and siltstone with very shallow cover in places (10 m) until finally reaching a section 1,750 m long in molasse with deep overburden. The rock strength of the molasse was subject to strong fluctuations, with a maximum uniaxial compressive strength value of 30 MN/m^2 being given. After the molasse followed a 1,150 m long section in Quaternary gravel/clay mixes above the groundwater and in the last section of 908 m before the west portal an inhomogeneous moraine with chaotic structure under water pressure, consisting of consolidated silts and clays with frequent boulders and blocks of rock as well as highly permeable gravel/sand sections.

Tunnelling and machine concept

The complex geological conditions posed very difficult problems for the machine concept. While relatively dry and stable rock could be assumed in the molasse section, normally requiring no support of the face, a tunnelling system with face support had to be provided near the portals. The decision was made to use a shield machine (Mixshield from Herrenknecht AG), which could work as a slurry shield with bentonite-supported face and slurry circuit and could also be converted in the tunnel for open mode with a conveyor belt. The separation plant for the slurry operation was installed on the backup in order to keep the pumping distance as short as possible to avoid any rapid saturation of the bentonite suspension with very fine material. For the continuation of material transport after the end of the backup, a transport conveyor belt was chosen (Figure 12-6).

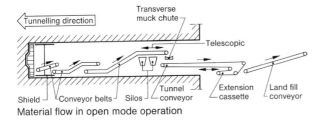

Material flow in open mode operation

Material flow in slurry support mode

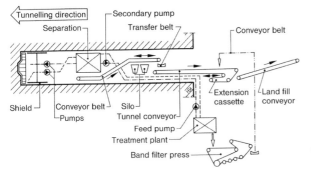

Figure 12-6 Material flow for open mode and slurry mode, Grauholz Tunnel [15]

The cutting wheel was constructed as a six-armed open star fitted with 46 scrapers for working in loose ground formations. In order to break up boulders and bore through the molasse, 65 17" disc cutters were provided as well. To break up larger lumps of rock, the machine was equipped with a crusher in the centre. Three process-controlled displacement cylinders enabled an exactly defined fully automatic overcut with twelve different positions and five different depths.

Project progress

Tunnelling work in the 5,548 m tunnel started in January 1990 and the works were successfully completed after a full three years of mechanised tunnelling at the end of May 1993. The most significant sections are now explained with the advance rates and problems encountered [274].

Table 12-1 shows the minimum, maximum and average daily advance rates for the sections described.

Table 12-1 Advance rates at the Grauholz Tunnel according to [274]

Station	Operating mode	Min. daily advance [m/d]	Max. daily advance [m/d]	Weekly average daily advance [m/d]
km 11 + 300 to 11 + 000	slurry shield	1.1	9.7	3.6
km 11 + 000 to 10 + 700	slurry shield	1.1	8.3	4.1 (incl. maintenance work)
km 10 + 700 to 10 + 000	slurry shield	4.0	7.7	5.8
km 10 + 000 to 9 + 600	open shield and slurry shield	7.7	18.5	15

Station	Operating mode	Min. daily advance [m/d]	Max. daily advance [m/d]	Weekly average daily advance [m/d]
km 9 + 600 to 7 + 850	open shield	10.8	28.8	18.5
km 7 + 850 to 7 + 660	slurry shield	1.8	11.5	6.7
km 7 + 660 to 6 + 500	slurry shield	5.0	18	12.6
km 6 + 500 to 5 + 750	slurry shield	1.1	9.7	3.6

Station km 11 + 300 to 11 + 000
The tunnel passed through very silty-sandy moraine material with a slurry-supported face and frequent necessary maintenance work under compressed air due to the high abrasiveness of the silt and the resulting tool wear. The laborious separation often slowed the advance rate.

Station km 11 + 000 to 10 + 700
After 413 m, there was a collapse while operating under compressed air. The jammed cutting wheel could only be freed after sinking a shaft from ground level. The cause of the collapse was presumed to be the lack of mechanical support of a 65 cm long section in front of the shield due to the cutting wheel projecting in front of the shield blade and being constructed as an open star (without perimeter rim). For this reason, protection canopies were formed by ground freezing every 50 m in order able to able to carry out maintenance work under compressed air.

Station km 10 + 700 to 10 + 000
The prevailing weathered rock with shallow overburden often caused problems by sticking to the cutting wheel and the excavation tools, which had to be laboriously removed by hand. The disc cutters combined with scrapers could not optimally excavate these soft ground formations. The machine operated in slurry mode, sometimes with partially lowered suspension level and compressed air application, as flowing sand under 2.5 bar water pressure was forecast in the crown at km 10 + 300.

Station km 10 + 000 to 9 + 600
During the rebuilding for open mode, a few technical modifications were made. A rapid closing system for the excavation chamber enabled rapid changes from open mode to slurry mode within a few hours. Where the cover was less than 4 m, it was planned to convert the machine back to slurry mode. At km 9 + 750, 50 m had to be operated with bentonite-supported face with an average daily advance of only 8 m.

Station km 9 + 600 to 7 + 850
Advance continued without problems in open mode with muck removal out of the excavation chamber by conveyor belt. The muck often consisted of up to a third of large chips, while the other two thirds was ground small by the excavation tools, which would explain the already described problems with sticking in slurry mode. The maximum monthly advance was 467 m.

Station km 7 + 850 to 7 + 660

Before entering the flood-transported moraine, some modifications were made to the cutting wheel. A continuous rim was intended to improve safety during maintenance work under compressed air and avoid the need for auxiliary measures. Nonetheless, there was another collapse during maintenance work under compressed air in June 1992. The cause was a sliding body above the cutting wheel, which extended into the lake clay strata about 6 m above the crown and jammed the cutting wheel. Grouting work and manual clearing of the excavation chamber took almost two months.

Station km 7 + 660 to 6 + 500

Excellent advance rates were achieved in continuous operation with slurry-supported face thanks to modifications to the separation plant and the support slurry. Almost a kilometre of tunnel was driven in only three months.

Station km 6 + 500 to 5 + 750

After a final general overhaul of the cutting wheel, the undisturbed advance continued with slurry-supported face until reaching the west portal.

Summary

The selected machine concept was proved absolutely correct in the subsequent evaluation. Using any other tunnelling method and without lowering the groundwater, the loose ground under high groundwater pressure could only have been tunnelled with compressed air. The high abrasiveness of the moraine material would have led to much worse wear problems for an EPB shield. The use of an EPB shield was ruled out in any case if only because of the very high permeability and gravelly grading distribution. Slurry transport in the freshwater molasse would have caused considerable sticking problems without face support due to the high fines content in the marl, considering the experience at station km 10 + 300 to 9 + 600.

Problems mostly occurred during maintenance work with lowered suspension level under compressed air. A cutting wheel capable of being fully retracted into the excavation chamber would have been very advantageous.

12.3.2 Zürich Thalwil contract 2.01

As the northern part of the Zimmerberg Base Tunnel, the two-track Zürich-Thalwil tunnel disentangles the regional and the long-distance traffic in the southern approach to the Zürich main station. Within the overall project, the 9.4 km long tunnel was split into two main contracts, of which the northern 2,620 m long section 2.01 between the Allmend Brunau installation site and the Lochergut portal was driven by a convertible shield machine with two conversions of operating mode to slurry operation and auxiliary construction measures under the densely built-up inner city.

Geological and hydrogeological conditions

The tunnel starts through the upper freshwater molasse, which consists of interbedded marls and sandstones, followed by a loose ground section through lake deposits and Sihl gravel, highly permeable gravels (10^{-3} to 10^{-4} m/s) and gravelly sands (Figure 12-7).

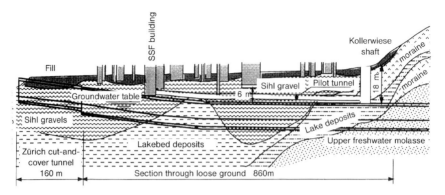

Figure 12-7 Vertical alignment and geology of Zürich-Thalwil 2.01, loose ground section

Tunnelling and machine concept

The convertible shield for boring molasse and loose ground at Zürich-Thalwil from Herrenknecht AG was capable of operating both as an open TBM and as a slurry shield. Compared to the predecessor of this combination for the Grauholz Tunnel, the shield designed almost ten years later showed various modifications:

– The flat cutterhead, diameter 12.39 m (Figure 12-8 a), was constructed as an eight-armed cutting wheel with a rim and equipped with 74 disc cutters. For the loose ground sections, 66 scrapers could be additionally fitted. The entire flow of material in the cutting wheel was optimised for working in one transport direction. The open spaces between the cutting wheel arms were closed with fixed steel plates for boring in hard rock. For slurry mode, the plates were removed from the cutting wheel and arranged on the submerged wall as classic breasting plates, which could be pushed forwards into the cutting wheel openings with hydraulic cylinders for compressed air interventions. The cutting wheel could thus meet the requirements of hard rock tunnelling with an almost completely close cutterhead with buckets at the perimeter and also the requirements of loose ground tunnelling with an open cutting wheel for a good flow of material in the excavation chamber filled with suspension.
– The rock crusher was designed as a jaw crusher to cope with blocks up to 800 mm and installed as usual in front of the screen.
– The shield had two personnel locks and one material lock.
– In the shield, drilling channels for investigation and grouting holes were arranged around the perimeter in between the double thrust cylinders.
– A two row brush seal with spring steel plates was used to seal the shield tail. The annular gap was grouted through the shield tail.
– The backup had a central gantry section over a length of 100 m in order to be able to install a concrete invert arch in blocks of 20 m, the longitudinal tunnel drainage and invert fill with asphalt paving. The working steps in each block were: 1. cleaning and isolation, 2. invert concrete, 3. side wall, 4. installation of drainage, gravel fill and asphalt paving (and 5. supplying the TBM).

Figure 12-8 a) eight-armed rimmed cutting wheel in the operating mode "hard rock cutterhead" ⌀ 12.39 m
b) loose ground cutting wheel ⌀ 12.36 m (Herrenknecht)

The tunnel was lined with rings of bolted segments 7 + 1 with a ring length of 1.7 m, 30 cm thick segments and an external diameter of 12.04 m.

Project progress

After the start in April 1999 with a short backup from the 35 m deep starting shaft and the step-by-step completion of the machinery, the daily advance as an open TBM in hard rock mode with conveyor belt to clear the excavated material was between 20 and 21 m in 16 h. Because of the shallow rock overburden, the first conversion was undertaken at tunnel metre 355 for a distance of 150 m in slurry mode with wet transport, entailing the installation in the tunnel of a slurry circuit, the fitting of the scrapers, changing the disc cutters and checking of the brush seals.

In the loose ground section 780 m long following the rock section, the advance rate was not determined by the slurry shield but by the separation capacity of the chamber filter presses used to separate very fine material. The highly permeable Sihl gravel section was driven using a suspension enriched with vermiculite (foliated silica similar to mica) and further auxiliary construction measures.

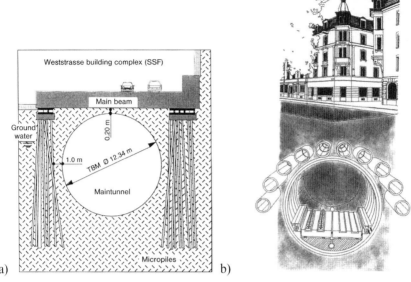

Figure 12-9 a) Tunnelling under an office building after underpinning the building with a beam grillage, micropiles and hydraulic cylinders
b) Tunnelling of the last 130 m in the protection of a pipe screen (10 x ⌀ 1.5m)

For the boring of a large tunnel under the inner city with shallow overburden, the tunnelling concept planned the use of a continuous grout body created in advance of the tunnel drive from a pilot tunnel. The pilot tunnel was also used during a five-week tunnelling break to improve the bedding of boulders embedded in the lake deposits, so that the boulders would be fixed better in the soil matrix and could be excavated without problems by the disc cutters.

Near the end of the loose ground section was a six-storey office building with the deepest basement that actually projected into the vertical alignment of the tunnel. This basement had to be removed and filled. Then the building was underpinned using micropiles and a prestressed reinforced concrete beam grillage and hydraulic cylinders for any necessary corrections (Figure 12-9 a). The shield passed the building at a distance of 0.2 m. The settlements at the surface and the settlement of the building were about 2 mm and no correction was necessary using the hydraulic presses. The last 130 m of the tunnel drive were bored in the protection of a pipe screen consisting of ten reinforced and concrete-filled pipes ⌀ 1.5 m (Figure 12-9 b) installed by pipe jacking. The breakthrough took place at the start of May 2001. The advance rates in this sensitive loose ground section under a heavily built-up inner city area with a large tunnel diameter and shallow overburden were about 7 to 30 m/week in the rock and lake deposits, while 60 m/week was achieved in the gravel under the pipe screen. The complete tunnelling machinery was dismantled after completion and relocated to the Oenzberg Tunnel.

12.3.3 Socatop

This 10.1 km long two-storey motorway tunnel was constructed in order to fill a gap in the motorway ring to the west of Paris. A second tunnel for trucks is planned for a later date. Two storeys, each with three carriageways, required an excavation diameter of 11.565 m.

Geological and hydrogeological conditions

The tunnel passed through the entire spectrum of geological formations to be found below Paris: chalk, marl, clay, limestone, gypsum and sand with three groundwater storeys (Figure 12-10).

Tunnelling and machine concept

In order to adapt to the geology as well as possible, a machine was developed that could be operated in various modes:

- as a slurry shield with suspension-supported face (Mixshield in slurry mode) (Figure 12-11 a),
- as an EPB shield with the face being supported by an earth slurry (Figure 12-11 b),
- in semi-open mode with earth slurry in the invert and muck transport through a screw conveyor (semi-EPB) and/or with compressed air applied to the upper part of the excavation chamber,
- in open mode (material removal through the screw conveyor, no compressed air in the excavation chamber).

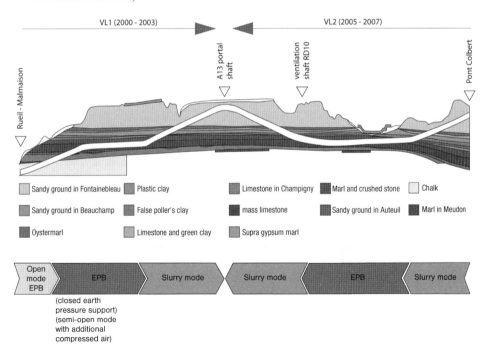

Figure 12-10 Vertical alignment of Socatop Tunnel

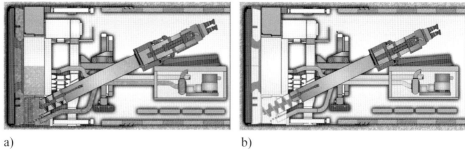

a) b)

Figure 12-11 Machine concept for the Mixshield at Socatop (Herrenknecht)
a) Slurry shield mode (left)
b) EPB shield mode (right)

The changes of operating mode were carried out in the tunnel:

- Shield and backup were equipped with the specific equipment for face support in each of the operating modes. For slurry mode, this was the hydraulic transport circuit components, but also the submerged wall/pressure bulkhead set-up typical for a Mixshield and the provision of a stone crusher. In EPB mode, the screw conveyor and machine conveyor should also be mentioned.
- The cutting wheel was designed so that it could be used in all operating modes without modification or rebuilding. A mostly closed cutting wheel was used, fitted with a combination of scrapers and 17" disc cutters for both directions of rotation.
- In slurry mode, the excavation chamber and the lower part of the working chamber were filled with suspension; the upper part of the working chamber contained the air bubble and the entire space was pressurised. In EPB mode, on the other hand, only the excavation chamber was pressurised, and the submerged wall became the pressure bulkhead. The pressurised space was accessible under atmospheric pressure and could be used as a safety chamber (both for regular excavation chamber entries and also as an additional escape chamber in the fire event in March 2002), and was only pressurised for access to the working face. For the change from EPB to slurry operation, the entire screw conveyor was retracted, leaving the opening through the submerged wall in the invert and the intake screen area free. A specially constructed stone crusher was swung out from its parked position into the operating position in front of the screen.
- Some of the equipment for slurry operation, like the compressed air regulation equipment or the bentonite circulation system, could also be used in EPB mode if required. The opportunity of having two systems available all the time offers a range of potential synergy benefits.

In addition to the capability of converting the operating mode, the convertible shield had the following key data and special features:

- In order to cope with the requirements of EPB mode, the cutting wheel drive power was 4,000 kW and the available torque of the hydraulic cutting wheel drive 35,000 kNm. The shield thrust force was designed for 150,000 kN, and the maximum advance rate 80 mm/min.
- The transport circuit was designed for 1,900 m³/h for an advance rate of 50 mm/min. The vertical alignment of the tunnel had a rising gradient and the shield was 160 m

above the separation plant by the end of the drive. In order that the transported soil-suspension mixture could be fed to the separation plant under almost no pressure, the resulting pressure difference of 16 bar was dissipated through the pipe friction losses in the pumped pipeline.
- A camera system developed specially for the excavation chamber was installed for the first time and successfully tested in semi-EPB/compressed air mode.
- As a result of the steep gradient of 4.5 %, the segments and grout were delivered using special wheeled vehicles (Figure 5-13). A semi-automatic loading station was set up at the tunnel portal, which could handle a complete ring of segments and grout at the same time in order to minimise the logistics cycle time, together with rapid unloading in the backup.
- About 200 m behind the tunnelling machinery, the precast invert units were installed, with the lower carriageway level being constructed of precast slabs as tunnelling continued.

Project progress

In November 2000, the convertible shield started drive VL1 in the open mode of EPB operation. The first 150 m were bored with the incomplete backup in chalk with flint inclusions in two-shift operation with screw conveyor, machine conveyor belt followed by a starting belt, with an advance rate of up to 80 mm/min. After the backup and the facilities at the portal had been completed, operation was changed to tunnel conveyor and three-shift operation. A stoppage of three months was caused by a fire in the following stretch of tunnel in March 2002. Operation continued in closed operating mode with earth pressure face support and semi-open mode with additional compressed air support of the face of 1 to 2 bar, which repelled the groundwater and enabled the excavated material to be kept dry. In October 2002, the machine was converted again to slurry mode to cope with the sands of Fontainebleau, achieving advance rates of up to 50 mm/min. The first tunnel drive was broken through in October 2003, after which the machine was dismantled and transported to the starting point at Pont Colbert.

The second drive VL2 started in Mixshield mode in June 2005. Immediately after the tunnel portal (10 m), the tunnel passed below a six-lane motorway with shallow cover. The restart of the machine and the crossing under the motorway were both achieved without incident in only nine days. After 1.2 km in slurry mode, the machine was converted for EPB operation. After passing through the access shaft RD10 at the deepest point of the tunnel, the machine was converted back for slurry mode and finally arrived at the underground target shaft at the junction with the A13 in August 2007.

13 Special shields and special processes

Most of the shield tunnelling systems that have actually been used can be allotted to one of the already described shield families, despite all the difficulties of categorisation. A few special types of shield machine, however, which are clearly different from other types in some of their constructional features, have not yet been properly dealt with and are therefore described in this chapter, "Special shields and special processes". Construction methods, which are associated with shield tunnelling and demonstrate the potential of shield tunnelling, also in combination with other underground construction processes, are described. This chapter also describes the newest, predominantly Japanese, ideas for mechanised tunnelling.

13.1 Blade shields

Blade shields or similar machines can also described as shuffle-shoe TBMs. Although the use of blade shields has declined in recent years, this development will continue to have its applications in the future. The particular characteristic of blade shields is the construction of the shield skin. In contrast to normal shields without joints in the skin, the shield skin is split into plank-like blades, which are mounted parallel to the tunnel centreline on support and arch frames and can be advanced individually or in small groups (Figure 13-1). The idea for this process goes right back to a patent from Brunel in 1806 (Chapter 1). They are still used today for sewer construction under urban areas.

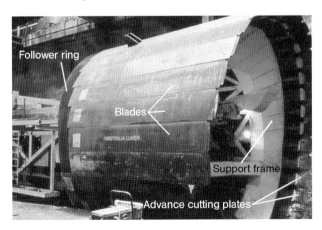

Figure 13-1 Blade shield, Frankfurt underground, 1975 (Westfalia Lünen)

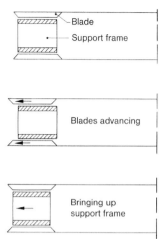

Figure 13-2 Principle of a blade shield [183]

This design of shield was developed to cope with the following requirements:

— In order to enable an in-situ concrete lining of continuously pumped or sprayed concrete and thus avoid the need to push the thrust cylinders against the support behind the machine, and thus to avoid having to design the tunnel lining to resist the loading case "thrust forces",
— improve working safety in the excavation area through the provision of "walking" or "shuffle-shoe" support to the tunnel sidewalls,
— decouple the excavation and support work and thus increase the advance rate compared to more conventional tunnelling methods,
— create possibilities to mechanise support work.

The working and advance principle of the blade shield is based on exploiting the friction between rock mass and shield skin. As only single blades or groups of blades are ever pushed forward, that is only a part of the shield skin, the friction forces between the blades that remain stationary and the rock mass can act to resist the comparatively small thrust forces required. Once all the blades have been pushed forward by one stroke one after another, then the support frame can be brought forward by one working stroke while the blades remain fixed in the rock mass (Figure 13-2).

The ground is normally excavated from only part of the face (excavator, roadheader), but full-face excavation by a cutting wheel has also been used (Figure 13-3).

Blade shields therefore have the following advantages in addition to the requirements listed above:

— Non-circular cross-sections are possible,
— they are easily adaptable when obstructions are encountered at the shield skin,
— obstructions can be removed under the protection of the advanced blades,
— setting off from a starting shaft without an expensive abutment,
— in some versions, the blades can be expanded to alter the cross-section (compensation of settlement, adaptation to ground convergence and yielding without damage) [36], [302],
— most of the equipment can be reused, for example to change the cross-sectional profile and size without heavy investment cost.

13.1 Blade shields

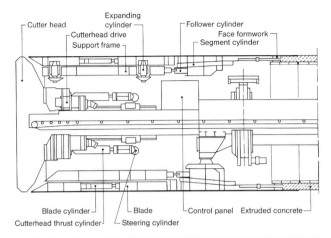

Figure 13-3 Expanding blade shield with compressed air support, pilot tunnel for the Freudenstein Tunnel, 1984 to 1986 (Westfalia Lünen) [38]

With regard to advance rates, blade shield cannot yet achieve the performance of similarly sized conventional shields, despite the decoupling of excavation and lining work. This is due to the time taken to advance the single blades or groups of blades and the time taken to place the support of in-situ or sprayed concrete and other factors.

Although the significance of blade shields as an alternative to simple shields has declined in recent years, it is certainly foreseeable that more will be seen of the idea of a walking support system, above all because of the cost-effectiveness resulting from the decoupling of excavation works.

13.1.1 Face support with blade shields

Support to the face in blade shield tunnelling is normally provided by natural support. The support frame in the front section can indeed be equipped with breasting plates or even with intermediate platforms in larger machines. The presence of groundwater can be countered by chemical soil consolidation [302].

The application of compressed air or support medium to support the face is problematic with a blade shield due to the difficulty of sealing the joints between the blades. Blades shields have only been used with compressed air on a few occasions and have not been successful. No blade shield has yet been used with a slurry-supported face. The problems of overcoming the use of compressed air were demonstrated, for example, by the construction of the North Collector sewer in Hamburg-Harburg (1978/79), where a blade shield operating under compressed air was used to drive a 3.54 m diameter circular section. 1.3 bar overpressure was necessary to hold the water. Due to the geological conditions, several blowouts occurred; the air losses through the surface of the shield could not be sufficiently compensated. The blade shield operating with compressed air was therefore replaced after about 250 m of the total length of about 1.400 m by a Hydroshield with extruded concrete lining [114], [36], [181]. There was also enormous consumption of compressed air during the driving of the pilot tunnel for the Freudenstein Tunnel on the route of the German Railways new Mannheim-Stuttgart main line (1984/86) in addition to numerous other difficulties. On contract 2, where an expanding blade shield was used in combination with the extruded concrete lining [38] for a tunnel with a clear diameter of 5.2 m, up to 435 m^3/min of compressed air was consumed at 1.9 bar. This was due to misinterpretation of the geology and hydrology [304].

13.1.2 Support types with blade shields

Because the excavation work is decoupled from the lining work, blade shields can be used with a wide range of lining systems. The lining system in turn has a decisive influence on the design of the shield, particularly of the backup. In order to illustrate the various blade shield concepts, tunnel projects are now described, where blade shields were used in combination with

– in-situ concrete,
– shotcrete and
– extruded concrete.

Blade shield with in-situ concrete lining

A blade shield for concrete placed in-situ was used from December 1976 for the construction of the Essen light rail link on contract 17a (Figure 13-4). A two-track tunnel about 400 m long with a horseshoe profile was driven under the very busy Rüttenscheider Straße. With a crown height of 7.48 m and an invert width of 9.82 m, this was the largest blade shield yet used in Germany. The support frame had a central platform combined with breasting as well as two working platforms, from which the excavated soil slid down chutes to the continuous conveyor installed at the bottom. Excavation was partly by hand, partly by continuous miner or hydraulic excavator and breaker. The skin of the blade shield consisted of 8 invert blades, 32 advance blades and 32 follower blades. Each advance blade could be extended forward hydraulically with a stroke of 600 mm. The follower blades were relatively thin, only 12 mm, supported at the front by the support frame and at the back on the last concrete section and served as external formwork and also provided a protected space for steel fixing and concreting. The waterproof concrete was placed in sections of 2.5 m. The three-part folding formwork was supported for concreting on the last and last two concrete sections. The annular gap, which was opened by bringing up the follower blades, was grouted to reduce any settlement.

13.1 Blade shields 313

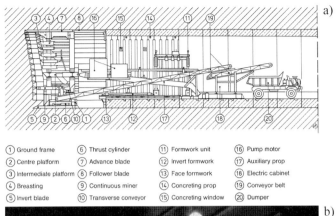

① Ground frame ⑥ Thrust cylinder ⑪ Formwork unit ⑯ Pump motor
② Centre platform ⑦ Advance blade ⑫ Invert formwork ⑰ Auxiliary prop
③ Intermediate platform ⑧ Follower blade ⑬ Face formwork ⑱ Electric cabinet
④ Breasting ⑨ Continuous miner ⑭ Concreting prop ⑲ Conveyor belt
⑤ Invert blade ⑩ Transverse conveyor ⑮ Concreting window ⑳ Dumper

Figure 13-4 Blade shield with in situ concrete lining Subway Essen, 1976 (Westfalia Lünen) [245]
a) Machine concept
b) Workshop assembly

A special feature of the blade shield at Essen was the provision of an auxiliary prop, which was intended to transfer forces through the support frame into the hardened invert concrete, especially under squeezing face conditions. The apparatus consisted of two support props, which were hinged from the support frame and supported from recesses integrated into the poured concrete invert. In order not to additionally increase the forces required to advance the support frame by having to pull the auxiliary props, the support apparatus was designed to "walk" itself.

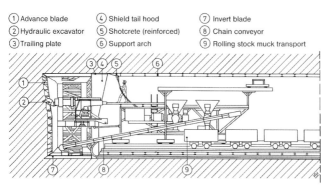

Figure 13-5 Blade shield with shotcrete support, Frankfurt underground, 1975 (Westfalia Lünen)

Blade shield with shotcrete support

Blade shields used with shotcrete support do not have any follower blades. Instead, the advance blades are fitted with trailing plates, so-called blade tails, which slide on a shield tail hood fixed to the support frame. Support arches and reinforcing mesh are installed under the protection of the hood. As soon as the sides of the tunnel are free after the shield has been advanced, they are sealed with shotcrete. The first blade shield of this type was used on contract 20 of the Frankfurt underground in 1975 [12]. The external diameter was 6.6 m. Two bores each 400 m long were driven (Figs. 13-1 and 13-5).

Inside the blade shield, a hydraulic excavator arm was installed on a platform. This could reach and excavate the entire face. As the ground to be excavated consisted of alternating soft soil beds and hard rock, the excavator arm had a crusher installed permanently with a special bucket fixed to its housing with a hinge. In order to excavate the hard dolomite, the bucket was folded back and a breaker point was inserted. In the shield invert, a chain conveyor removed the piled muck and transferred it to railway trucks. The average advance rate was 8 m per day with peaks of 12 m. These rates were only made possible by simultaneously excavating and applying shotcrete support in the shield tail.

The use of fibre-reinforced shotcrete would have avoided the need to install steel arches and mesh and this simplification could have led to a further increase of the advance rate.

Blade shield combined with the extruded steel fibre concrete

A blade shield of this type was first used for the construction of the underground in Frankfurt, contract 36, 1980 to 1982. The equipment consisted of the actual blade shield and a following shield to produce the extruded concrete (Chapter 6). The blade tails hinged to the advance blades lay on the following shield, which was fitted with a shield tail hood and joined to the blade shield with hydraulic cylinders. The face formwork for the support concrete with elastic shield tail seal was integrated into the following shield. The inner formwork was constructed similar to segments and also supported with hydraulic cylinders hinged on the following shield. The following shield could thus be pulled forward at the frame of the blade shield and pressed forward from the steel segments (Figure 13-6).

Blades, support frame, following shield with face formwork and formwork segments were advanced independently of each other, which enabled the entire apparatus to be driven forward.

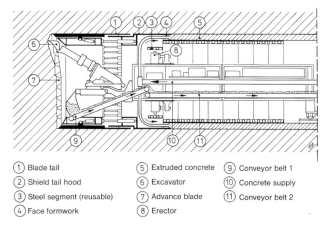

① Blade tail	⑤ Extruded concrete	⑨ Conveyor belt 1
② Shield tail hood	⑥ Excavator	⑩ Concrete supply
③ Steel segment (reusable)	⑦ Advance blade	⑪ Conveyor belt 2
④ Face formwork	⑧ Erector	

Figure 13-6 Blade shield combined with extruded concrete, Frankfurt underground, 1980/82 (Westfalia Lünen) [36]

The contract mentioned above passed through clay with limestone banks in the protection of a groundwater pressure relief system using vacuum wells. The external diameter was 6.86 m, the length 1,650 m. As this was in the nature of a trial site, the steel fibre reinforced extruded concrete was intended as temporary support and an in-situ inner lining was concreted later. A further example of the use of a blade shield together with extruded concrete was the construction of the pilot tunnel for the Freudenstein Tunnel from 1984 to 1986 (see Figure 13-3).

13.2 Multi-face shields

The special feature of multi-face shields is the profile of the excavated cross-section, which is formed of two or more circles. The idea behind the design of multi-face shields resulted from the demand for the construction of transport tunnels under open ground as far as possible in order to save time and cost. This would typically be beneath public roads. However, there is often insufficient space for two adjacent tunnels, which would normally have at least a half diameter between them for structural reasons. The only alternatives are the boring of a large two-track tunnel or the use of a multi-face shield, which avoids the conventional principle of a circular tube (Figure 13-7) [191].

It is not surprising that this generally applicable demand first led to definite developments in Japan. A multi-face shield was used for the first time for the construction of the twin-bore Kyobashi Tunnel in Tokyo, a section of the Keiyo Metropolitan Line, where a shield tunnelling machine with two cutting wheels was used (Figure 13-8) [191].

Despite the greater technological complication, there is a range of technical and economic arguments for the use of a multi-face shield. One important advantage is the flat shape of the profile. Figure 13-9 shows a comparison of a multi-face cross-section with an alternative single-bore solution.

In this example, the flattened shape produced by the multi-face shield reduces the excavation area by about 13 %.

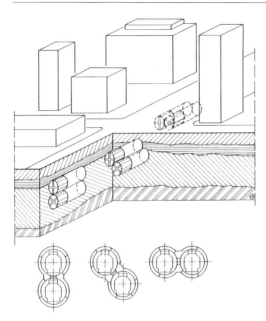

Figure 13-7 Possible layout variations for a shield with two cutting wheels [119]

Figure 13-8 Multi-face shield, Kyobashi Tunnel, Tokyo, 1988 [86]

Another advantage derives from the shape of the cross-section, which means that the alignment can be nearer to the surface than possible with a circular profile, with all resulting advantages for inner-city tunnel.

13.2.1 Arrangement of the cutting wheels in multi-face shields

Three different layouts of the cutting mechanism are possible for the construction of a multi-face shield (Figure 13-10).

In multi-face shield type A, the cutting wheels are in the same plane, meaning they have to be star-shaped and cannot rotate independently of each other. As an example of this principle, a DOT shield is shown in Figure 13-11.

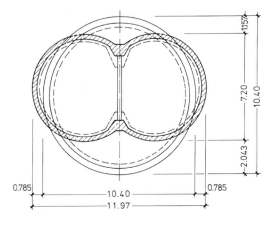

	Unit	Circular cross-section	Multi-face cross-section	Ratio
Excavated area	m²	87.90	76.10	0.87
Outer lining	m³	12.60	10.30	0.82
Inner lining	m³	6.30	5.60	0.89
Annular gap	m³	4.40	5.50	1.25
Invert concrete	m³	9.70	1.80	0.19
(Quantities per meter tunnel length)				

Figure 13-9 Multi-face cross-section in comparison to a conventional profile, Kyobashi Tunnel, Tokio [214]

With its offset cutting wheels, the cutting wheels in shield type B can occupy any part of their perimeter. The front and back shields can rotate independently of each other and have a common chamber.

Type C functions similarly to type B except the front and back shields have separate chambers. This means that the longitudinal offset has to be larger as there has to be a seal between the part of the face cut by the front wheel and that cut by the back wheel.

The constructional principles of multi-face shields therefore open up a range of new possible applications (Figure 1-1). An extension to three, four or even more cutting wheels is certainly conceivable. This means that, for example, underground stations, multi-lane road tunnels or the underground shopping malls that are increasingly built in Japan can be driven in one pass. The high cost of such shields should of course be borne in mind, which demands a certain length of tunnel to write off the cost. It is also possible to operate a two-wheel shield vertically. For the construction of the Subway Line No. 12 in Tokyo, a multi-face shield with three cutting wheel was used (Figure 13-12) [306].

13.2.2 Tunnel support with multi-face shields

As a result of the special cross-section, the tunnel support has a few special characteristics. Multi-face sections require a few additional types of segments compared to circular sections. In additional to the normal arc elements for the outer parts of the tubes, special K-elements are required in the overlap area and columns for central stiffening. Figure 13-13 shows two different multi-face cross-sections with the associated segment type.

318 13 Special shields and special processes

Type A

Cutting wheels in one
plane and star-shaped,
one chamber

Type B

Cutting wheels offset
and full-face, one
chamber

Type C

Cutting wheels offset
and full-face, indepen-
dent chambers

Figure 13-10 Construction principles of multi-face shields [214]

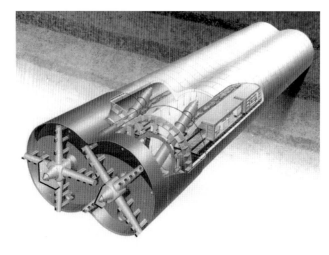

Figure 13-11 DOT shield [79]

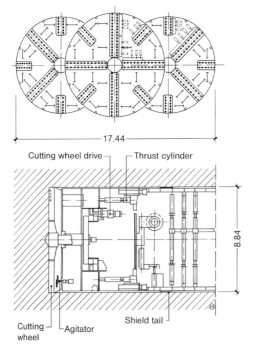

Figure 13-12 Multi-face shield with three cutting wheels, Subway Line No. 12, Tokyo [306]

The stability of a multi-face cross-section depends greatly on the central column and the material around the segmental lining. When the rings are assembled, great care has to be taken with precision to enable the central column to be placed correctly. For example, for the construction of the Kyobashi Tunnel in Tokyo, first the lower K-element, then the arc elements and then the upper K-element were assembled. The bolts between the elements, which were only used to simplify assembly, were not fully tightened. In order to install the central column, the upper K-element was raised slightly with the hydraulic erector and the central column, a welded hollow box construction, was installed with the second erector. At the end of the process, all bolts were tightened and the annular gap was grouted in order to ensure good contact with the surrounding ground. A special grout mix was used, which hardened immediately after injection [191].

Horizontal and vertical curves are problematic due to the complex assembly of the lining.

13.3 Enlargement of shield tunnels

The enlargement for various purposes of the section bored by a shield poses particular challenges for the process technology in mechanised shield tunnelling. Such enlargements can be necessary, for the example for the construction of underground stations or underground crossings and points. As the use of a special shield for these enlarged locations is normally ruled out as uneconomic, they are often constructed using a combination of shield tunnelling and other underground construction methods. The shield-driven tunnel than has the character of a pilot tunnel.

The idea of widening a tunnel driven by a shield machine to form a station structure is not new. Szechy already gave various examples in his book "Tunnelbau" [277].

As can be seen from the illustration of the individual phases of construction of the station at Gants Hill in London (Figure 13-14), three auxiliary shields of small diameter and four station shields of larger diameter were used. For the construction of the central section, a so-called roof shield was used, whose shield skin only forms the upper part of a circle (Figure 13-15).

This construction method is still in use today in London. In recent years, the stations have been supported with shotcrete.

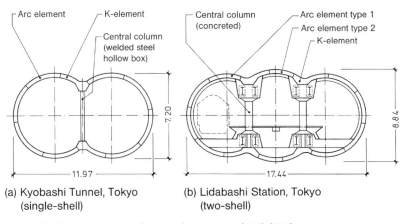

(a) Kyobashi Tunnel, Tokyo (single-shell)

(b) Lidabashi Station, Tokyo (two-shell)

Figure 13-13 Segment types for multi-face sections [191], [306]

A roof shield was also used for the driving of the Nagatacho Station on the Yurakucho Line in Japan. Figure 13-15 shows a modern roof shield from the machine manufacturer Mitsubishi.

The excavation of the ground between the two shield-driven tunnels under the protection of these shields is undertaken when the geological conditions rule out open excavation. If the use of a roof shield is intended for the excavation of the centre, then special segments have to be used for the construction of the two running tunnels in the area to be enlarged. The roof shield then slides along guides in the special segments and is kept in position. Figure 13-16 shows the construction sequence for Nagatacho Station, Tokyo.

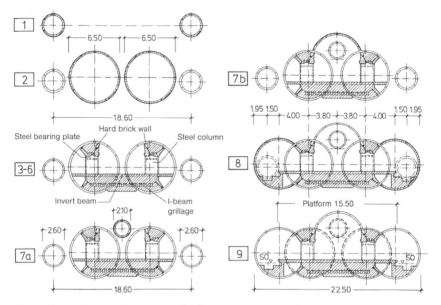

Figure 13-14 Construction sequence for the station at Gants Hill in London (1950) [277]

Figure 13-15 Modern roof shield (Mitsubishi)

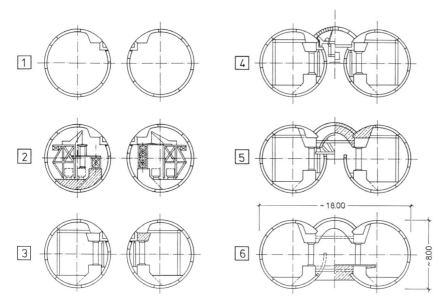

Figure 13-16 Construction sequence at Nagatacho Station on the Yurakucho Line, Tokyo, 1974

A particularly simple and often used method of constructing an enlargement was used for the construction of the light rail link in Essen, contract 32 (northern section). After the completion of the shield tunnel, the lining was partially removed and the enlargement required for the platforms was excavated and supported with shotcrete. The transition from the running tunnel to the enlarged area was designed so that the lining of the running tunnel and the enlarged structure met each other tangentially (Figure 13-17).

Because of their dimensions, an excavated section 29 m wide and 21 m high, the crossovers in the Channel Tunnel between France and England were a particularly interesting construction task. For the English crossover in relatively favourable geological conditions, two side headings were first advanced and supported with shotcrete. According to the multiple heading method, the top headings were then excavated followed by the invert and supported with shotcrete [223] (Figure 13-18).

On the French side, a special if rather antiquated method was used. Because of the geological situation with water-permeable chalks, overlapping headings with a cross-section of about 5.5 × 4.8 m arranged in a ring around the perimeter of the future structure were first excavated with roadheaders. These were filled with concrete and finally the centre was excavated (Figure 13-19).

The illustrated examples of enlargement structures, which certainly only reflect a small part of the possible construction sequences and variants, show that a functional collaboration between shield tunnelling and other methods of underground construction offers almost unlimited scope for the construction of underground cavities.

Figure 13-17 Enlargement, light railway Essen, northern section, 1990

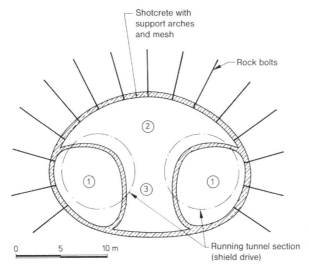

Figure 13-18 Excavation sequence for the crossover on the English side of the Channel Tunnel, 1988 to 1991 [223]

13.4 Pipe jacking

Pipe jacking can also be categorised as mechanised shield tunnelling. The essential difference to the shield tunnelling methods already described in this book is the location of the thrust cylinders and that a prefabricated pipe or box is jacked forward. While the thrust cylinders in a shield tunnelling machine are located in the machine and travel with it by pushing against the already installed lining, the thrust cylinders in pipe jacking remain in the launching shaft, which can also be called a thrust pit in this case. They push forward the entire pipe (or lining) including cutting shoe or tunnelling shield in the direction of advance. New pipes (or lining sections) are added in the launching shaft as the pipe jack advances.

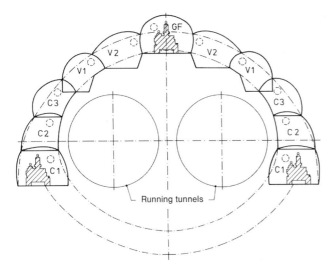

Figure 13-19 Excavation sequence for the crossover on the French side of the Channel Tunnel, 1988 to 1991 [202]

Progressing development of the process has led to continually longer lengths and larger diameters being possible (Figure 13-20). The use of intermediate jacking stations and bentonite lubrication of the outer face has now made pipe jacking possible over 1,000 m long with curves (hot water transport pipe in Berlin, 1986/87). The setting up of field production plants near the construction site has also eased the former limits on cross-section resulting from the limitation of the size of precast elements that could practically be transported. Pipe sections of more than 5 m diameter have now been jacked (diversion of the River Alte Emscher, Duisburg 1978) [225] and the in jacking of complete structures called box jacking, box sections of up to 300 m² (Section 13.4.2).

No end of this development is in sight, and the use of the jacking process for the construction of longer transport routes is under consideration. A 170 m long urban rail tunnel consisting of precast sections has already been jacked (City S-Bahn Hamburg, contract 4/5) [278].

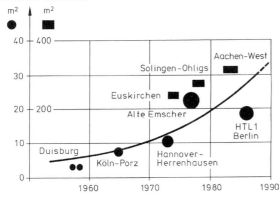

Figure 13-20 Development of excavated cross-sections for pipe jacking and box jacking [225]

13.4.1 Pipe jacking

The classic field of application for pipe jacking is the installation of sewers and drainage pipes. The first known application of the process in Germany took place in 1935 in Nuremberg. Since the start of the 1950s, the technique has been used increasingly [178]. According to a survey by STUVA, 60 % of pipes between 1 m and 5.5 m diameter are installed by pipe jacking [225]. Figure 13-21 shows a diagram of the principle of pipe jacking.

The thrust wall, thrust jacks, thrust ring and in general one length of pipe with tunnelling shield or cutting shoe are assembled in the thrust pit (starting shaft). The pipe section in the thrust pit is jacked forward while at the same time the ground is excavated at the working face, either by hand or with an tunnelling machine suitable for the prevailing geology. The pipes are either delivered as precast elements or produced on site in a field production plant.

The relocation of the thrust unit into the starting shaft and the fact that equipment for installing segments is no longer needed are the decisive advantages of the pipe jacking process over conventional shield tunnelling. The extra space freed up in the tunnelling area makes the process very flexible when the ground conditions are changeable.

Despite all the further technical developments, the process remains limited by the maximum pipe jacking length, which is limited by the increasing wall friction forces between the pipeline and the surrounding ground as the advance continues, and also by the size of the cross-section that can practically be jacked, which is limited by the weight and size of the pipe or box elements. In addition, the settlements, which have been measured due to pipe jacking in loosely consolidated, non-cohesive soils, tend to be rather larger than with conventionally shield-driven tunnels [225]. Caution is also demanded by longer pipe jacks in jointed or friable hard rock.

A wide range of types of cutting shoe or tunnelling shield is available, starting with a simple steel ring formed as a blade and mounted in front of the first pipe, and extending to include all the developments in shield tunnelling, which have already been described. Regarding measures to hold back groundwater and support the working face, all known principles like compressed air support, slurry support and earth pressure balance support have already been used.

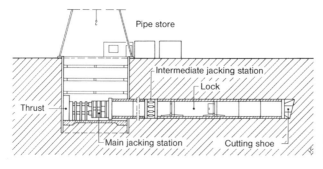

Figure 13-21 Principle of pipe jacking [183]

The provision of intermediate jacking stations enables longer headings to be jacked. The intermediate stations divide the pipeline into sections, each of which is jacked forward, and thus limit the required thrust forces.

Horizontal and vertical curves can be pipe jacked through the provision of steering jacks behind the tunnelling shield or cutting shoe.

13.4.2 Box jacking

A further development of the pipe jacking process is the jacking of entire structures with large cross-sections, called box jacking or tunnel jacking. This technique has mostly been used until now for the construction of road tunnels under railway embankments. The German Railways always applies stringent requirements concerning disturbance of rail operation for such crossing structures. Any disturbance of rail operation should be restricted to the shortest possible time and the disturbance should be as slight as possible, depending on the status of the line.

Compared to all other construction methods like cut-and-cover or tunnelling, box jacking offers decisive advantages over the conventional methods. As the structure can be constructed and prepared for jacking next to the railway line, railway operation is only disturbed in the relatively short time taken to jack it through. As the jacked box is pushed under the railway line as a complete, load-bearing system, railway operation is even less severely disturbed during the actual jacking process. Jacked structures have now reached enormous dimensions. In Aachen, a rectangular cross-section of over 300 m^2 was jacked a distance of 126 m under the west station as part of the building of the western ring road L 260 [22]. In Bochum in 1988, a box-section 30 m wide, 7 m high and 50 m long was jacked under a railway embankment as part of the Westtangente Bochum (Figs. 13-22 and 13-23).

There is a wide range of special problems and considerations regarding the construction of the box for jacking, the layout of the cutting shoe and the overall construction process, and a detailed description of these would exceed the bounds of this book. For further information, reference is made to the extensive literature available [14]. In order to offer an impression of the construction process, the box jacking for the Westtangente Bochum is described as an example, which still illustrates the current state of technological development.

The construction problem in this case was: the Westtangente required two lanes in each direction to pass under two high-speed ICE lines, two urban rail tracks and a buffer track with a traffic density of about 350 trains per day without any severe limitation of rail traffic. Further problems were caused by the shallow cover and the very variable and unfavourable soil formations.

As the result of an alternative proposal (Dyckerhoff and Widmann) to construct the underpass by box jacking, rail operation remained completely undisturbed until the actual jacking operation. During the five-week jacking period, the trains were restricted to 80 km/h.

The structure was constructed in a field to the south of the railway embankment on three tracks beneath the centre of the longitudinal walls.

Large pipes with an external diameter of 2.32 and 2.78 m were jacked through from three thrust pits to form the tracks for the main box jacking. Due to the rock strength at the level of these tracks, the excavation at the face required some blasting, with careful calculation of the charges. The actual tracks for the box jacking consisted of continuously welded crane rails, which were levelled on steel beams in the large-diameter pipes. Then the lower half of the pipes below the rails was concreted.

22 main jacks with a total thrust force of 66,000 kN were installed for the jacking of the structure. These were divided into three main groups of jacks (Figure 13-24). Half of the jack group in the middle was assigned to each of the outer groups, which made it possible to correct the direction of the structure during jacking.

For the jacking operation, precast concrete blade planks were installed on the ceiling and side walls of the structure. The side blade planks were jacked with fixed jacks with a compressive force of 3,660 kN, while four jacking frames each with a thrust force of 4,630 kN were provided for the ceiling blade planks. The jacking frames were withdrawn after the completion of a stroke and transported to the next plank along an overhead track. The control and monitoring of all jacks and the jacking process was undertaken from a central control position in the tunnel.

The jacking operation started in a cycled sequence with the narrow blade planks (Figure 13-25) on the ceiling and side walls of the structure being jacked forward by 25 cm. The working face was then excavated under the protection of the blade planks. After completion of the excavation and clearing of the soil, the structure was jacked forward 25 cm by the main jacks.

This procedure and an excavation device specially developed for the zone at the ceiling and the blades ensured an advance with little settlement. What proved particularly problematic was the excavation of the soil between the underside of the tunnel ceiling slab and the steel blades of the blade planks, as there was only a clear thickness of 16 or 32 cm along a length of 2.50 m. This problem was solved by the use of a specially developed telescopic and slewing narrow clamshell bucket, which was mounted on a hydraulic excavator.

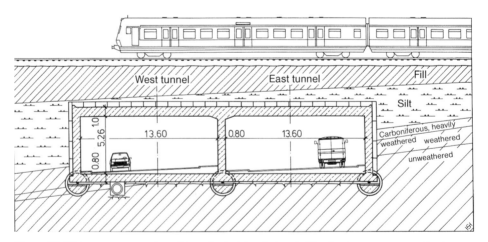

Figure 13-22 Tunnel cross-section with geological details, box jacking for the Westtangente Bochum, 1988

Figure 13-23 Prefabricated box section, Westtangente, Bochum [85]

Figure 13-24 Layout of the three jack groups for box jacking, Westtangente, Bochum [85]

In order to ensure sufficient stability of the working face, the structure was divided horizontally by an intermediate slab. Additional stability was provided by vertical walls, which supported the ceiling slab and the intermediate slab half way between the side walls. The leading edges of the ceiling slab, the intermediate slab and the side walls were formed with a raking angle as "blades" (Figure 13-26). The high reaction forces acting on the blades during jacking required a heavy concentration of reinforcement in this area and an appropriate concrete mix.

For the structure in the serviceability state, it was necessary to fully support the underside of the structure which was only supported on the sliding tracks during jacking, from the subsoil. In order to fill any gap underneath the structure and for additional waterproofing, a suspension of ground limestone, cement and bentonite was injected into the gap between structure and blade planks.

An extensive monitoring system including a warning system for railway traffic provided continuous control of the progress of construction. The average advance rate per 24 h was 1.60 m, so the structure reached its final position in 31 days [259].

Figure 13-25 Laying the blade planks, box jacking for the Westtangente, Bochum [85]

Figure 13-26 Inside view of the box section, box jacking, Westtangente, Bochum [85]

13.5 New concepts in mechanised shield tunnelling

Above all in Japan, interesting new ideas intended to solve the current deficits in the technology of mechanised shield tunnelling are being produced constantly. These developments lead to opportunities for the even wider application of shield tunnelling, as already with multi-face shields.

13.5.1 Shield machines for flexible cross-sections

One significant failing of the shield machines used until now has been the normally applied limitation to a circular excavation cross-section. Above all elliptical, egg-shaped or vaulted cross-sections, which can be advantageous for structural reasons or to improve the relationship between excavated area and usable area, have so far seldom been possible. Shield machines capable of producing non-circular cross-sections have recently been developed and tested in Japan.

The shield machine shown in Figure 13-27 can excavate the part of the working face not reached by the main cutting wheel through the provision of a number of planet cutters, which are mounted on swinging arms on the main cutting wheel and turn with it. The swinging arms are controlled by their own actuating cylinders. The cross-sectional form excavated is determined by the changing position of the planet cutters as the main cutting wheel rotates. The system can excavate, according to information from the manufacturer, not only of elliptical sections but also the egg shapes and horseshoe sections etc. shown Figure 13-28.

As part of the improvement of the drainage system in Tokyo, the machine shown in Figure 13-27 was used successfully as an EPB shield to drive a 50 m trial section [159].

It should be mentioned that the cutting system of the machine described above is based on the Wohlmeyer principle developed in Germany (Figure 4-26). A machine built by the Austrian Alpine Montangesellschaft according to this principle was used as long ago as 1958 in Grandenberg (Styria) in brown coal and limestone.

The concept of a shield machine shown in Figure 13-29 is based on a combination of circular cross-sections, partially with a pivoted mounting, and sliding cutters, which enable an elliptical cross-section. This machine has not yet been tried out.

Figure 13-27 Shield machine for flexible cross-sections [154]

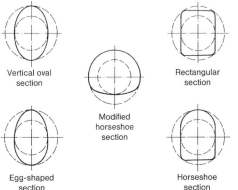

Figure 13-28 Examples of possible excavation shapes [154]

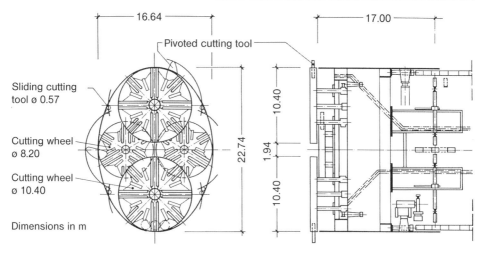

Figure 13-29 Concept of a machine for elliptical cross-sections [142]

13.5.2 Ultra-flexible shield

Particularly in inner-city areas, there is a demand for curves of ever-tighter radius in tunnel alignments. This makes it necessary not only to enable a large overcut but also to provide a special hinge in the shield skin, to enable the shield to follow tight curves without leading to mechanised constraints. The sealing of this joint against the ingress of water, soil and any bypassing support suspension is a major problem.

In the shield machine shown in Figure 13-30, this sealing problem is overcome by the inclusion of a foam rubber ring and a two-layer seal in the articulation joint.

13.5.3 Horizontal and vertical shield machines

Horizontal and vertical face machines are a further development of the multi-face shield. They enable a flexible change of the cutting wheels between a horizontal and a vertical position through the exact control of the roll angle.

Figure 13-30 Ultra-flexible shield (Mitsubishi)

The shields consist of two entirely independent part machines, which can also work separately from each other. The possibility of separating the two tunnelling machines and perhaps bringing them back together opens up new possibilities for the design of tunnels to be driven by shield machines [154]. No shield machine of this type has yet been tried out.

13.5.4 Enlargement shields

The idea of the enlargement shield, also called a reamer machine, is to enlarge the size of existing shield-driven tunnels to adapt them for new use requirements by making use of the advantages of shield technology. Another use of such machines is the partial enlargement of new tunnels, for example for underground stations (Section 13.3). The principle of enlarging a tunnel in stages has already been used for some decades in order to reduce the necessary thrust forces (Wirth system).

The essential advantage of this technology is the possibility of performing all works inside the protection of the shield skin of the tunnelling machine. No practical application of this concept (Figure 13-31) is yet known.

13.5.5 Rotation shields

Further reduction of the disruption of surface infrastructure, particularly for the construction of launching and reception shafts, is promised by the rotation shield, an idea from Japan. This type of shield features the ability to rotate the shield so that the machine can bore the starting shaft, the horizontal tunnel itself and emerge from the reception shaft (Figure 13-32).

The following advantages are stated for this shield design [154]:

- Reduction of the cross-section of the shafts to 25 %,
- reduction of the time taken to construct the shaft by 50 %,
- reduction of manufacturing costs,
- improved safety, particularly for deep shafts, as no compressed air working is necessary and the completed shaft does not have to be cut to install the shield machine,
- In longer, horizontal stretches of tunnel, it is relatively simple to change the cutting wheel by turning the cutting face inwards,
- improved navigation possibilities.

Figure 13-31 Concept of an enlargement shield [154]

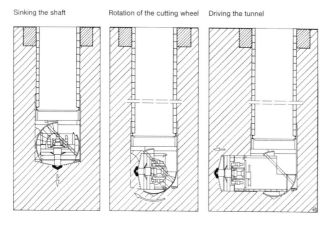

Figure 13-32 Working principle of a rotation shield [154]

Figure 13-33 Rotation shield, 1993 (Ishikawajima-Harima)

No practical application of this type of machine was known at the time the first edition of this book was published. However, in summer 1993, the Tokyo Sewer Bureau placed an order with the Japnese manufacturer Ishikawajima-Harima, who had been experimenting with rotation shields since 1983 [239], for the first rotation shield (Figure 13-33). The trial operation of the 7.3 m long shield machine with an external diameter of 5.5 m has already taken place. The change of direction while tunnelling is achieved by the integration of a second, smaller shield of 3.7 m external diameter inside the main shield. The second shield is supported on a ball joint moved by four hydraulic cylinders. The ancillary equipment of the main shield is mostly used by the smaller shield during a change of tunnelling direction (e.g. screw conveyor) [240].

13.5.6 Shield docking method

Longer tunnels can be driven by two shield machines, which start from the two portals and meet in the course of the tunnel. Despite the technical problems posed by the connection, this method offers above all cost advantages through the reduction of the construction time. A typical example is the Channel Tunnel, which was driven simultaneously from the English and the French sides (Chapter 8).

13.5 New concepts in mechanised shield tunnelling   333

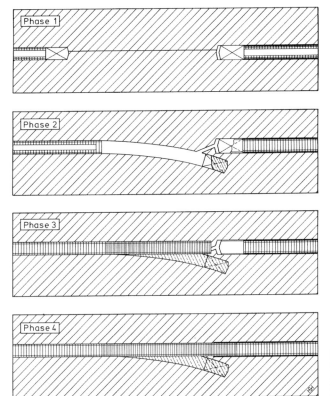

Figure 13-34 Four stages of the breakthrough in the middle of the service tunnel under the English Channel [202]

Figure 13-34 shows the procedure used for breaking through the service tunnel at the end of 1990.

– The two shields were stopped. An investigation hole was drilled through the cutting wheel of one shield. Then the other shield drove forward to the investigation hole, which was the final position. The shield was than dismantled and brought back to the starting shaft except for the shield skin, which remains in the service tunnel as lining.
– The other shield continued forward in a curve until next to the dismantled machine. Finally, this machine and the final curve were filled with concrete.
– Finally, the two tunnels were connected with a side cut.

The procedure is more complicated when the geological conditions are unfavourable or there is groundwater. This would make additional measures like chemical grouting or ground freezing necessary.

In order to simplify the connection of two tunnels driven by shields from both ends, a method has been developed in Japan to connect two shields (Shield Docking Method).

In this case, the two machines approach each other directly until they are almost in contact (Figure 13-35 a). The cutting wheels of the two shields are fitted with retractable tools, which are now retracted. Then a outer cylindrical skin, which was integrated in one of the machines, is pushed forward into a groove provided in the other. The connection is sealed by a rubber gasket. The cutting wheels and all the other equipment can now be dismantled and transported back to the launching shaft. The shield skins of both machines remain in the tunnel as lining. The connection is completed with a concrete inner layer (Figure 13-35 c).

The connection procedure does not demand two identical shield types, e.g. a slurry shield and an earth pressure shield can be connected. The first project where two shields were docked in this fashion took place in 1992 in Tokyo, Japan. A collector of about 3.5 km length was to be driven in cohesive soil. The external diameter of both shields was 3.43 m; on this project, one was a slurry shield and the other was an earth pressure shield (Figure 13-36) [141].

For the construction of the "Trans-Tokyo Bay Highway" (start of tunnelling in 1994), four connections of altogether eight slurry shield machines were planned. The connections under Tokyo Bay were to take place using ground freezing from the shield.

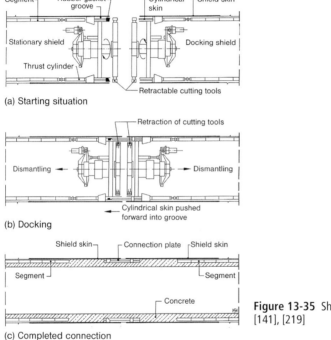

Figure 13-35 Shield Docking Method [129], [141], [219]

13.5 New concepts in mechanised shield tunnelling

Figure 13-36 Shields for the Shield Docking Method, Tokyo Metropolis 1992 [141]
a) Shield with groove for the docking procedure (slurry shield)
b) Shield with projecting cylindrical joining skin (earth pressure shield)

14 Guided microtunnelling processes

Microtunnelling (Table 14-1) like pipe-jacking is a trenchless or no-dig technology, which has been growing steadily for some years and is now well established. Microtunnelling machines have reached a high level of reliability and are capable of boring long distances, even under the groundwater table, and application in almost any geological formation [195]. Appropriate monitoring and control units ensure high position precision. Mobile, quickly installed tunnelling units, which are available in containers as a building block system, permit simple design and construction [187].

These developments are being increasingly used in the construction of underground sewers, water pipes and pipelines, so that the disruptive and environmentally damaging effects of trenching like dug-up roads, traffic congestion, diversions, increased noise and fumes, the ecological impact of groundwater lowering and settlement damage to roads and buildings can be reduced to a minimum.

The development of microtunnelling in recent years has shown that this method of underground construction offers a clear alternative for the laying of non-accessible supply and waste pipes and with increasing acceptance is also an economical method. This applies particularly when all costs associated with the laying of new pipes in trenches are included, for example the cost of traffic obstruction and similar problems resulting from trenching are also considered. Some construction works should indeed only be undertaken by microtunnelling, a fact that is already being implemented systematically, particularly in Berlin und Hamburg.

The main feature of microtunnelling is the ability to control the position, that is to steer the tunnelling machine. There is a difference between guided and unguided processes. Only the guided variant is handled further in this book.

Table 14-1 Guided microtunnelling processes and working principles

Pilot tube process	Guided jacking of a pilot tube either by removing or displacing the soil: followed by the jacking of pipes enlarging the bore behind an uncontrolled reamer while pushing out the pilot tube sections
Guided auger microtunnelling	Jacking of a pipe while the soil is removed from the face by a cutterhead; continuous soil removal with an auger
Shield microtunnelling	Jacking of a pipe while the soil is removed by a cutterhead at the mechanically- or slurry-supported face; continuous hydraulic soil removal

Guided processes are used today in almost all areas of pipe laying. In the case of foul, surface water or combined sewers, the prefabricated product pipes are jacked forward, while in the case of district heating pipes, gas pipes, water mains or electric cables, the jacked pipe is used as a casing for the actual product pipe or cable. Guided pipe jacking can be carried out in cohesive and non-cohesive soils, in loose ground and in rock.

14.1 Pilot tube process

The pilot tube process first became possible with the guidance of the ground displacement process.

The first pilot tube is jacked through displacing the soil with the direction and position being controlled by surveying a target installed on the steering head with a theodolite. The steering in the example shown in Figure 14-1 is achieved using a rotating steering head with a slant to one side, although other processes are also possible for the production of the pilot bore. Then the casing or product pipe is jacked through behind a reamer head by unguided pipe jacking (Section 14.2) simultaneously displacing the pilot tubes. According to information from the manufacturer, pipes of up to 400 mm diameter can be installed by the illustrated process with an accuracy of millimetres over 35 m.

The geological and hydrogeological range of application of this process is essentially limited by the soil displacement required to install the pilot tube. Ground that cannot be displaced like soil containing obstructions or rock are unsuitable, as are changeable ground conditions. Soil formations with groundwater can be overcome with auxiliary measures like groundwater lowering. The main application of this process are sewer connections.

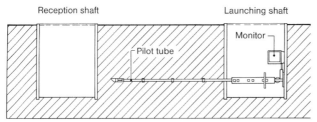

Phase 1: Jacking of the pilot tube while checking direction with a theodolite, tube displaces soil

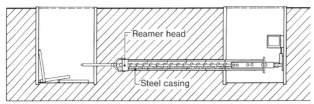

Phase 2: Jacking of the steel casing

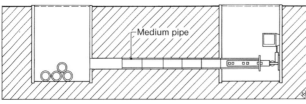

Phase 3: Jacking the product or medium pipe

Figure 14-1 Principle of the pilot tube process

14.2 Auger microtunnelling

Guided auger microtunnelling (Figure 14-2) is a direct further development of the equivalent unguided system, and differs mainly in the integrated steering and surveying equipment.

The material excavated by the closed cutting wheel is transported by an auger to the drive pit and generally removed with a crane. The product pipes are jacked forward simultaneously. Auger microtunnelling is predestined for driving shorter distances due to the relatively low cost of the machinery. The excavated material can be tipped without further treatment, in contrast to shield microtunnelling with slurry support (Section 14.3). Figure 14-5 shows the range of application of the process.

The direction and position are checked with a laser beam, which illuminates a target on the cutting wheel. The measured data are continuously transmitted to the control position and displayed (Figure 14-3). A log of all measured data is produced automatically for the operator of the equipment and can be produced for supervisory authorities [124].

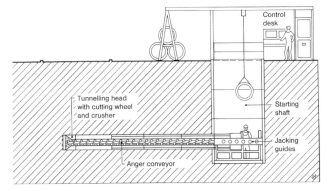

Figure 14-2 Principle of guided auger microtunnelling

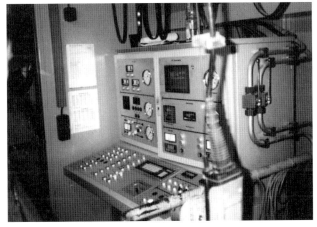

Figure 14-3 Control panel of a microtunnelling system (Herrenknecht)

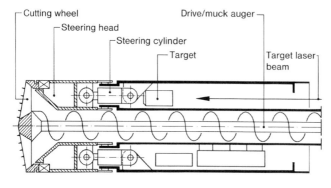

Figure 14-4 System diagram of a tunnelling machine for guided auger microtunnelling

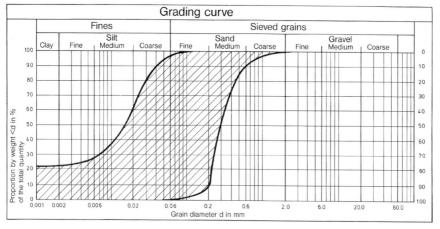

Figure 14-5 Range of application of guided auger microtunnelling [221]

Any deviation of the tunnelling machine (Figure 14-4) from the correct alignment is automatically corrected in modern systems, although manual interventions remain possible. Steering cylinders in the backup of the cutting head actuate the corrections. It is difficult with conventional pipe jacking systems to correct rolling, because the cutting wheel and auger can only turn in one direction.

14.3 Shield microtunnelling

In shield microtunnelling (Figure 14-6), the full face is supported mechanically or by pressurised slurry and excavated by a cutting wheel with the muck being removed continuously in the slurry circuit from a suspension chamber immediately behind the cutting wheel. The necessary pumps are located in the launching pit and on the surface, where the excavated soil is also separated from the support and transport medium, which returns to the circulation. A separating plant is particularly necessary in clay and loam soil. The shield machines used in microtunnelling are built according to the systems known from large-scale pipe jacking and tunnelling and have been further developed for the special requirements of microtunnelling [195].

14.3 Shield microtunnelling 341

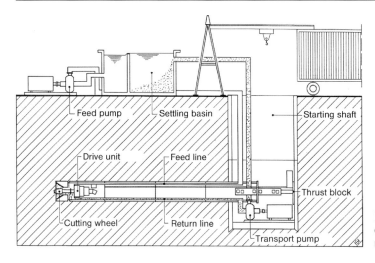

Figure 14-6 Principle of the shield microtunnelling process

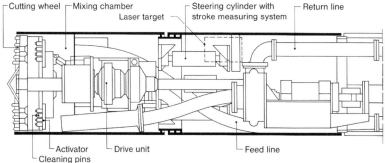

Figure 14-7 System diagram of a shield microtunnelling machine (Maschinen- und Gerätefabrik)

The control of direction and position and the steering essentially work in the same way as in pipe jacking, with a laser beam and a target installed in the tunnelling head and steering cylinders in the backup of the shield.

Heaving and settlement in shield microtunnelling are largely avoided by the automatic adaptation of the pressure of the support and transport medium (water or bentonite suspension) and the contact pressure of the cutting wheel, which is held between the active and the passive ground pressure (Figure 14-7).

According to the manufacturers, distances of up to 250 m can be driven in the inaccessible diameters and up to 400 m in the accessible diameters by microshields, even in aquiferous formations. Nominal diameters greater than 150 mm can be driven by a microshield, and the geological range of operation is shown in Figure 14-8.

As with auger microtunnelling, various excavation tools are available for shield microtunnelling. These have to be selected according to the soil conditions (Figure 14-9).

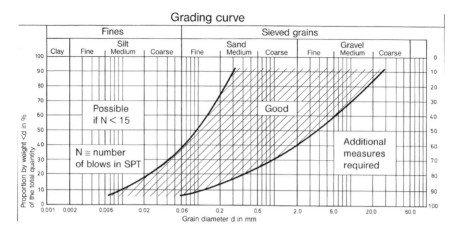

Figure 14-8 Range of application of microshields [307]

The integration of all site facilities including a crane rail and a diesel generator into a container has become standard practice for setting up sites in Germany. The containers are placed directly above the thrust pit and this enables short setting up times, independence from external sources of energy and continued working through the winter (Figure 14-10). Tunnelling machinery constructed as a modular system enables quick changing from auger mucking to a hydraulic slurry circuit and the changing of the cutting wheel, which all enables rapid adaptation to the prevailing ground and groundwater conditions.

The use of precast rings for the lining of launching and reception shafts (Figure 14-11), which are installed as caissons as part of a complete microtunnelling system, and prefabricated thrust constructions reduce the equipment time and enable simple alteration of the direction of tunnelling.

14.4 English Mini Tunnel system

This process for construction tunnels in diameters from 1 to 2 m is very economical. The entire process is highly standardised with an impressively clear concept. A simply equipped minishield is pressed forwards against a simple segment ring. The annular gap is filled with pea gravel grade 3 to 6 mm and then grouted with cement mortar. The launching shaft has a diameter of 2.5 m and is thus considerably smaller than would be required for pipe jacking, as no thrust block or lengths of pipe have to be accommodated. After the completion of tunnelling work, this shaft can be used as an inspection shaft or manhole. Further information about the English Mini Tunnel system can be found in the "Handbuch für Tunnel- und Stollenbau", Volume I [183].

14.4 English Mini Tunnel system

Figure 14-9 Cutting wheels for shield microtunnelling (Herrenknecht, Maschinen- und Gerätefabrik)

Figure 14-10 Containerised construction site

Figure 14-11 System launching shaft

14.5 New developments

The installation of crushers in the tunnelling head can extend the range of application of microshields to include loose ground with coarse grading. Maximum grain sizes up to about 40 % of the cutterhead diameter are possible according to information from the manufacturers (Figure 14-12).

High-pressure water jetting in the cutterhead can also enable boring through inhomogeneous ground with rock outcrops. The jets can also be used to clean blockages and material sticking to the mucking openings in the rock cutterhead used in this case when boring though cohesive soil. Another effective measure is the integration of cleaning pins behind the cutter disc, which can be extended if the mucking openings are blocked (Figure 14-7).

Hydraulic mucking is increasingly used instead of auger mucking, particularly for longer distances. More recently, systems using pneumatic soil transport have also become available and applications of these have been successfully completed.

Figure 14-12 Cutterhead with crusher

Figure 14-13 EPB shield, nominal diameter 1,200 mm (Markham)

Because the control and steering technology has now reached a reliable standard, driving microtunnels with diameters of 1,000 m and more around curves for longer distances is becoming more interesting. This has already been achieved.

The principle of earth pressure support (Chapter 11) has also been applied successfully in microtunnelling. Earth pressure support offers advantages for boring through very changeable soil and non-cohesive soil in the groundwater. Settlement can be reduced still further than would be possible with slurry-supported microshields, because uncontrolled entry of soil into the shield cannot occur in any phase of the drive, even in very variable soil formations. Using this process, there can be no destabilisation of the face due to heavy loss of support slurry or compressed air blowouts. The earth pressure balance shield shown in Figure 14-13 was used in Rochdale (UK) to construct a 570 m long sewer of 1,200 mm diameter in four sections.

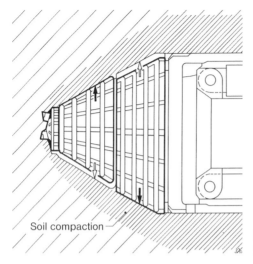

Figure 14-14 Schematic illustration of a ground displacement cutterhead, Perimole System (Iseki)

Another interesting development is the Perimole (Iseki) tunnelling system (Figure 14-14), which works on the soil displacement principle. Steered by laser, distance of up to 80 m can be achieved in nominal diameters ≤ 250 mm. The displacement cutterhead is divided into two contra-rotating parts and with steering cylinders installed in the tunnelling head is capable of accuracy of 25 mm vertically and horizontally according to the manufacturer.

In addition to the new installation of utility pipes, guided microtunnelling has also moved into the field of repair and renovation. Worth mentioning in this respect is the Pipe Eating process (Figure 14-15), which can install new pipes along the alignment of pipes that are defective or have to be replaced. The old pipe run is driven through, crushed (eaten), removed and replaced by a new pipe run. The new pipes are normally of larger diameter.

In this context, the opportunity for simplifying and redesigning the drainage system should also be pointed out, which is closely associated with the technical development of microtunnelling machinery. One example is the "Berlin construction method" (Figure 14-16), in which the collectors are installed by underground microtunnelling and jacking from a starting shaft to a reception shaft and then the house connections are bored in a star shape from the shafts. The advantage of this method is that the main collector pipe runs are not disturbed by incessant house connections and can be installed by an underground (trenchless) method.

Figure 14-15 Cutterhead with crusher for "Pipe Eating" (Soltau)

14.5 New developments

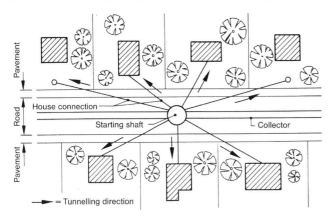

Figure 14-16 Berlin construction method [221], [273]

The idea of an accessible (by walking or crawling) pipe as a communal duct to house a number of different utilities will also increase in significance in the future [273]. The advantages of communal installation and ease of inspection as well as the improved repair capability have long been recognised and practised in the industry. In municipal engineering, developments in this direction are already being encouraged due to high level of development of shield microtunnelling and the possible cost savings.

One speciality in the field of communal utility ducts is a 710 mm external diameter jacked pipe developed in Japan, in which six cable ducts are preinstalled (Figure 14-17). The pipes are delivered to the site in prefabricated form and installed [158].

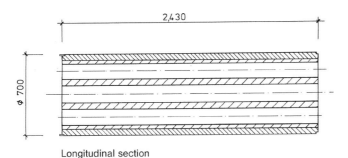

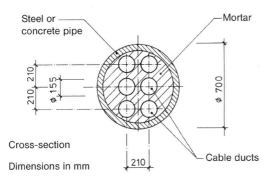

Figure 14-17 Schematic illustration of a prefabricated communal utility duct according to [158]

15 Surveying and steering

The surveying and steering of a shield are of increasing importance due to the increasingly stringent requirements for positional accuracy of shield drives and for the monitoring of the machine during tunnelling. Ever more complicated alignments and the use of machines with high advance rates also contribute to the high demands placed on surveying and control. The fundamentals of surveying and control technology are discussed in the book "Handbuch des Tunnel- und Stollenbau", Volume II [184].

Tunnelling projects can, depending on their magnitude, be driven from a number of starting points or prepared in advance with specialised civil engineering works at various locations. This requires exact knowledge of the current position of each tunnel boring machine working on the project. In order to maintain the construction tolerances and to avoid extra costs, modern tunnel drives are required to break through with great precision and small deviations from the designed tunnel axis. Particularly for traffic tunnels, the maintenance of the tolerances from the given axis is also important for the planned vehicle dynamics. Later alterations of the alignment or recalculation can cause high additional costs, which should be avoided.

The use of automatic navigation systems is now standard for tunnelling machines in order to comply with these requirements. The different types of tunnelling machine often pose new challenges for the instrumentation and the calculation algorithms, particularly when multiple axes of different TBM parts have to be determined, like for example with gripper TBMs, double shield TBMs or machines with inclined thrust cylinders. Collaboration with the design department of the manufacturer in this case extends far beyond simply keeping sight lines free for the surveying system, as extensive understanding of the working of the tunnelling machine is required.

Solutions are adapted to suit the requirements and offer machine drivers constantly updated information about the position of the machine, the current tendencies and chainage, related to the designed tunnel axis. At the currently usual advance rates of some centimetres per minute, the TBM operator must receive a precise response to show the effects of control movements carried out in order to steer the shield as near as possible along the given alignment.

The use of automatic navigation systems enables the surveying specialists to concentrate on the gathering of data for risk management in addition to the regular checks and grid measurements.

Additional measurements are also taken in the tunnel, for example the monitoring of the ring convergence of each ring or the measurement of the shield tail gaps, although these measurements can be made automatically or semi-automatically once they have been set up.

The documentation requirements are increasing in extent and have to be produced ever sooner, with the result that computer-supported solutions have become common here as well. This serves to investigate the causes of any deviation, but also delivers valuable data for invoicing and the management of variations and claims. For the example of pipe jacking, EN 12 889 (formerly DWA-A 125) requires complete documentation of the measured position every 10 cm or 90 s, whichever occurs sooner.

These extensive tasks from navigation to documentation and the use of very varied surveying systems demand from the tunnel surveyor not only very good expert knowledge but also good capabilities in data management.

15.1 Surveying

In order to be able to characterise the spatial position of the TBM, reference points on the main axes of the tunnel boring machine are determined in a global system as well as related to the designed tunnel centreline. Not only the coordinates have to be determined but also the roll, longitudinal pitch, yaw angle and the tendencies (of pitch or yaw relative to the alignment) (Figure 15-1).

The basis of each navigation system are the local coordinates of the system components like target, prisms, inclinometer and other points with relevance for surveying in relation to the machine coordinate system. For navigation purposes, these points are related to known global coordinate points so that the position of the tunnel boring machine can be calculated in relation to the alignment.

Standard practice for tunnelling machine navigation systems is the use of laser tachymeters in combination with automatic laser target units in the TBM. Alternatively, there are variants with prisms or with fixed laser and target for straight tunnels. Systems based on a gyroscope with electronic hose water level are used particularly for small diameters where it is often not possible to provide a clear line of sight or because they are too small to be accessible. Gyroscope-based systems are, however, inherently more laborious to operate, and for longer tunnels can lead to less precise breakthrough.

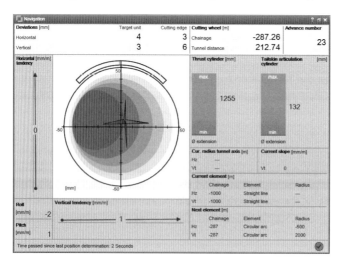

Figure 15-1 Display of machine position and TBM tendencies for the TBM operator in the control cabin of a tunnelling machine

15.1 Surveying

The spatial position of the machine centreline is continuously displayed at the control stand, to ensure that the TBM operator is informed about the position, heading and tendencies of the TBM and can react to this information with fine steering movements. According to the tunnelling method, the control cabin may be in the tunnelling machine or on the surface.

Fully automatic control systems have not yet become established for tunnel boring machines due to the multitude of process parameters and the quick reaction of individuals to the current excavation situation, but some progress has been achieved in specialised areas.

15.1.1 Navigation with tunnel laser and automatic target unit

The challenge for the navigation of pipe jacking is that the entire structure is constantly moving. This makes it impossible to set-out fixed points along the section that has already been driven.

When pipes are jacked along straight drives, a fixed laser in the launching shaft can be used with its beam aimed at a target mounted in the shield. A dual axis inclinometer integrated into the target can also measure roll and longitudinal pitch. The chainage is normally measured by a measuring wheel in the shaft. Supplemented by the determination of the cylinder strokes and the known machine dimensions, the position of the shield blade can be calculated relative to the designed tunnel axis. This process can be used for drives up to about 250 m long (Figure 15-2).

In order to be able to correct the influence of refraction, particularly at the entrance of the tunnel, a hose water level is additionally used. This enables the tunnel laser system to be used for distances up to 400 m. Nonetheless, refraction still causes inaccuracies in the determination of position and level, which have to be minimised with regular control measurements.

15.1.2 Navigation with gyroscope system and hose water level

Gyroscopic systems are well suited, supplemented by a hose water level, for curved tunnels with tight radii and small diameters (Figure 15-3). But this method also has advantages for straight drives, as refraction effects at the start of the tunnel are insignificant and no special free sight lines are required.

The level is measured as described in the last section with an electronic hose water level and the chainage is measured with a measuring wheel. Roll and pitch are measured with a dual axis inclinometer.

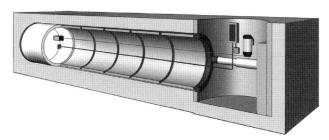

Figure 15-2 Tunnel laser with automatic target unit and hose water level in pipe jacking (VMT)

The gyroscope measures the angle against north. As its local coordinates are known with reference to the machine centreline, the azimuth of the machine can be determined. The principle of positional determination with a gyroscope is based on dead reckoning, i.e. the alteration of the azimuth after a defined distance is recorded and used for the calculation of the new position.

Drifting of the machine cannot be recorded by the gyroscope. The deviation resulting from drifting increases with the distance travelled. It is therefore necessary to survey the position of the machine and make corrections if required. These control measurements should be done every 30 to 40 m of advance.

15.1.3 Navigation with total station and automatic target unit

The most commonly used navigation system is the use of a total station with integrated laser on the tunnel wall, a target in the shield and at least one prism as back sight reference (Figure 15-4). The total station is oriented against known coordinates and the laser is automatically aimed at the target. The target unit determines the position of the impact point of the laser and the angle of incidence. All values are transmitted to the navigation computer. The distance between total station and target unit is determined with a prism on the target or without a reflector. Alternatively, the chainage can also be determined from the measured stroke of the thrust cylinders and the addition of the installed segment rings. In this way, the chainage of the TBM can be calculated. The combination of all the measured values can be used to calculate the exact position of the TBM in three dimensions. According to the type of machine, the position of the machine can be determined at two or three reference points of the shield centreline and their distance from the correct alignment in position and level and the horizontal and vertical tendencies can be displayed.

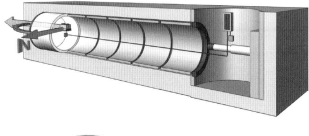

Figure 15-3 Gyroscopic system with hose water level in pipe jacking

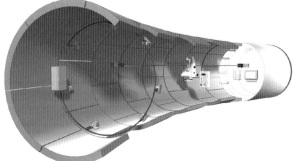

Figure 15-4 Total station with target unit and back sight prism as well as Ring Convergence Measurement System in segmental lining

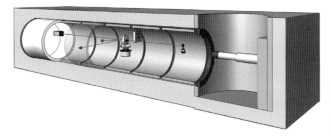

Figure 15-5 Total station with target unit, reference point prisms and back sight prism in pipe jacking

When positioning a total station in a tunnel drive with segmental lining, it s important to fix the instrument to a stable area of the installed tunnel lining. Considering atmospheric conditions and refraction, this should not be too far behind the tunnelling machine.

Regular automatic direction checks to the back sight reference prism checks the accuracy of position. As inaccuracies can be caused by refraction, a tolerance range is specified.

It is important to make sure the laser window is free to enable a clear view to the target unit.

This navigation method is the most common for mechanised shield tunnelling through loose ground due to the lower level of vibration compared to other tunnelling methods.

A similar process of navigation with a total station is used in pipe jacking (Figure 15-5). The difference is that the entire pipeline is in movement and it is not possible to set-out fixed points along the completed section. For this reason, the coordinates of the back sight reference prism and the total station have to be constantly recalculated with reference to the centreline of the already completed tunnel and the current chainage. At a configured interval of time or distance, the system automatically orients itself against the back sight reference prism. Afterwards it immediately determines the position of the reference pipe using the reference targets and then the actual machine position. In order to compensate rolling of the laser station, an automatic tribrach is used. This ensures a permanent horizontal setting of the total station. Lateral deviations caused by rolling are recorded by an inclinometer and included in the calculations. In comparison to gyroscope based systems, this has the advantage that the position can be determined continuously and drift tendencies can be detected.

15.1.4 Navigation with total station and prisms

Analogously to the method described above, two or more prisms in the shield can be used instead of an automatic target unit, although this does of course accordingly increase the requirements for a clear view. A dual axis inclinometer is required with this system to determine roll and pitch.

The prisms are measured one after the other in a sequence. As the TBM is advancing during this cycle, a position correction has to be applied to the measurements of the prisms. Then the global coordinates of the reference points on the shield centreline are determined and their horizontal and vertical deviations and tendencies are calculated with reference to the designed tunnel axis. The accuracy of this method is not so good and it takes up more space, but it is more resistant to vibration and is therefore the best method for tunnel boring machines operating in hard rock.

15.2 Ring design and calculation of the ring installation sequence

In addition to navigation tasks, there is often a demand in tunnels with segmental linings to calculate the optimal ring installation sequence, at the design phase and during construction, and then to document the installation of the rings.

For the designers of tunnels, questions have to be considered about fitting the ring design into the planned tunnel alignment, particularly for example the ring taper and the planned minimum curve radius, calculating the deviations to be expected from the ideal alignment or the design of cross-passage locations. Additional limitations or compulsory positions of individual rings or impermissible keystone locations can also be significant.

If long-term calculations for various ring designs, e.g. universal or left and right rings, are compared, this can be used to make well-founded decisions for the selection of a ring design with the best technical parameters.

During the tunnel advance, ring assembly calculations are carried out considering the current position and tendencies, the planned centreline, the shield tail gap and other parameters, also particularly the subsequent rings. These values can be differently weighted according to the requirements of the individual parameters, which result from the precision specified for the later use.

The measurement of the shield tail gap, the calculation of the subsequent ring sequence and the documentation of the completed rings can all be performed manually, semi-automatically or automatically.

Advance calculation of the subsequent sequence of ring installation also permits the information to be forwarded to the segment store to load the required ring types in good time and transport them into the tunnel in the right order.

15.3 Ring convergence measurement

The greatest effect on the initial deformation of a segment ring directly after its installation is caused by the thrust cylinders pushing the next stroke. Subsequently, further variable loads and the loading from ground and water pressure act on the ring. Ring convergence measurement systems are used in order to monitor the stability of the ring and thus the structural safety of the tunnel.

Systems based on the principle of inclinometers arranged in a measuring ring have great advantages compared to other methods. They work automatically, are very precise, do not require free sight lines like visual methods and can thus measure continuously. One inclinometer is installed in each segment, and the data is transmitted by WLAN. The initial assumption is that the individual segments are stiff in bending and that even deformed rings can be considered as a chain of stiff elements with the longitudinal joints represented as hinges. Evaluations of convergence measurements during stoppages have shown that convergences above 1 mm can be measured reliably.

If absolute deflections, particularly crown settlement or invert heave values, have to be determined instead of just relative deflections, then the inclinometer results are combined with the observation of a fixed reference prism in the observed section. If the absolute changes of level in the crown are combined with the known relative vertical convergence, these can then be split into the components crown settlement and invert heave.

15.4 Steering

The data gained by surveying is displayed in numeric and graphic form for the TBM operator at the control cabin for purposes of steering the shield (Figure 15-1). The operator can then set different pressure levels in the thrust cylinders, which are collected into cylinder groups, to correct the position of the tunnelling machine. Because a shield is heavier at the head, the thrust forces are highest in the invert and conversely the lowest in the crown. Driving around curves requires additional steering forces at the side to ensure the machine follows the intended three-dimensional curve.

In addition to the geodetic survey data described in Section 15.1, which is used to check the direction and position of the shield, various machine and tunnelling data is significant for the steering and monitoring (observation?) of the shield, according to the type of shield (Table 15-1). This is provided by sensors, which convert the physical measurements of most important control elements of the tunnelling machine (distances, pressure, switch and valve positions, angles) into analog signals (current/voltage). The relevant values or information are then displayed on a screen or can also be given as analog values on the operating panel (Figure 15-6).

Table 15-1 Machine parameters shown on the screen in the control cabin

Thrust cylinders	forces, stroke, extension speed, ring assembly pressures
Cutting wheel drive	revolution speed, rotation direction, torque, breakdown torque, overcut, electricity consumption, pressures or position of the cutting wheel displacement
Face support	support pressure, air consumption
Shield tail seal	pressure, sealant quantities
Annular gap grouting	grout quantities, grouting pressures, ideal values
Material transport (according to system/ type of shield)	**Slurry transport** pump revolutions; flow quantities, densities, gate valve positions **EPB operation/conveyor belt** screw conveyor revolutions and torque, gate valve position, **tonnages**, ideal values, electricity consumption, surveillance pictures of material flow or material discharges
Other measurements	temperatures, level monitors, releases/bypasses or error messages

The operators cabin can be located on the surface or in the tunnelling machine itself (Figure 15-7). As manual control should remain possible, for example in case of a breakdown of an automatic control system, current experience shows that the best location is on the backup. For smaller, inaccessible shield diameters (microtunnels), the control position is housed in a container on the surface.

For electronic data logging, further signals are recorded, digitised and transmitted through an interface to a PC in addition to the already listed machine parameters. The PC has software to save the data. In addition to the electronic storage of the tunnelling data, the most important data is normally logged in numeric or graphical form as a ring number-based tunnel log.

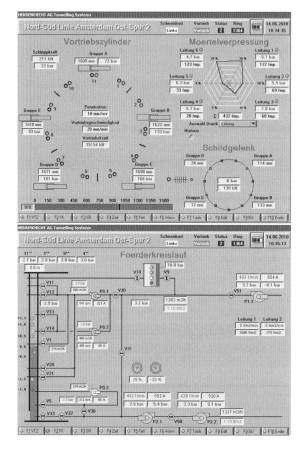

Figure 15-6 Screen display of thrust cylinders and annular gap grouting (top), and hydraulic transport circuit of a Mixshield (bottom)

Figure 15-7 Operator cabin of a Mixshield, Socatop Tunnel, Paris 2006 (Herrenknecht AG)

15.5 Further surveying and data logging tasks

On tunnelling projects, safety and profitability are of great importance. Construction projects should also have a minimal impact on the surroundings despite their magnitude and now have to consider environmental requirements more than ever. Decision makers are therefore faced with various conflicting ideas and methods.

Information systems play a central role in the assessment of the overall situation – the overall context of all past events, presently occurring events or future events with a holistic approach. As systems based on databases with recording, analysis and evaluation functions, these platforms integrate the fields of surveying, geology, geotechnics, mechanical engineering and civil engineering. Data collected at various times and in various places is collected by one system, which can provide the user with up-to-date, non-redundant and secure data. This can be accessed at any time through a uniform structure and ergonomic and multi-language graphical user interface; see also Chapter 18 "Process Controlling".

Tunnelling information systems (e.g. IRIS, 2doc, TPC, SISO, PROCON) offer data administration and monitoring suitable for the relevant user group (site managers, TBM operators) based on parameters, which are either defined or derived from the previous construction process (Figure 15-8). Comprehensive reporting (numerical and graphical display) makes it possible to immediately recognise deviations from the drawings or reactions at the surface. Virtual sensors can be used to generate performance figures for the progress of the projects, which are available in the entire system. Important information is provided in real time – within the system. The responsible parties receive messages on a screen, by email or text message to ensure a safe and cost-effective construction process. At the same time, the information gained can also be useful for continued project design work or machine control. Information systems can supply information for the management of variations (claim management) and help to reduce risks.

Position and time provide the framework for the data and enable combined evaluations. The navigation data of the TBM in combination with the machine data provide the basis for documentation and a wide range of evaluations.

The application of modern measurement technology is becoming ever more significant in tunnelling. New sensors provide improved precision and replace conventional time-consuming monitoring concepts. Surveying tasks include underground and above ground measurements. Underground, not only the position of the TBM is determined but also its spatial orientation (roll, pitch, yaw angle) and additionally the position and orientation of the rings forming the tunnel structure. Deformation measurements are carried out above ground as well as below ground. In addition to manual measurements, automatic systems are also available. The results should be rapidly and transparently available together with other facts like for example machine data, in order to enable constant effective action on the site.

The linking of various professions with one information system helps to ensure low-settlement tunnelling. The geological ground conditions can be displayed visually by advance investigation systems.

15 Surveying and steering

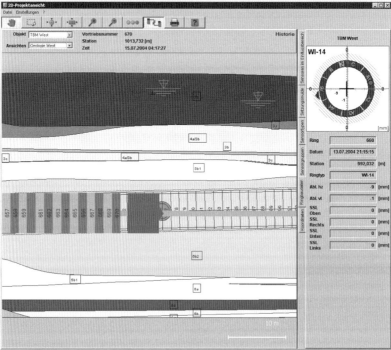

Figure 15-8 Tunnelling information system IRIS

16 Workplace safety

Mechanised tunnelling is inherently more suitable than conventional tunnelling in meeting requirements for ergonomic and humanitarian workplaces. The need to consider profitability criteria has been known for many years, and now quality assurance requirements have to be considered as well.

There are currently no generally valid national laws, standards or guidelines for workplace safety that are specifically applicable to mechanised tunnelling. The regulations applicable to tunnelling should be complied with, particularly the accident prevention regulations and the safety rules of the occupational accident insurers. These are specifically:

- Accident prevention regulations "Construction works" [27],
- Accident prevention regulations "Principles of prevention" [28],
- BG (occupational accident insurer) regulations "Underground works" [24],
- Compressed air regulations [25],
- EN 12 336 "Tunnelling machines – Shield machines etc." [73],
- EN 12 110 "Air locks" [72],
- Evaluation of risk in mechanised tunnelling BG BAU [26].

The brochure from the former civil engineering accident insurer (TBG), today BG BAU, "Practical instructions for safe working in tunnelling" should also be mentioned [23].

The primary source for occupational safety in mechanised tunnelling is the standard EN 12 336 [73] from the European Committee for Standardisation which will combined together with the EN 815 to the new prEN 16 191. This serves as the basis for this chapter and is quoted on the important points, in places with slight formal modification. This should illustrate the stringent requirements for workplace safety in mechanised tunnelling. For the study of general passages or to clarify specialised questions, the interested reader may also have to obtain the original version.

The standard has been prepared by the working group 4 "Pipe jacking and tunnelling machines" of the technical committee CEN TC 151, which is responsible for the standards for construction and construction materials machinery. Delegations from all EU member countries and Switzerland were involved in its production (as of 2005). The secretariat of the technical committee is held by Germany, as is the secretariat of the directly responsible working group. There are three official versions in German, English and French.

The standard applies to all types of shield and associated equipment, horizontal pipe jacking machines and auger boring machines or parts of them. It specifies the essential safety requirements for the design, construction and operation of such machines including methods of verification. Shields include shields for hand and mechanised excavation, tunnel boring machines and microtunnel boring machines. Auxiliary equipment includes towed backup systems, segment installation equipment and pipe lifting equipment (microtunnelling).

Additional equipment, which can form an integral part of a shield or the auxiliary equipment and is used for working under compressed air, is not covered by this standard.

The standard designates significant dangers caused by shields and the associated backup equipment, pipe jacking machines and auger boring machines, when these are used as intended and under the conditions intended by the manufacturer, and gives details of their limitations. When there is a danger of explosive air conditions, as in some mining applications, then special rules apply.

16.1 General safety requirements

This chapter states the requirements and/or measures against such dangers, which seem to demand special handling. If applicable requirements/measures are already included in other standards – particularly in EN 292 [90], in Note A or in Type B standards – then reference is made to these or to relevant sections and/or the applicable work category.

In the design of the machines, the following measures should be considered to achieve safety:

– Description of dangers and estimation of risk,
– elimination of dangers or limitation of risk,
– precautionary measures against identified dangers, which cannot be fully eliminated,
– necessary level of training for the operating personnel of the machine.

Material

Material used in the production or use of the machine should be selected so that danger to the safety of the exposed personnel is reduced. Plastic materials should be flame-resistant or self-extinguishing.

Note from the authors: particular care is required when using compressed air due to the increased fire risk [25].

Contact surfaces

Accessible parts of the machine should be designed and constructed so that contact of an exposed person with sharp edges, angles or rough surfaces, which could possibly cause injury, is avoided [90]. The same applies to hot surfaces [90], [44].

Protection against bursting hoses and pipes

Hoses and pipes, which could burst and cause injury to people, should if possible be permanently secured and protected against external damage and attack. There should be reasonable shielding for personal protection according to ISO 3457 section 4.9 [146].

Cutter head of shields

If it is necessary to gain access through a bulkhead to the area behind the cutter head and similarly through a cutter head to the forward area, then reasonably large access holes should be provided.

In case the shield is used in unstable ground conditions, then appropriate support measures, like supporting or closing measures and/or compressed air should be provided in order to enable safe access for checking and repair works to the cutter head.

The cutter head should be equipped with a braking system in order to avoid unintended movement of the cutting head. This system should be in operation when the cutter head is stopped for other than normal reasons in its working cycle.

Handling heavy loads

When the weight, size or form of parts prevents their manual lifting, then the parts should be fitted with attachments for lifting devices or be designed so that they can be fitted with such attachments, or they should be formed so that standard lifting apparatus can be simply fixed. There should be equipment for transport and lifting of such parts, which require regular replacement, like excavation tools.

If the tunnel lining requires the lifting of parts heavier than 50 kg, than a handling and erecting device should be provided. Its winches and drive motors should be fitted with mechanical brakes, which are switched off during operation.

Certain types of flat-surfaced segments can be handled with vacuum attachment pads. Vacuum-operated lifting devices for segments should be fitted with audible and visual alarms activated by the loss of more than 20 % of the minimum vacuum.

All parts of the equipment used for lifting operations should be designed and constructed in accordance with EN 292-2 [90] if applicable. All hydraulic and pneumatic rams and motors and vacuum attachment pads, which carry load, shall be constructed so that in the event of a circuit or power failure the load is held by a directly coupled, controllable control lock or a brake, which is switched off while in operation [149], [150].

Loss of stability

All shields serve as temporary support during the excavation of a tunnel. They should therefore be designed to resist all loading acting from the surrounding ground together with any loads acting through the process of advancing the shield.

Machine designers and manufacturers should ensure compliance with these requirements for as long as the machine is used at the location, for which it was designed. It is known to happen that machines may be used at locations, which were not intended at the time of their design and construction. In this case, it is the responsibility of the operator of the machine to comply with these requirements. All information belonging to the structural construction of the shield should either be included with the maintenance book or be available from the manufacturer for the lifetime of the shield or for at least 10 years.

When grippers are provided in tunnel boring machines and are in use, the cutter head drive should only be started and the thrust force should only be applied when the machine reaches the minimum necessary gripping pressure to prevent rotation or slipping backwards. Should the gripping pressure fall below its minimum, the rotation of the cutter head and the thrust force should switch off automatically.

All shields can be subject to slow rotation (rolling) due to unequal distribution of forces. Attention should therefore paid in the design and manufacture of shield machines and their auxiliary equipment to avoid eccentric loading, and the shield should be equipped with an effective system for roll correction in order to bring it and the auxiliary equipment back into the right position.

Rapid rotation of a shield can occur when a cutter head or copy cutter is embedded in the working face. All shields should therefore be fitted with a protection device, which cuts off power to the drive motor in the event of rapid rotation of the shield.

Access to operating locations and for maintenance work

The manufacturer should provide means of access (stairs, ladders, walkways etc.) in order to enable safe access to and exit from all working areas. Areas of a machine, in which people work, should be designed and constructed so that the risk of falling is minimised and collisions are avoided. The areas of the machine and auxiliary equipment, in which the handling of material to and from the tunnel transport system takes place, shall be designed and constructed so that safe access and exit for the personnel is guaranteed. The instruction handbook should include detailed dimensions of the intended tunnel transport system and the support system.

Walkways inside the machine or between the machine and the tunnel wall should have a minimum cross-sectional area of $0.5\ m^2$, within which there should be a free rectangular body opening of at least 700 mm height and at least 450 mm width. The walking surface of the walkway should be at least 300 mm wide. Handrails and toeboards should always be provided when the walking surface is less that 300 mm wide. Walkways should not be closed by access openings or storage areas for materials or equipment. Alterations of the level or direction of the walkway should be avoided wherever possible. Where rampsteps or stairways are necessary, handrail or handholds should be fitted on at least one side of the walkway.

If the machine is too small to fit stairs or ladders, it should be equipped with handholds and steps, non-slip surfaces and fixing points for safety lines. Reference can be made to ISO 2867 sections 6 and 7 [145] and ISO 2860 [144] as instructions, but the requirements mentioned above have precedence.

Access openings, e.g. in the bulkhead or cutter head, should have a free body opening with a diameter of 400 mm within an area of $0.2\ m^2$. Wherever feasible, these dimensions shall be increased.

Protection against falling objects, face collapse and flood

The control stations of machines should be provided with protective roofs in order to provide reasonable protection for the operating personnel. This does not apply when a control station is in an underground area inside the tunnel lining or under a reasonably protected slab. The types of intended lining and slab support should be given exactly in the operating handbook. Exposed people, who undertake support work, should be protected in a similar way. The test of the adequacy of a protective roof should be verified by calculation. The criteria for impact protection according to EN 23 449 section 5.4 [89] should be complied with.

All sensitive components like cables, transformers, hydraulic pipes and hoses, which are exposed to a risk from falling objects, should be protected where this can be implemented within reason.

In tunnels with an open working face, there is always a danger of a collapse of the face. All shields, in which excavation can be undertaken with an open face, should be equipped with a face support system, which can cope with the foreseeable ground conditions. This could be hydraulically operated breasting plates, platforms or in simple shields, timber breasting.

There is a risk to be taken seriously of injury or drowning when the tunnel or shaft is flooded. All shields should therefore be designed so that appropriate pumping equipment for all foreseeable conditions can be installed. Shield machines and auxiliary equipment should be equipped with a clear and effective means of escape for the entire operating personnel.

The authors would like to suggest here that a submerged wall should be provided in every shield. In case of unexpected water inflow, this can offer a safe room for a certain time.

16.2   Control stations

The control station should be placed in a cabin or protected in order to prevent the operating personnel being exposed to mechanical hazards, harmful dust, gases, vapours or excessive noise emissions. The cabin should be constructed, built and/or fitted out so that safe working conditions (e.g. ventilation, view, noise reduction, lack of vibration, protection from falling objects, heating or cooling) are provided for the operators and that rapid escape is possible. The materials used for the control position should, where possible, be fireproof and not develop toxic vapours in case of a fire.

The control station should be ergonomically constructed in order to hinder tiredness and stress of the operators. This should include consideration of the fact that operators may be wearing heavy gloves, boots and other personal protective clothing. When the construction permits, the dimensions according to EN 23 411 [88] should be used. The seating should, after appropriate adjustment, offer the operator a constant stable and comfortable position at all times. The seating should be constructed so vibrations transferred to the TBM operator are reduced to a reasonably low extent.

The view from the control station should be so that the TBM operator can operate the machine under the intended conditions of use in complete safety and with complete safety of other exposed personnel. Where necessary, optical aids including closed-circuit television should be provided in order to assist in cases where the direct view is inadequate. In case the entire machine is not visible from the main control position, indicator lights or similar devices should be installed in the control station to show that the machine is ready for operation, that all auxiliary equipment required for operation is ready or started, that the monitoring system is working and that no emergency switches have been activated.

16.3 Electrical cut-out and safety devices

All movable parts of the machine should generally be fitted with electrical cut-out and protection devices in order to avoid the risk of contact between exposed people and movable parts, which could lead to accidents. If this is not practical for technical or economic reasons, appropriate warning signs should be displayed on the machine, which are easily visible, or the movable parts should be painted in warning colours [148]. Reference is made to [90] for the description of suitable safety protection.

Access to the cutter head

Access to the area where there can be contact with the cutter head, should only be possible after the cutting head has been brought to a complete stop. Access to the cutting head should be protected with interlocked doors and guards [90]. This requirement can be eased for machines during creep or stepwise operation (Section 16.4).

Transport equipment

Drive and guide rollers of a conveyor should be fitted with safety devices to prevent parts of the body coming between the moving belt and the idlers. The idlers of the upper and lower belts should be equipped with safety devices, except when the conveyor system is completely enclosed. Safety devices should be effective when the conveyor is equipped with a loading system. For requirements concerning emergency switches, see section 16.4.

If a conveyor belt serves as access to the cutting head, it must be possible to stop the belt or in case it is a conveyor at less that 6° to the horizontal, operate it at creep speed controlled from an additional control position near the conveyor.

16.4 Control devices and control systems

Control devices should:

- be clearly visible and identifiable and if appropriately marked where necessary,
- be situated for safe operation, for example so that unintended activation of nearby operating devices is avoided,
- be located close to each other, when the start and stop switch functions are operated by the same control device,
- be equipped with safety devices if dangerous movements could be caused by unintended activation,
- be positioned so that frequently used operating elements are within easy reach of the TBM operator,
- be laid out so that the arrangement and the movement agree with the action to be carried out and be ergonomically constructed,
- provide a "hold-to-run" function for creep or jog operation of a cutter head or conveyor belt,
- be of robust construction appropriate for the operating conditions.

If an operating device is designed and built to perform a number of processes, then the process that is to be performed should be clearly displayed and if necessary confirmed.

Each machine should be equipped with such display devices (scales, signals etc.) as are necessary for safe operation. The TBM operator should be in a position to read these from the control position.

The control system should be constructed and built so that it is extremely reliable in operation in an underground environment and that dangerous situations are avoided in case it is defective. It should resist rough handling and heavy loading and vibration.

In particular:

- the switching on of drive motors for hydraulic pumps should not in any way lead to uncontrollable movements, which could be dangerous,
- no dangerous operating conditions should occur in case of a power failure,
- the failure of hydraulic or electrical operating switches should not lead to any unexpected or unintended movement of any part of the machine,
- it should be possible to individually start and stop all different operating functions, which are normally started or stopped in sequence, during maintenance work.

Each control station should be equipped with a switch controlled by a key, which can switch off and prevent the operation of all systems controlled from the station. This key switch should function so that all systems, which can be operated from the control station, can be stopped automatically in a safe manner. A shield for full-face excavation should be provided with an auxiliary operating panel, which is connected directly to the cutter head at a connection box. This auxiliary operating panel should have precedence over all other operating functions, which control the cutter head, and only enable the positioning of the cutter head in creep or jog operation. The connection of the auxiliary operating position should be locked with a key.

If creep or jog operation is activated from the auxiliary operating panel, a warning signal should sound as long as the creep operation continues.

Switching on and off

Machines should only be started by intended activation of the operating device provided for the purpose. There should only be one switch, which should be situated on the main control board and any switch in the auxiliary equipment should be overridden by this switch. After the machine has been stopped, it should only be restated by one switch.

An audible maintenance warning system should be installed on quickly moving machinery like open conveyor belts. The start signal should be electrically connected to ensure a minimum advance warning time of 15 s. In shields, which permit open access to the cutter head, a similar audible warning system should be installed and electrically connected in order to provide a minimum advance warning time of 10 s before operation. Whenever the backup equipment moves independently of the shield, a warning system should be fitted, which provides a minimum advance warning time of 10 s.

Machines should be equipped with an operating device, which can be used to bring them safely to a complete stop. Each control station should be equipped with an operating device, to stop one or all of the moving parts independent of the type of danger. The stop controls should have priority over the start controls.

Emergency stop

Electrical or electrically controlled hydraulic equipment, including auxiliary equipment, should be equipped with emergency stop devices, which can include trip wires [90]. Emergency stop devices should be mounted where they can be used to reduce danger, particularly at the main control position and at auxiliary control stations.

Emergency stop devices should comply with EN 418 [92] and particularly:

- be operated in separate electrical circuits,
- be simply and safely accessible, at least at the workplace of the main operator and in each potential danger zone or at the access point to the cutter head,
- lock mechanically in the off position and only be released intentionally at the switch, which was activated,
- be situated near the high-voltage transformer, where they are fixed in order to break the high-voltage circuit.

If central hydraulic or pneumatic operating devices have no emergency stop device, they should return automatically to the neutral position when they are not used.

Electric power failure

The interruption, resumption of supply after an interruption or the flow of any type of current to the machine shall not lead to a hazardous situation. The following should be particularly noted:

- After the re-establishment of the power supply, the machine shall only start after intentional actuations of the start controls,
- the machine shall not be prevented from stopping if the command to stop has already been given,
- the safety measures, which controls access to the cutter head, shall remain fully effective.

16.5 Towing connections

All towing connections between the shield machine and the backup should be designed and constructed so that they can bear the entire tension load with the following safety factors:

- Chains: 4 times maximum breaking load,
- bars etc.: 2 times maximum yield stress.

When two connections are used, each should be capable of bearing the entire tension load. The following friction factors should be applied for the calculation of the tension force:

- Sledge mounted backups: $\mu = 1.0$,
- Backups supported on rails: $\mu = 0.3$.

For applications in tunnels and inclined shafts with more than 15° angle to the horizontal, a second brake should be provided.

Steel cables should not be used as towing connections, with the exception of between backup equipment and auxiliary equipment, like for example California switches.

If the auxiliary equipment is not moved simultaneously with the machine, than an acoustic or optical signal should be activated, before the auxiliary equipment is moved.

16.6 Laser guidance

When a machine is guided by laser, the laser window should be positioned so that the exposure of the eyes of personnel to the laser beam is minimised. Appropriate warning signs shall be provided.

16.7 Ventilation and the control of dust and gas

All shields should be designed to include adequate ventilation, dust suppression and collection equipment. Details of the size and type of the equipment provided should be stated in the operation handbook.

Exhaust gases

Whenever possible, no internal combustion motors should be used permanently on the shield. Should an internal combustion motor be used as part of backup equipment, screw conveyor or pipe jacking equipment, this must be driven by diesel or LPG. All reasonably practical precautions should be taken to prevent the emission of toxic exhaust gases into the tunnel. Diesel-operated, hydraulic power supplies for pipe jacking equipment, microtunnelling machines and screw boring machines should be stationed outside the tunnel or shaft opening. They should be positioned so that the emission of toxic exhaust gases inside the tunnel is minimised.

Alteration of the air conditions

Gas ingress from the surrounding ground can represent a physiological danger in tunnelling. These gases can be toxic or flammable or simply reduce the oxygen concentration in the normal air to a dangerous level.

All shields (except simple shields) should be equipped with air monitoring equipment, which is capable of detecting low oxygen concentrations and flammable gases. Precautions should be taken to provide room for additional equipment to monitor toxic or radioactive gases (e.g. radon), which could occur under the given ground conditions. In simple shields, measures should be provided for the installation of a portable air monitoring system.

In all machines, the sensors of the monitoring equipment should be mounted as near as possible to the face. On large machines, additional sensors can be necessary, distributed over the machine and the backup. In microtunnelling machines, sensors should be installed on the pipe jacking equipment and in operating containers on the surface, whenever these are sited over the working shaft. If air locks are used, precautions should be taken to provide portable measurement equipment inside the bulkhead for inspections, and when reasonable inside the pressurised working chamber.

All air monitoring equipment should emit a visible and audible warning of dangerous concentrations of toxic or flammable gases or in case of oxygen deficit to a dangerous degree. If high concentrations of flammable gases can be expected, then the monitoring equipment should be coupled to the machine and all electrical and mechanical devices, which are not explosion-proof, should automatically switch off when the concentration reaches 1.5 %. In these circumstances, equipment for dust removal, additional ventilation, emergency lighting and communications should be designed to be explosion-proof.

Explosive air conditions

If the concentration of flammable gases cannot be kept below 0.25 % by positive-pressure ventilation, the shield and auxiliary equipment should have explosion-proof electrical equipment [93].

All machines, which are built and used for these operating conditions should comply with the appropriate standards for equipment in an explosive atmosphere and coal mines.

16.8 Fire protection

The outbreak of a fire in a tunnel represents another serious risk. All shields should be equipped with clearly signed and effective means of escape from all workplaces. They should be designed as far as possible so that no combustible materials are used in their construction. Cushioning and insulation should be made of fire-resistant material, which permits a linear propagation of flames of a maximum of 250 mm/min, tested according to ISO 3795 [147].

All mineral hydraulic fluids should contain a vivid dye so that leaks are quickly visible. Tanks, which contain mineral hydraulic fluids, should be equipped with warning systems to indicate low and excessively high oil levels. All hydraulic systems, which contain mineral oil, should be designed so that the loss of oil is minimised in case of a breakage of a part and the breakage is indicated early.

Shields and auxiliary equipment should be equipped with fire extinguishing systems or alternatively with fire extinguishers with a mass of an appropriate extinguishing agent of not less than 6 kg. The fire-extinguishing medium should be suitable for all classes of fire. The fire extinguishing system should comply with the requirements of EN 3 [91].

Fire extinguishing systems should be distributed on the shield and its associated equipment, particularly at danger points like the workplace of the TBM operator, near the main drive motor for the cutting head, at hydraulic drive motors, at electric cabinets and transformers. Fire extinguishers should be positioned between the personnel and the areas where fire could break out. They should be easily accessible and the access should never be blocked. Fire extinguishers should be fixed so that no tools are required to remove them from their holders.

Note from the authors: when compressed air is used, this causes an increased fire risk (see compressed air regulations [25]).

16.9 Storage of safety equipment for the personnel

If there is enough room, stores should be provided on the backup for the safety equipment of the crew, e.g. for blankets, stretchers, breathing apparatus etc. The store should be clean and protected from dust and damp. It should be clearly marked.

16.10 Maintenance

There is a danger of injury to maintenance personnel in the restricted space inside the shield and during work to the cutter head. All shield machines and auxiliary equipment should therefore be built so that adjustment, lubrication and maintenance work can be carried out without danger for the affected personnel. Whenever possible, such machines should be built so that adjustment, maintenance, repair, cleaning and supply work can be carried out while the machine is stopped.

When access to the cutter head of a machine is necessary, access routes and bulkheads should be provided. Wherever it is possible, maintenance should be carried out in the protection of the shield.

When it is necessary to enter the area between the cutter head and the face (excavation chamber), suitable ground support measures should be undertaken to support the face and the crown. Before personnel enter this space, and while they are in the space, the entire power supply to the drive motors of the machine shall be switched off.

16.11 Content of handbook

Manufacturers of shields should provide operating instructions with information for safe operation and maintenance of the machine as part of the delivery. The operating instructions are part of the shield and are important documents. The text should be simple, reasonable and complete. The formulations should correspond to the level of training and knowledge of the personnel. The information should be comprehensive and clear. All information, which is relevant to personal safety, should be in a prominent typeface, different from the rest of the text. These instructions should give extensive information about all matters, which are covered in EN 292-2 section 5.5.1 [90] and all dangers together with all information relevant to the reduction of risk to personnel. They should also contain a description of the ground conditions and loading conditions, for which the machinery is designed.

When reasonable, the operating instructions should also include suggestions regarding operating conditions other than those for which the machine was built. If the instructions assume a certain level of competence and experience from the operator, then this should be clearly defined.

The operating instructions should include a section for a machine maintenance logbook, in which details of significant modifications can be entered.

The details regarding occupational safety in the operating information are detailed in CEN TC 151/WG4 N8 [46].

16.12 Evaluation of risk in mechanised tunnelling [26]

The Evaluation of risk in mechanised tunnelling [26] was issued in April 2009 as a cooperative effort of the main association of the German construction industry and the accident insurance body responsible for construction BG BAU and is intended to avoid any accidents. The following are required:

– Recognition,
– evaluation,
– remediation and
– steps for an evaluation of the danger on each project.

The index quoted on the next pages repeats the entire content of the Evaluation of risk in mechanised tunnelling [26].

As an example, an EPB drive is quoted, furthermore the special conditions for compressed air working for interventions in the excavation chamber under compressed air.

The authors of this book consider this systematic and expertly grounded procedure as an important basis for risk analysis.

The pages quoted here as excerpts are intended to show the level of expert competence applied in the production of the text regarding danger evaluation. The complete work is available on a CD-ROM from the BG BAU.

16.12 Evaluation of risk in mechanised tunnelling [26]

 BFA Underground Construction

Evaluation of Risk
Mechanised Tunnelling

 BG BAU Berufsgenossenschaft der Bauwirtschaft

Index

Module	Catalogue content	Pages
	Preamble	1
	Index	2
TVM 1	Construction implementation	2
TVM 2	Construction site works	
2.1	General site facilities	6
2.2	Lone workers	2
TVM 3	Infrastructure / specialised site facilities	
3.1	Infrastructure / traffic routes	2
3.2	Electrical installations	5
3.3	Machinery	2
TVM 4	Surface production locations	
4.1	Separation	1
4.2	Batching plant	3
4.3	Materials handling	3
4.4	Workshop / stores	4
4.5	Segments – reinforcement	2
4.6	Segments – concreting	1
4.7	Segments – formwork setting and striking	2
4.8	Segments – fitting out	1
4.9	Segments – intermediate stores	1
TVM 5	Regular tunnelling	
5.1	Assembly, conversion and dismantling	10
5.2	Shield start, shield exit	2
5.3	Slurry shield tunnelling	1
5.4	EPB shield tunnelling	2
5.5	Compressed air shield tunnelling	2
5.6	Hard rock shield tunnelling	4
5.7	Ring assembly	2
5.8	Annular gap filling	2
5.9	Materials handling	3
5.10	Pipe extension	3
5.11	Track extension	1
5.12	Rail operation and rebuilding	3

| | BFA Underground Construction | Evaluation of Risk Mechanised Tunnelling | | BG BAU Berufsgenossenschaft der Bauwirtschaft |

Index

Module	Catalogue content	Pages
TVM 6	Special tunnelling events	
6.1	Changing the shield tail seals	2
6.2	Maintenance incl. tool changing	4
6.3	Compressed air working	7
6.4	Forward probe drilling	1
6.5	Cross-passages	2
6.6	Making good of concrete linings	2
TVM 7	Pipe jacking	
7.1	Advance	8
7.2	Pipe installation	1
7.3	Pipe stores	4
7.4	Compressed air working, working in the excavation chamber	8
8	Important application notes	1
9	Empty hazard list (sample)	2
10	Imprint	1

16.12 Evaluation of risk in mechanised tunnelling [26]

	BFA Underground Construction	Evaluation of Risk **Regular Tunnelling**	BG BAU Berufsgenossenschaft der Bauwirtschaft
Page 1 of 2		**TVM 5.4**	Company:
Issue 04/2009		**EPB tunnelling**	Site:

Hazard group	Hazard	Measures	Checked
Auger	Squeezing/trapping	☐ When working on the screw (e.g. to remove obstructions), switch off motor and secure against unintended restarting	
	Suffocation	☐ When welding or cutting work is carried out to the screw, ensure adequate ventilation and/or extraction	
	Injury through uncontrolled air escape with soil	☐ Ensure sufficient filling of the excavation chamber	
Conveyor belt	Impact Entanglement	☐ Provide protection against material falling off, e.g. trough ☐ Provide belt scraper ☐ Secure guidance idlers against direct contact ☐ Protect idlers against direct contact	
	Slipping/falling	☐ Clean access routes regularly	
	Dust	☐ Reduce dust by extraction or water jetting ☐ Clean belt sweeper	
	Fire	☐ Use fire-resistant belt material ☐ Maintain conveyor idlers	
Loading soil	Vehicle impact slipping/falling	☐ Secure loading area against entry ☐ Provide falling protection ☐ Regularly clean transport routes	
Foam/ polymers/ tensides	Eye damage/ skin damage	☐ Observe safety data sheets ☐ Use protective goggles ☐ Avoid skin contact, e.g. disposable overall	
Stators/rotors	Trapping/impact Falling	☐ Clean excavation space from top to bottom on entering ☐ Use platform and ladder access	

⟁ BFA Underground Construction	Evaluation of Risk **Regular Tunnelling**	BG BAU Berufsgenossenschaft der Bauwirtschaft
Page 2 of 2	**TVM 5.4**	Company:
Issue 04/2009	**EPB tunnelling**	Site:

Hazard group	Hazard		Measures		Checked
	Date	taken by	Company / representative of the company		
Creation			.... Place, date		
Implementation					
Instruction			.... Signature		
Checked			.... Name in capitals		

16.12 Evaluation of risk in mechanised tunnelling [26]

	BFA Underground Construction	Evaluation of Risk Special Tunnelling Events	BG BAU Berufsgenossenschaft der Bauwirtschaft
Page 1 of 7		TVM 6.3	Company:
Issue 04/2009		Compressed air working	Site:

Hazard group	Hazard		Measures	Checked
Compressed air lock operation and working in the excavation chamber				
Compressed air support	Unstable working face	☐	Investigate and secure the alignment with regard to natural and artificial fault zones (e.g. peat gullies, wells, boreholes, sewers etc.)	
		☐	Support pressure calculation for air, slurry support and if appropriate additional mechanical support, shotcrete	
		☐	....	
	Blowouts	☐	No entry if there is a risk of blowouts	
		☐	Support pressure calculation for air, slurry support and verification of safety against collapse	
		☐	Additional measures, e.g. surcharge blanket on the surface, ground improvement measures (e.g. reduction of the pores volume)	
		☐	....	
	Compressed air loss	☐	Monitoring of air quantity to detect unusual demand	
		☐	Sealing of the face	
		☐	Check the pressure bulkhead, closing gates, pipework etc. for air leaks	
		☐	....	
	Breakdown of compressed air supply (mechanical, power failure)	☐	Provide a redundant system in compliance with the German compressed air regulations	
		☐	....	
Personnel lock: workplace of the lock attendant	Posture injury forced posture	☐	Provide an ergonomic workplace	
		☐	....	
	Inadequate lighting/ eye damage	☐	Provide adequate lighting	
		☐	....	
	Lack of view into the lock	☐	Suitable location of the workplaces at the lock	
		☐	Use camera, provide window	
		☐	....	
Personnel lock: compressed air worker	Forced sitting posture	☐	Maintain structure gauge	
		☐	Provide ergonomic seating for dynamic sitting	
		☐	....	

▲ BFA Underground Construction	Evaluation of Risk **Special Tunnelling Events**	BG BAU	Berufsgenossenschaft der Bauwirtschaft
Page 2 of 7	**TVM 6.3**	Company:	
Issue 04/2009	**Compressed air working**	Site:	

Hazard group	Hazard	Measures	Checked
Working in the excavation chamber	Hitting head	☐ Maintain free space above seats ☐ Keep clear height of lock free of obstructions ☐	
	Compressed air sickness	☐ Reduce the frequency of entry through technical solutions e.g. use of wear resistant materials, wear dedection, maintenance free/low-maintenance systems ☐ Provide suitable tools for work ☐ Improve the repair features of the machinery ☐	
	Impact by material from the face, blow-outs	☐ Mechanical protection with „closed cutting wheel", safety plates or similar ☐ Constant observation of the face, stopping work and refilling the excavation chamber with bentonite suspension or earth slurry if necessary, ☐ When support measures are used, separate evaluation of danger for the individual case ☐	
	Drowning	☐ Constant observation of the face and the suspension/water level ☐	
	Lack of escape route or escape route blocked	☐ Access for all workers into a lock chamber at all times ☐ Escape routes must be kept clear ☐ All doors must be immediately lockable ☐ Sealing gaskets must be kept clean ☐ No pipes and cables in lockable doors/openings ☐	
	Lack of access for helpers	☐ Ensure free access for assistance workers at all times ☐	
	Falling	☐ Use working platforms including mobile platforms if necessary ☐ Use ladders ☐	

BFA Underground Construction	Evaluation of Risk **Special Tunnelling Events**	BG BAU Berufsgenossenschaft der Bauwirtschaft
Page 3 of 7	**TVM 6.3**	Company:
Issue 04/2009	**Compressed air working**	Site:

Hazard group	Hazard	Measures	Checked
Working in the excavation chamber	Fall/ falling/ slipping/ drowning	☐ Clean access with water in advance ☐ Use working platforms including mobile platforms at workplaces ☐ Fall protection ☐ Provide safe working conditions for assembly/dismantling works with mobile platforms ☐	
	Lifting excessive loads/ bodily injury	☐ Use lifting devices and connection points ☐ Test and practice before putting into use ☐	
	Uncontrolled movement of the cutting wheel, excavator, rock crusher or screw conveyor	☐ Operation only possible from the excavation chamber with good view ☐ No parallel working ☐	
	Injuries while using high-pressure lances	☐ Instruction / training ☐ Personal protection equipment ☐ No working in front of the device ☐	
	Eye damage (metal splinters)	☐ Eye protection ☐	
	Chilling through standing in the bentonite suspension	☐ Suitable waterproof, warm protective clothing ☐ Limitation of working time ☐	
	Inadequate air quality, temperature, humidity	☐ Provide adequate ventilation / cooling ☐ Limitation of working time if required ☐ If appropriate, additional medical ("hot working") ☐ Monitor the temperature of the excavation chamber ☐	

BFA Underground Construction	Evaluation of Risk Special Tunnelling Events	BG BAU Berufsgenossenschaft der Bauwirtschaft
Page 4 of 7	TVM 6.3	Company:
Issue 04/2009	Compressed air working	Site:

Hazard group	Hazard	Measures	Checked
Environmental conditions	Air pollution through oil vapour	☐ Regular maintenance of pneumatic devices ☐	
	Poisoning	☐ Check toxicity of substances (e.g. polymers) ☐	
	Drying of the body	☐ Provide warm drinks ☐	
	Chilling of the body	☐ Provide warm clothing ☐ Avoid contact with cold parts (e.g. steel) ☐	
	Forced posture	☐ Avoid kneeling work ☐	
Passing out through lock	Forced posture	☐ Encourage movement ☐ Dynamic sitting ☐ Prevent sleeping ☐	
Transport of injured persons	Lack of rescue plan	☐ Produce escape and rescue plan (fire services / emergency services)	
	Lack of transport for injured persons	☐ Provide a stretcher for the personnel lock ☐ Provide personnel lock with sufficient space (length, width, door openings etc.) ☐	
	Lack of access for rescue services	☐ Ensure access of doctor/first aider through personnel lock ☐	
	Insufficient space for passing uninjured through lock	☐ Provide adequate capacity of the lock(s) ☐	

BFA Underground Construction	Evaluation of Risk **Special Tunnelling Events**	BG BAU Berufsgenossenschaft der Bauwirtschaft
Page 5 of 7	**TVM 6.3**	Company:
Issue 04/2009	**Compressed air working**	Site:

Hazard group	Hazard	Measures		Checked
Welding and cutting work under compressed air				
Welding and cutting work under compressed air: fire	Poisoning by toxic gases (nitrous gases)	☐	Ensure provision of adequate air for breathing	
		☐	Ensure additional extraction of smoke gases	
		☐	....	
	Burns	☐	Provide suitable protective clothing for welders and assistants	
	Electric shock	☐	Wear protective gloves	
		☐	Provide isolating layers	
		☐	Wear dry protective clothing	
		☐	Switch off before changing electrodes	
		☐	Training of personnel	
		☐	....	
Welding and cutting work under compressed air: general	Hydrogen detonation	☐	Provide ventilation	
		☐	....	
	Scalding by hot water	☐	Relief drillings	
		☐	Provide suitable protective clothing against scalding for welders and assistants	
		☐	....	
	Lack of escape routes or escape routes blocked	☐	Keep escape routes free of equipment and materials	
		☐	Do not lay pipes or cables in door access	
		☐	....	
	Explosion/burns	☐	Store gas bottles outside the pressurised area if possible	
		☐	Limit gas bottles to the necessary quantity (no storage, no reserve bottles)	
		☐	Safety post at the bottles	
		☐	Fire watch provided with extinguisher	
		☐	Personnel in danger zone must wear fire-resistant clothing	
		☐	....	

	BFA Underground Construction	Evaluation of Risk Special Tunnelling Events	BG BAU Berufsgenossenschaft der Bauwirtschaft	
Page 6 of 7		TVM 6.3	Company:	
Issue 04/2009		Compressed air working	Site:	

Hazard group	Hazard		Measures	Checked
Diving work under compressed air (diving in bentonite suspension)				
Divers working in suspension: general		☐ ☐	Diving work should only be carried out under the conditions that air support cannot be verified but slurry support can be verified See module M22 (CD "Compendium of occupational safety" from BG Bau)	
Separation of compressed air working and diving work	Legal uncertainty, which regulations are applicable (diving or compressed air)	☐ ☐	Report the works to the responsible authority and accident insurer with statement under which rules is being worked	
Diving work in bentonite suspension	General	☐	Produce dedicated risk analysis	
	Lack of connection to the diver (air, communication, working materials etc.)	☐ ☐	Provide and check diver supply flange	
	Injury through inadequate face stability	☐ ☐	Produce structural verification of face stability	
	Lack of escape route or escape route blocked	☐ ☐ ☐	Analyse escape routes Train divers	
	Suffocation through sticking of the breathing out valve (bentonite)	☐ ☐	Use water flushing of the valve	
	Skin injury	☐ ☐ ☐	Use dry diving suit Regular cleaning and care of the diving suit	
	Lack of visibility	☐ ☐	"Dry run" before starting work	

Notes:
- Tool changing, see Section TVM 6.2
- Resources for the analysis of risk in "General compressed air working" is available in "Module 19" of the "Compendium of occupational safety" from BG Bau
- Diving with mixed breathing gas and saturation diving are not included in this evaluation of risk and have to be dealt with separately

BFA Underground Construction	Evaluation of Risk **Special Tunnelling Events**	BG BAU Berufsgenossenschaft der Bauwirtschaft
Page 7 of 7	**TVM 6.3**	Company:
Issue 04/2009	**Compressed air working**	Site:

Hazard group	Hazard		Measures	Checked
	Date	taken by	Company / representative of the company	
Creation			.... Place, date	
Implementation				
Instruction			.... Signature	
Checked			.... Name in capitals	

17 Partnering contract models and construction

17.1 Introduction

Difficult competitive conditions, tenders based on performance specifications or with gaps and the interactions between construction process and surrounding ground, which are difficult to estimate, often lead to under-priced bids, extra costs and contractual disputes in highly mechanised tunnel construction. Risk transfer to the other party, aggressive management of variations and the lack of systems of payment incentives can sustainably restrict the optimisation of the construction process. The geological conditions described by the employer are often a source of disputes. An international outlook confirms that the German contract model is no longer appropriate and requires urgent revision.

In public sector tunnelling, pure partnering models or alliance contracts, in which employer and contractor form a virtual project company, are scarcely conceivable in the middle term. This is due to the current complicated process of design, approval and tendering. Important characteristic features of the partnering model can, however, be integrated into the existing contract model. After the completion of the approvals process, the most suitable bidder can be determined through a competitive dialogue.

Partnering models require contract documents with a much higher level of detail based on an exhaustive and contractually unambiguous description of the works, complete bill of quantities and specification and a fair and contractually backed-up distribution of risks. In the future, the demand for a detailed breakdown of work and costs could hinder under-pricing and improve the common control of costs during the construction period (open books principle).

The difficulty of forecasting the geological conditions demands a particularly high degree of technical and contractual flexibility in tunnel construction. The Austrian and Swiss contract models include detailed procedures for tunnelling and have already been implemented in codes. In contrast to the German contract model – but also to the pure partnering model – the design is normally part of the sphere of responsibility of the employer in Switzerland and Austria. If the two contract partners work together from an early stage of the design in the future, then the performance and invoicing basics can be produced in collaboration for the specific project. Innovative payment systems are certainly suitable for performance- and time-dependant bill items in mechanised shield tunnelling. The existing methods of the Swiss and Austrian models can serve as a starting point and the details worked out for the specific process.

Cooperation in the future will demand the active involvement of the employer in all decision-making processes. A high degree of expert knowledge on the employer's side and transparency are imperative preconditions. Constant collaborative analysis and checking of the system behaviour (the interaction of the construction process with the surrounding ground) should therefore be included in the existing contract framework. Information asymmetries can be relieved through the implementation of a system-oriented knowledge/information management system. The collaboratively produced target-actual performance comparison (mutual project controlling; Chapter 18) of all processes at the level of cost, time and safety in real time enables the optimisation of the construction progress and the clarification of the sphere of risks in terms of the partnering philosophy in mechanised tunnelling.

17.2 Requirements for the contract model

Tunnelling in the public sector, but also in the private sector, is significantly different from conventional civil engineering and requires specialised contract models. In recent decades, hundreds of kilometres of tunnels have been bored through countless geological and hydrological combinations all over the world. Highly mechanised shield tunnelling today is equivalent to an industrial production process in many aspects with a constantly repeating sequence of operations. This standardisation of the construction process and today's technical methods for the precise monitoring of construction progress offer an opportunity for the production of unified contract conditions and a chance for innovative contract models based on the partnering philosophy.

In the development of new contract models, the special features of tunnel construction should be taken into account, as also the weak points of existing contract models. In the past, considerable cost and time overruns have occurred with both performance specifications and tenders including the detailed description of the works with specification and bill of quantities. One regular source of disputes between employer and contractor was and still is above all the unambiguous and exhaustive description of the works required by contract law according to VOB/A as well as the distribution of risk based on it. The ground surrounding a tunnel is at the same time a construction material, whose behaviour is often difficult to estimate. The behaviour of the ground is of decisive importance for tunnel projects, as not only the selection of a construction process but also numerous other contractual aspects depend on it [55]. This includes unit prices, estimated quantities as well as extra work. Deviations of the properties of the ground often lead to an alteration of the expected performance and thus to a justified claim for extra payment by the contractor. Changes of ground conditions are also fairly often used to justify increasing the invoice sum or the assertion of an unfounded alteration of performance according to VOB/A.

The next section first discusses the weak points of the current model of VOB/A (German construction contract procedures - award). Numerous disputes between employer and contractor have their origins in structural under-pricing of tenders. The specific chances and opportunities of partnering for tunnel projects, based on trust and learning are explained as well as the possibilities for practical implementation in the existing contract framework.

17.3 Contract model according to VOB

According to § 9 VOB/A, the works can be described either as "description of the works with bill of quantities and specification" or as "description of works with performance specification". A performance specification should, however, only be used in exceptional cases. This is the case when the design of the works is to be subject to competition in addition to the construction [58].

In this context, the geological risk proves particularly problematic in contract law. According to the VOB/A § 9, the ground belongs fundamentally to the risk of the employer. This means that the employer has to investigate the ground conditions as extensively as possible [56]. This applies particularly to a performance specification, as this assumes an exhaustive description of the ground. In practice, however, it is often the case that the use of a performance specification rather tempts the employer to reduce the costs of site investigation [77].

The use of a performance specification therefore tends to be advised against for projects using highly mechanised tunnelling machinery. The reasons are: the complex interaction between ground and construction process, lack of transparency in the evaluation and comparison of tenders and disputes when the expected conditions are not encountered.

In the past, numerous tunnel projects have nonetheless been tendered with a full or partial performance specification. Experience shows that this procedure can indeed be successful as long as a number of preconditions are satisfied:

1. The price is right.
2. The ground and system behaviour is according to expectations.
3. Employer and contractor are competent.
4. There is a good relationship between employer and contractor.

It will naturally be noted that under these conditions, any contract model would probably work.

An alternative or possible extension of the performance specification is the unit price contract. Experience shows that this entails considerable risk, as standard bills of quantities and a catalogue of unit prices that are not specific to the project can in no way replace detailed analysis and estimation of the project.

The main reason for the disturbed relationship is the awarding of contracts to under-priced bids in the market situation, which has been difficult for years. Insufficient bids sometimes even negotiated lower still in the award negotiations and risks are intentionally or unintentionally transferred into the sphere of the contract partner. This makes conflicts of interest inevitable during the construction phase. The current market situation in the construction industry and the influence of globalisation, but also the revision and introduction of European laws governing the awarding of contacts demand a fundamental modification of the contract models.

17.4 Time and cost drivers

Independent of the contract framework, time and cost overruns of over 10 % are not rare in mechanised tunnelling. In order to assert a claim for extra costs, the contractor normally cites the VOB/A. This states that the work should be described clearly and completely so that the contractor is not burdened with any unusual risk. If, however, there are no basic errors, defects or gaps in the description of the works and the bill of quantities, deviations of time and cost from the planned figures are assigned to the geological conditions on the exclusion principle. The following quote [94] characterises the contractual positioning of the contractor:

> *"If the question is in dispute on a tunnel construction project, what was the cause of extra costs or a schedule overrun, then the contractor has to demonstrate and prove the construction of the works for the achievement of their performance according to the regulations and the state of the technology. If they succeed in demonstrating this, for example through an expert report, which is based in particular on the available construction documentation, then any contradictory presumption (prima facie evidence) argues that the cause of the inadequacy of the performance was the ground conditions provided by the employer. The employer then has to refute this prima facie evidence by substantiating the reasons why the ground could not have been the cause of the inadequacy of the performance."*

The employer then takes up a contradictory position and refers to the system risk. According to VOB/C (German construction contract procedures - construction) § 3.1.1 DIN 18 312 "Underground construction work", the choice of construction sequence and the choice of machinery are matters for the contractor. Even if the construction process is prescribed in detail in the description of the works, the contractor has to prove that the cause of the inadequacy of the performance lay in the ground not being as described in the contract.

Without a contract law appraisal, the extreme positions of the contract parties described above show that there is no incentive for collaborative optimisation from under-pricing or risk transfer and the intended win-win situation cannot be achieved. When the cost estimates are exceeded, there is no basis for a win-win situation and a legal dispute is often the outcome.

Cost overruns are often not solely caused by the ground. In addition to unambiguous contract parameters to establish the intended works, numerous other factors play a decisive role and not least the relationship between the contract parties. The localisation of cost and time drivers is extremely complex and affects all phases of the project. The essential drivers of cost and time assigned to project phases are:

Production of tenders:

- the selection of an unsuitable tunnelling machine,
- specification and bill of quantities are incomplete or defective (definition of the intended works),
- incomplete listing of risks,
- missing approvals,
- lack of standardised bill texts.

Estimating:

- information asymmetries (state of knowledge) employer-contractor,
- under-pricing and incomplete estimation,
- too little time for the processing of a bid.

Award:

- award to the cheapest bidder,
- self-interested award negotiations with the intention of a subsequent improvement of the contractual position,
- complicated and non-transparent objection procedure.

Contract:

- contractual and technical contradictions,
- lack of risk distribution.

Construction:

- lack of payment regulations for works outside the contract,
- low-budget construction particularly of the functional and non-functional works,
- decision-making conflicts due to asymmetry of information and interests,
- coordination of technical experts.

17.5 Under-pricing as a performance killer

The award of a contract based on an under-priced tender often only leads to a short-term award success for the employer. Insufficient prices favour counterproductive cost-saving measures at the cost of reduced quality. Instead of optimising the construction progress, the contractor looks for gaps in the contract and claims whenever possible. The resulting additional costs finally reduce the overall construction costs, slow construction progress and often result in a lose-lose situation for both contract parties.

In Austria, bidders must completely lay open their estimation calculation with the tender in the so-called K-7 form. Speculation is, in contrast to Germany, impermissible and bidders with under-priced bids are excluded under federal law. In Spain, the Netherlands and the USA, bids are extensively evaluated and assessed regarding their technical merits in addition to the price. The contract is awarded to the economically and technically best bid. There are even models, which consistently exclude the cheapest bidder from the award.

In Germany, the estimation calculations that are handed in are often not suitable for checking due to the lack of standardisation and are not usually considered in the evaluation of the bids. As speculation regarding quantity and price is not ruled out, under-pricing cannot be recognised. The lack of transparency of the tenders also hinders the later checking of variations and the pricing of extra-contractual works.

17.6 Chances and risks of partnering

Alliance contracts and partnering models have proved successful in the GB and US markets, particularly for PPP, BOT and BOOT projects. In an alliance contract, the employer and the contractor form a virtual project company with an innovative risk distribution and payment model for the implementation of the contract [105]. Characteristics of an alliance contract are the representation of the virtual project company by waiving boards, disputes and legal claims, a unanimity rule and the determination of indicators, which are used to calculate the distribution of profit and loss between the members. The partnering model does not require a permanent commitment to a project company. The usual works contract relationship is developed into a partnering relationship. The foundations are the mutual determination of the detailed design and the specification and bill of quantities, an open-books method of calculating payments and a bonus-malus payment system. A future development is the GMP (guaranteed maximum price) approach. The contractor guarantees a maximum price based on the preliminary design. The contract also proceeds according to the open books process, and the employer has a view of all contract processes [279]. There are no standardised contracts for the various forms of contract, but a similar approach. The early involvement of the contractor in the design and cost estimation is intended to exploit their experience and lower material and personnel costs.

One way of involving the contractor at an early stage would be to enter a partnership at the design stage (design phase 3) through a "competitive dialogue" according to § 3a point 4 VOB/A [300]. This is, however, much more difficult, expensive, time-consuming and possibly riskier for employers in public sector tunnel projects than, for example, in private sector building. The following reasons could be mentioned:

– highly detailed design as early as the planning approval stage,
– the interests of authorities and third parties,
– complicated and tedious approval process,
– complicated tendering and award guidelines.

Any thorough revision of the planning approval process, but also of contract and tendering models, would have to be carried out at EU level and is not to be expected in the short or medium term. There is not any urgent reason for this either. The existing contract models also contain proven components.

The following sections are concerned with the possibilities of integrating the partnering philosophy into the existing contract framework according to § 9 No. 11 VOB/A, but can also be applied to the competitive dialogue after the completion of approvals (design phase 4).

The aims, premises and methods of the partnering model and alliance contracts have been described in detail in [32], [47], [232]. For tunnel construction, the following aspects are considered significant:

Use and added value:

– shorter decision processes,
– optimisation of costs, time and quality,
– lower administrative costs,
– value-engineering and encouragement of innovation,
– win-win situation.

Principles:

- mutual processes, no simple relationship,
- common resources,
- open communication,
- intention of constant improvement,
- start at an earlier stage of the project (design phase 3 or earlier),
- mutual overcoming of risk instead of self-interested risk transfers,
- internal conflict resolution strategies,
- involvement of third parties only in case of intentional severe infringements.

Elements:

- mutual management,
- common project management team.

Investments:

- training and workshops,
- experienced seniors,
- monitoring and audits.

Difficulties:

- Partner finding and competitive conditions,
- tendering and award guidelines,
- conflicts of aims and interests between the partners,
- price fixing,
- mutual cost determination (fees),
- target price determination.

17.7 Partnering – contractual implementation

As already mentioned, the complete and thorough implementation of the partnering model in the short or middle term has to be regarded as difficult due to the predominant contract culture in Germany, the guidelines for the awarding of contracts and the complicated planning approval process in tunnelling (exception: operator/concession projects). The aspects mentioned above can, however, be at least partially integrated into the existing contract award system based on the VOB/A. A summary of all aspects leads to the following core demands for the improvement of the existing contract models:

1. mutual optimisation of the construction sequence,
2. mutual optimisation of costs,
3. mutual risk management,
4. mutual decision making.

It would therefore be sufficient to integrate the essential aspect of the partnership model, the "mutual cooperation" into our contract culture and the existing models. The amalgamation of both parties into a jointly and severally liable project company as early as design phase 3 (design), as is intended in alliance contracts [105], is not imperative.

The following preconditions must be present:

- complete, technically optimised but flexible tender design (employer),
- realistic cost determination for the construction target (employer),
- cooperative risk distribution (employer),
- measures to prevent under-pricing (employer /contractor),
- mutual target/actual comparison of the construction progress at all levels (costs, time and safety),
- mutual remedies in case of target/actual deviations,
- mutual resolution of risks.

A look at events abroad shows that essential components are already included in the contract models or are dealt with successfully in other ways than in Germany. For example in Switzerland and in Austria, the employer is normally responsible for the detailed design and undertakes the design risk, in complete contrast to the partnering philosophy. If the local conditions alter with an effect on the detailed design (for example through different ground conditions) and this results in additional costs, these are also paid be the employer. According to the Swiss standard for underground structures [264], works are paid in a way helpful to the company according to the actual consumption of materials and for mechanised tunnelling after mutual measurement of the actual progress of boring.

17.8 Partnering – mutual process optimisation

The existing models can be further developed towards motivating payment systems – bonus/malus approach, GMP approach (guaranteed maximum price) – or the profit distribution model based on KPIs (Key Performance Indicators). The contractor guarantees a tunnelling performance that is accepted by the employer as realistic, with a tunnelling process defined in detail, in a tunnelling class also defined in detail, at a transparently itemised price. If the tunnelling performance is not achieved or if additional works are required, payment is made on the mutually agreed cost basis. If the performance is better or if work is not required, the contract parties share the profit.

The "Austrian model" [227], [228], [257] denotes the calculation of payment for conventional tunnelling in the form of a matrix. The evaluation factors used in the matrix are not subject to competition but are defined and fixed. Adaptability to changing conditions is the first priority. The standard provides for the identification of the time-related costs split into construction implementation and machinery costs as well as a very detailed breakdown of site overheads. For unusual circumstances and stoppages, there is a simplified form of daywork payment calculation, which is also included in the tender comparison and is thus subject to competition [264]. The daywork payment rate and optimisation through value engineering could also be developed further according to the partnering philosophy.

The mutual aim of sharing risks, profit and loss and optimising construction processes presumes that both partners are prepared to learn from mistakes and trust each other. In [104], openness is discussed and it is emphasised how important it is to be open about mistakes and maintain transparency. This is the basis of the load planner and lean management approaches, which are used to transport increased motivation, cooperation, transparency, reliability and the fulfilled undertakings from the project level into the project.

17.8 Partnering – mutual process optimisation

In tunnelling, partnering is only possible when both partners communicate at the same eye level and are in possession of the same knowledge.

Figure 17-1 shows an organisational method of implementing the optimisation process.

Trust and the mutual readiness to learn from mistakes require fairness, expert knowledge and transparency. For highly mechanised tunnelling, this requires a powerful knowledge and information management system. The open-books approach of the partnering model should not be restricted to cost control but should incorporate the entire technical and commercial construction progress. Only when both partners are subject to a continuous target/actual analysis at all levels can mistakes be recognised early and preventative measures introduced mutually and in good time.

Deviations from the schedule, failure to achieve performance rates and in particular the cost driver and source of contractual disputes "deviations or geological conditions" as well as the overcoming of unforeseen interruptions should be clarified at the technical level. The primary aim is to optimise the construction progress and avoid "disturbances to construction progress". The central approach is the knowledge-based and prompt target-actual comparison of all process parameters as well as the construction costs and resources. This makes possible a technical dialogue between the partners on the same eye level with the common aim of continuous construction optimisation and enables a win-win situation for both contract partners.

In mechanised shield tunnelling, it is not sufficient to carry out the target-actual comparison separately for the geological conditions and the machine data. The system behaviour (the interaction between geology and construction process) is also of significant importance. In mechanised shield tunnelling, the following procedure has proved successful [116]:

– Description of the geological conditions including consideration of the construction process,
– analysis of the geological conditions including consideration of the construction process, for example through a finite element simulation,

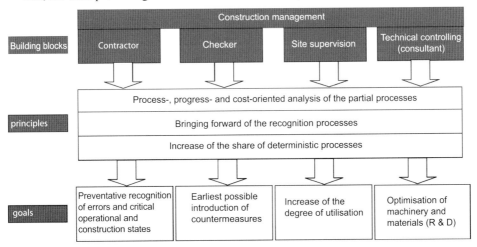

Figure 17-1 Flow diagram of the optimisation of construction processes between partners [208a]

- real-time analysis of the advance and the interaction of the machine with the surrounding ground,
- decision-making based on the results and the introduction of preventative measures in good time.

Important remarks about the points listed above are given in Chapter 18 (Process Controlling). The implementation of computer-based data management systems enables the construction progress to be electronically analysed and the knowledge gained made available to all project parties in real time. Knowledge-based methods will be able to incorporate experience form other projects in the future. Using mutually agreed priced bills of quantities, it will be possible to integrate the development of construction costs into the cybernetic system of process control. Negative target-actual deviations will not only be detected but also avoided through active countermeasures.

18 Process controlling and data management

18.1 Introduction

Tunnels driven with shields can be regarded as construction measures with a high degree of difficulty, with pronounced interaction between the structure, materials and surroundings. They are also subject to particular risk because the principal construction material, the surrounding ground, is difficult to recognise and describe. The safety, also the effectiveness of the use of resources, of highly technical shield tunnelling can be improved considerably through the thorough analysis of the process data. The aim of process controlling is to analyse the system behaviour – in situ and if possible in real time – under consideration of all interactions between ground and construction process.

The interactions of structure, construction material and surroundings are also described as system behaviour in DIN 1054 thus following the Austrian standard ÖNORM B 2203-2 "Underground works – Works contract – Part 2: Continuous advance" [228]. This term includes the behaviour of the overall system resulting from ground and tunnelling process. The new versions of ÖNORM EN 1997-1 (Eurocode 7) [226] and DIN 1054 [74] prescribe the observation method for complex geotechnical structures. The purpose is to verify measures, which were decided before the start of construction, during the construction phase with measuring systems. Prognoses are to be checked or the calculation model is to be adapted when the behaviour of ground and structure do not settle down as expected. If the serviceability or even the structural safety is endangered, then countermeasures should be taken.

18.2 Procedure

The method of process controlling [209] described below is based on successful methods of optimising production processes in mechanical engineering and originally comes from system engineering. The production process of the shield tunnelling machine is for this purpose split into partial processes. Figure 18-1 shows the partial processes and their interaction for the example of an earth pressure balance shield.

Controlling is a description for the summary of the target-actual comparisons of the essential process parameters and quality criteria. Only when it is clear which quality criteria a process has to fulfil can the process sequence be sensibly developed and monitored. This means that suitable methods have to be laid down for process controlling, which can ensure that the processes fulfil the expectations placed on them. Suitable methods therefore have to be laid down for process controlling, which can be used to ensure that the processes fulfil expectations. A well thought-out concept of process controlling not only serves as an instrument for evaluating whether the intended targets have been reached but also offers the longer-term advantage of delivering the necessary indicators for continuous improvement of the process. This requires the setting of objective quality criteria (Key Performance Indicators or KPIs). The objective quality criteria can be used to check whether processes are running or there is a need for improvement. Therefore the overall targets have to be fixed before the decision of suitable KPIs. With these targets in mind, suitable indicators for the evaluation of the successful execution of a process can be determined. Which KPIs are finally selected depends on factors including the tunnelling process, individual experience and the available possibilities for determining the values. In the ideal case, the indicators can be calculated automatically, for example using a numerical simulation (support pressure specification) or mathematical algorithms (tool wear). The measurement procedures defined here therefore also represent at the same time requirements for the system to be implemented [303].

In process controlling, the aim is not to specify as many KPIs as possible. In practice, it has often been found that an excessively complex structure of indicators leads to a disproportionately large amount of work, is little accepted and is therefore not used any more after a short time. It is recommended instead to define only a few but significant indicators so that the amount of work involved in determining the indicators and reporting remain within reasonable bounds.

18.3   Data management

Data management is the most important element of a process controlling system. It includes the entire reporting and thus includes all essential indicators for design and construction. Highly mechanised shield tunnelling is particularly suitable for the implementation of computer-based data management systems. Every one to ten seconds, between 200 and 400

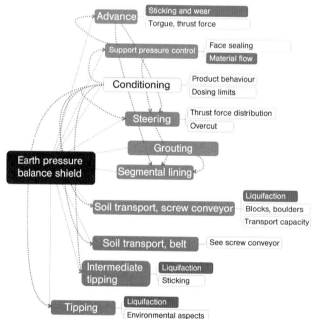

Figure 18-1 Process sequence scheme for an earth pressure balance shield

items of machine data, called momentary values, are recorded. This amounts to between 1.7 and 3.5 million items of data per day, which can be collected automatically for each advance cycle in average and final value files. Separately from this, a great amount of geodetic and geotechnical data is usually logged during the advance. While this was formerly done manually and thus at very varied intervals, measurement robots are now used, which automatically record defined survey points and transmit the data digitally.

The special challenge of the observation method is the real-time processing of the information and data as the advance progresses. Fig. 18-2 shows the database structure for the implementation of a controlling system based on construction process. The database contains process-related data related to the ground, the tunnelling machine and the results of geotechnical measurements. In addition to the machine data and the geodetic survey data, data can be recorded at other process interfaces (separating plant, segment stores etc.), which is normally only available in analogue form in shift or daily logs. The implementation of construction operations data, for example time-dependent costs, cost of materials and electricity consumption, is also possible. All information relevant to the advance is transmitted in real time over a VPN link and analysed by a programme developed specially for shield tunnelling according to Key Performance Indicators (KPIs), like for example face stability, advance rate, specific electricity consumption, foam injection and expansion rates. Operational interruptions, changes of ground conditions and their consequences for the process can be followed in real time by remote analysis or subsequently.

18.4 Target-actual comparison

The target-actual comparison considers all significant process-related and geotechnical parameters. The computer-based results are compared with the design specification (reference values) and the results are made available to the responsible geotechnical specialists (regulator-control unit) in evaluated and visualised form. The target-actual comparison is then interpreted by experienced users and instructions are issued and implemented by the site management.

The design specification (reference values) is derived from the geotechnical calculations and forecasts (Section 18.5) carried out before the start of construction. In accordance with the philosophy of the observation method, the observation data from the surveying and monitoring system is included. As shown in Figure 18-3 the behaviour to be expected is forecast using detailed analyses, back analyses and simulations.

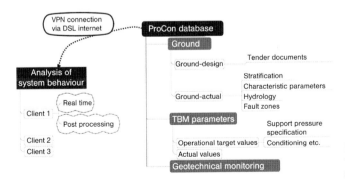

Figure 18-2 Database structure for process controlling in shield tunnelling [209]

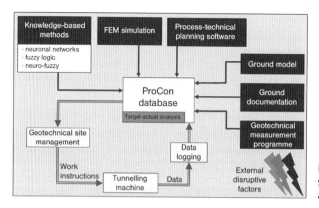

Figure 18-3 The cybernetic system as part of an expert target-actual comparison [209]

As the classic cycle of control regulation through the site management and the geotechnical specialists would not be capable of making real time decisions due to the flood of data, a future method shown in Figure 18-3 would be to implement so-called reaction programmes or expert systems (knowledge-based methods) to aid decision making. For shield tunnelling, the following methods are particularly suitable:

– analytic mechanical, fluid mechanical and geotechnical calculations,
– expert systems,
– statistics, stochastics,
– neuronal networks,
– fuzzy logic,
– neuro-fuzzy.

Fuzzy Logic makes is possible to consider engineering know-how (expert knowledge) and intuition in data analysis and process controlling. The excellent compatibility and integration possibilities in standardised regulation and PLC systems (PLC = programmable logic control) have proved to be particularly suitable for use in shield tunnelling.

Neuronal networks (NN) are mostly used when the complexity of a problem being considered is too great or the knowledge is too unstructured. In contrast to conventional deterministic methods, closed mathematical modelling of the system is not used in this case, i.e. no statements are required about the significance of the individual variables or their interaction.

Fuzzy logic or neuro-fuzzy in particular are interesting possibilities for mechanised tunnelling as they make it possible, as shown, to implement expert knowledge gained through practice into automatic evaluation and analysis processes. As the resulting systems are based on linguistic descriptions and rules corresponding to human thinking, they can be understood and subsequently manually optimised. They also offer the advantage that they can be easily integrated into the standard hardware and software of control circuits. In addition to the analysis of data for the specification of guideline values, this makes further automation of partial processes leading to integral process control systems conceivable in the future.

Further tools, which contribute to the gaining of knowledge, can be integrated into the control circuit. The automatic advance investigation of the ground (e.g. the SSP system from Herrenknecht AG) could make an important contribution in the future. The more reliable and unambiguous the information is about the ground conditions, the easier it is to improve and calibrate the simulation model.

18.5 Target process structure

How should the decision be reached, whether a process is "running well" or not? How can the target performance be quantitatively or qualitatively defined for the individual partial processes? Different methods are used for the various partial processes as shown in Table 18-1.

FEM simulation of the shield advance is an important tool for the determination of the design specification and also serves to verify the actual system behaviour through " back analyses". With the assistance of FEM simulation, it is possible with the current state of the technology to display the extremely complicated interactions between ground and tunnelling machine. The FEM simulation can be structured into the following phases:

– Geometrical and process-technical modelling with realistic discretisation of all machine elements (face support, shield gap grouting, shield tail grouting etc.),
– material modelling with the selection of a suitable constitutive law (consideration of positive pore water pressures, creep effects, non-linear-elastic behaviour etc.),
– step-by-step analysis to consider the effects of different phases of construction on the stress state in the ground,
– verification of the results and plausibility checking.

Table 18-1 Procedure for the determination of the KPIs/Key Performance Indicators

Partial process	Intended process analysis	KPIs
Ground excavation (tool wear)	analytic calculation model of cuttability (balance of forces), structural calculation model based on laboratory tests	contact force, torque, penetration, skin friction, material density, cuttability index
Face support	FEM simulation, kinematic support pressure calculation, analytic calculation model for volume control	support pressure, suspension losses, foam injection pressure, density distribution
Soil conditioning	FEM simulation of the hydraulic processes, analytical calculation models	support pressure distribution, foam injection pressure, foam parameters
Annular gap grouting	FEM simulation, analytical calculation models	Grout volume under consideration of the overbreak, grouting pressure
Soil tipping	analytical calculation model (consistency development, hazardous materials calculations)	liquidity index, carbon content,-tenside/polymer concentration
Soil separation	analytical calculation models	Suspension densities, solids content, viscosity measurements, yield strength

Figure 18-4 shows the 3D modelling of a shield tunnelling machine in the surrounding ground. The excavation chamber, shield area, annular gap grouting and segmental lining can be recognised. The model is designed so that all measured technical machine parameters, for example the support pressure at the working face, support pressure in the annular gap or the "volume loss" through overcut at the shield skin are taken into account. With this procedure, it would theoretically be possible to import measured machine data without delay directly into the model and receive selected calculation results, for example surface settlements, ground stresses, loading on the tunnel lining etc.

Optimisation of the process requires further knowledge about the fluid mechanics processes in the excavation chamber. When active face support is provided, the rheological properties of the soil in the excavation chamber are important for optimised active support pressure control, in addition to the pressure values P1, P2, P3 according to Figure 18-5.

The material flow in the excavation chamber can also be realistically simulated using numerical models.

When active conditioning is not used, the material flow in the excavation chamber concentrates in the immediate vicinity of the screw conveyor; in the other parts of the excavation chamber, there is scarcely any flow of material. There is a danger that the soil consolidates in these areas and thus loses its flowing properties. Through process-controlled soil conditioning, the flow pattern can be reconstructed and actively manipulated or optimised. For this reason, all sensors for the control of pressure and volume are included in the analysis of the flow behaviour.

In summary, it can be stated that decisive interactions between ground and tunnelling machine can be at least qualitatively investigated through numerical simulations. The knowledge gained is used as part of the process controlling to optimise the advance. The shield tunnelling processes can thus be represented with their complex interactions between ground and machine and analysed using efficient system tools. The optimisation of the key parameters in real time and subsequently can be done through the implementation of the know-how of the experts and through knowledge-based methods like fuzzy logic and neuronal networks in the process controlling system, which makes the system capable of learning.

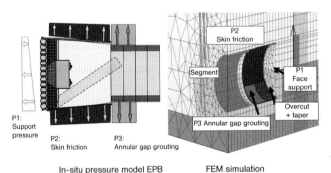

Figure 18-4 FEM model for the analysis of system behaviour [209]

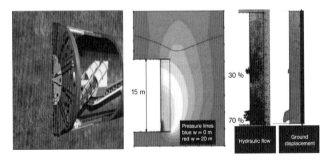

Figure 18-5 FEM model for the analysis of the hydraulic behaviour in the excavation chamber [209] (Herrenknecht EPB; Madrid M-30; D = 15.16 m)

18.6 Analysis of the actual process

The process sequence is now archived and analysed in real time by the computer according to the procedure described above. Processes with sharp criteria, for example the control of support pressure, can be clearly visualised with target curves and coloured bandwidths. Particularly advantageous is the display of the geological longitudinal section and layout plans including the display of measured settlements.

Experience shows that in case of complex interactions, the sole consideration of single KPIs (Table 18-1) does not normally lead to an unambiguous result. It is not normally the quantitative value of a single indicator that is significant but rather the relationships and the interaction of the single quality criteria. In this case, the expert knowledge of the user is indispensable.

For example, if the yield strength is too low in a shield advance with slurry-supported face, this does not necessarily represent an impairment of the stability of the face as long as the suspension has sufficient density and the suspension losses are low. The KPI yield strength must therefore always be considered in relation to the soil permeability and suspension density (Figure 18-6). The suspension losses in turn can be verified through the development with time of the bentonite level during stoppages.

The results of knowledge-based process analyses can be provided for the responsible parties on the project in real time. As a part of the collection of evidence, the completely analysed progress of the advance is permanently archived and can be called up at any time according to time or position.

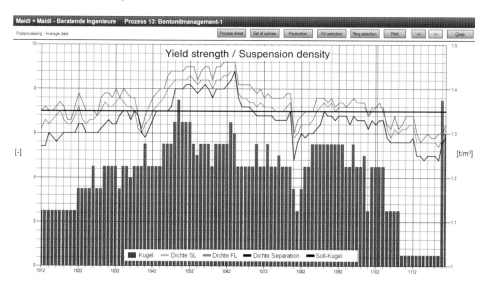

Figure 18-6 Visualisation of the KPI yield strength/suspension density [209]

Special Deep Foundation

■ The methods and equipment technology employed in the deep foundation industry have improved rapidly in recent years. The ingenuity of civil engineers, the results of new scientific research and the ongoing and new developments in machine technology have all led to the acceleration of this process. Applying technologies that have become very complex, and selecting the suitable machinery and equipment, demand ever more specialized knowledge and practical experience. It has become very difficult for users and manufacturers of special deep foundation machinery to maintain an overview of the level of technology in the sector. Both volumes provide a comprehensive overview of the special deep foundation applications, equipment and processes. They are intended as an aid to planning and implementation, and aim to help practitioners, public authorities, engineering companies and students to broaden and complete their level of knowledge. They are targeted primarily at occupational engineers and applications in the field.

The individual chapters discuss manufacturing techniques and potential applications, along with the associated machine components. The specifics of each method and machine technology are examined in detail. Since the first volume of the compendium on Special Deep Foundation was published in March 2008, it has become a standard reference book.

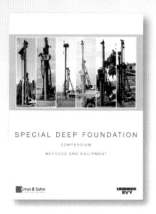

LIEBHERR-WERK NENZING GMBH (ED.)
Special Deep Foundation
Compendium Methods and Equipment.
Volume I: Piling and Drilling Rigs
(LRB Series)

2008. 370 pages, 300 figures, Hardcover.

€ 89,–*
ISBN 978-3-433-02905-3

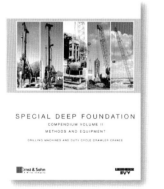

LIEBHERR-WERK NENZING GMBH (ED.)
Special Deep Foundation
Compendium Methods and Equipment.
Volume II: Drilling machines and hydraulic crawler cranes

2009. 336 pages, 297 figures, 43 tables, Hardcover.

€ 89,–*
ISBN 978-3-433-02932-9

Package-Price
€ 129,–*
instead of € 178,–*
ISBN 978-3-433-02935-0

*€ Prices are valid in Germany, exclusively, and subject to alterations. Prices incl. VAT. Books excl. shipping. Journals incl. shipping. 0234100006_dp

Online-Order: www.ernst-und-sohn.de

Ernst & Sohn
A Wiley Company

Ernst & Sohn
Verlag für Architektur und technische
Wissenschaften GmbH & Co. KG

Customer Service: Wiley-VCH
Boschstraße 12
D-69469 Weinheim

Tel. +49 (0)6201 606-400
Fax +49 (0)6201 606-184
service@wiley-vch.de

40 years experience, more than 150 tunnels
Since 2010 as a new consulting firm

9 m EPB Emisor Oriente, Mexico City

Metro Caracas, Venezuela

Planning • Verification • Consultancy • Controlling

Process Controlling with MAIDL-PROCON
Knowledge-based TBM analysis software

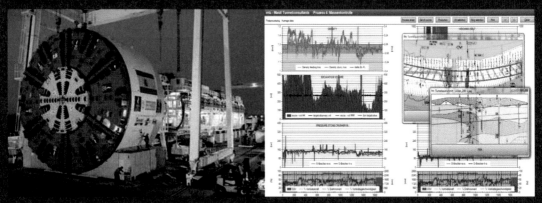
15 m EPB Madrid M-30   13 m Slurry Brenner Lower Inn Valley

mtc
Maidl Tunnelconsultants GmbH & Co. KG
Dr.-Ing. U. Maidl – Prof. Dr.-Ing. B. Maidl
Of Counsel: Prof. Dr.-Ing. M. Thewes

Contact: office@maidl-tc.de   www.maidl-tc.de

19 DAUB recommendations for the selection of tunnelling machines

Chapter 19 now repeats a translation of the recommendations of the German Tunnelling Committee DAUB[1] for the selection of tunnelling machines.

19.1 Preliminary notes

The purpose of these recommendations is to provide a decision-making basis for the selection of tunnelling machines for use in rock and soft ground based on process-technical and geotechnical criteria. This takes into account the prevailing technical, local and environmental conditions, also process and machine technology. The recommendations are intended as an additional decision-making aid for the engineer, but cannot replace project-related analysis, which will remain the most important basis for the selection of a tunnelling machine.

These recommendations particularly supplement the existing recommendations of DIN 18 312 "Underground Construction Work"; VOB Part C. For the selection of tunnelling machines for pipe jacking, reference is also made to guideline DWA-A 125 of the German Association for Water, Wastewater and Waste DWA e. V.

---

[1] The recommendations were produced by the working group „Recommendations for the selection of tunnelling machines" of the German Tunnelling Committee (DAUB).

Members of the working group:
Dr.-Ing. Ulrich Maidl, mtc – Maidl Tunnelconsultants, Duisburg (leader of the working group)
Dipl.-Ing. Wolfgang Frietzsche, Wayss & Freytag Ingenieurbau AG, Frankfurt/Main
Prof. Dipl.-Ing. Fritz Grübl, PSP Consulting Engineers GmbH, Munich
Prof. Dr.-Ing. Dieter Kirschke, consulting engineer for rock mechanics and tunnelling, Ettlingen
Dipl.-Ing. Gebhard Lehmann, Herrenknecht AG, Schwanau
Dr.-Ing. Roland Leucker, Research Association for Underground Transport Facilities – STUVA e. V., Cologne
Prof. Dr.-Ing. Dietmar Placzek, ELE Beratende Ingenieure GmbH, Earthworks Laboratory, Essen
Dipl.-Ing. Dr. techn. Klaus Rieker, Wayss & Freytag Ingenieurbau AG, Frankfurt/Main
Dipl.-Ing. Dieter Stephan, Ing.-Büro Dipl.-Ing. H. Vössing GmbH, Düsseldorf
Prof. Dr.-Ing. Markus Thewes, Chair of Tunnelling, Utility Engineering and Construction Management, Ruhr University, Bochum
Dipl.-Ing. Helmut Wanner, Spiekermann AG, Düsseldorf

External collaborators with the working group:
Dipl.-Ing. Winfried Schuck, Deutsche Bahn AG
Dipl.-Ing. Jörg Wingmann, mtc – Maidl Tunnelconsultants, Duisburg

19.2 Regulatory works

The following documents were used for the production of these recommendations:

19.2.1 National regulations

- Supplementary Technical Conditions of Contract and Guidelines for Engineering Structures (ZTV-ING) of the BMVBS from December 2007:
  - Part 5: Tunnel Construction, section 3 "Mechanised Shield Tunnelling".
- Guideline 853 "Design, construction and maintenance of rail tunnels" of DB Netz AG from 01/12/2008:
  - Module 853.2001 "Structural Stability Calculations" (including regulations concerning the actions from thrust cylinders of tunnelling machines),
  - Module 853.4001 "General rules for tunnelling, support and lining",
  - Module 853.4005 "Segmental lining" (including regulations concerning annular gap grouting),
  - Module 853.6001 "Construction, construction documents and documentation" (with regulations concerning the control of shield tunnelling works),
- Worksheet DWA-A 125: Pipe jacking and associated processes,
- Regulations for working under compressed air (Compressed Air regulations),
- Regulations for health and safety on construction sites (RAB 25): further details to the Compressed Air Regulations,
- Guideline "A code of practice for risk management of tunnel works" of the International Tunnelling Insurance Group (ITIG).

Laws and regulations concerning the use of conditioning agents

- General administrative regulations to the water supply law with the categorisation of potential water pollutants into water risk classes (VwVwS), 1999,
- General administrative regulations for the revision of the administrative regulations concerning potential water pollutants, 2005,
- Law concerning the environmental acceptability of washing and cleaning agents (WRMG), 2007.

Laws and regulations concerning landfill

- Interstate waste committee (LAGA); Note M20; Requirements for the material recycling or mineral residues/wastes – technical rules (version 6 November 2003),
- Federal Ministry for the Environment, Nature Conservation and Nuclear Safety: Regulations for the simplification of landfill law-draft, 2008,
- European Union: decision of the Council from 19/12/2002 to lay down criteria and procedures for the acceptance of waste on landfill sites according to Article 16 and Annex II of the directive 1999/31/EG.

19.2.2 International standards

- DIN EN 815/A2: Safety of unshielded tunnel boring machines and rodless shaft boring machines for rock – Safety requirements; English version EN 815: 1996/prA2: 2008,
- DIN EN 12110/A1: Tunnelling machines. Air locks – Safety requirements; English version EN 12110:2002/prA1:2008,
- DIN EN 12336:2010-03 (D): Tunnelling machines – Shield machines, thrust boring machines, screw boring machines, lining erection equipment – Safety requirements; German version EN 12336:2005,
- Code of practice for the planning and implementation of a health and safety concept for underground construction sites. Issued by: DACH; DAUB; FSV; SIA/FGU,
- SIA 198 (SN 531 198) Underground structures. Construction; Swiss Engineers and Architects Association, issue 10/2004,
- ÖNORM B 2203 Underground works – Works contract, issue 1994
- ÖNORM B 2203-2 Underground works – Works contract – Part 2: Continuous driving, issue 2005,
- Guideline for shield tunnelling from the Austrian associations B; FSV, OIAV,
- Recommendations and guidelines for tunnel boring machines (TBMs), working group No 14 Mechanized Tunnelling ITA,
- BS 6164, Code of practice for safety in tunnelling in the construction industry,
- Detergent regulations, Regulation (EC) 648/2004 of the European Parliament and Council (2004),
- Organisation for Economic Co-operation and Development (OECD), Guidelines 201–203; 301 B and 302 B: Freshwater alga and cyanobacteria, growth inhibition test, 2006; Daphnia sp. Acute Immobilisation Test, 2004; Fish, Acute Toxicity Test, 1992; Ready Biodegradability, CO_2 evolution test, 1992; Inherent Biodegradability: Zahn-Wellens/EMPA test, 1992).

19.2.3 Standards and other regulatory works

- DIN 4020: Geotechnical investigations for civil engineering purposes,
- DIN EN ISO 14688-1 (2003): Geotechnical investigation and testing – Identification and classification of soil – Part 1: Identification and description,
- DIN EN ISO 14689-2 (2004): Geotechnical investigation and testing – Identification and classification of soil – Part 2: Principles for a classification,
- DIN 18122: Soil, investigation and testing – Consistency limits,
- DIN 18130: Soil, investigation and testing – Determination of the coefficient of water permeability,
- DIN 18196: Earthworks and foundations – Soil classification for civil engineering purposes,
- DIN 1054: Subsoil – Verification of the safety of earthworks and foundations,
- DIN 18312: German construction contract procedures (VOB) Part C: General technical specifications in construction contracts (ATV) – Underground construction work
- "Minimum measures for the avoidance of injury in case of significant dangers from fire, gas ingress, water ingress and rockfall/collapse (Appendix A to the code of practice for the planning and implementation of a health and safety plan for underground construction sites, produced by the DAUB working group "Incident plans").

19.3 Definitions and abbreviations

19.3.1 Definitions

Abrasiveness The abrasiveness describes the influences resulting from the geology on the wear of tools. The abrasiveness of hard rock is often characterised with the CAI test (Cerchar Abrasivity Index) and that of soft ground often with the LCPC test (Test of the "Laboratoire Central des Ponts et Chaussées"), in addition to the mineralogical composition and strength parameters.

Active face support Measured and monitored support of the face by a suitable medium (for example slurry or remoulded earth) based on a support pressure calculation.

Air pressurisation The excavation chamber is pressurised with compressed air to hold back groundwater. Support against ground pressure is only possible in almost impermeable soil or if the face is sealed, e.g. by a filter cake.

Annular gap Cavity between the sides of the excavated cavity and the outer face of the segments.

Articulated shield Shield machine with more than one shield section, which are articulated with active steering cylinders or passive hydraulic cylinders in order to improve steering around curves.

Blowout Uncontrolled escape of compressed air to the ground surface or riverbed associated with loss of support effect.

Breasting (plates) Additional mechanical support to the face with extendable plates.

CAI (Cerchar Abrasivity Index) Value from the test of the same name for the characterisation of the abrasiveness of solid/hard rock.

Clamping units Side-mounted bracing apparatus in double shield machines intended to transfer the thrust forces radially into the surrounding rock mass, resist rolling and stabilise the tunnelling machinery.

Closed mode In closed mode, the excavation chamber of a tunnelling machine is held under a measured and monitored positive pressure. The pressure is applied by slurry, remoulded earth or also compressed air.

Cuttability The facility of excavating rock with a tunnelling machine depending on the rock properties. The most important process technology parameters for the quantitative description of the ease of boring are the cutterhead penetration and the contact pressure.

Cutterhead A tool carrier in hard rock tunnel boring machines fitted with disc cutters for full-face excavation. In soft ground, the term cutting wheel is normally used.

Cutting wheel Mechanism for the full-face excavation of a tunnel cross-section in soft ground. The ground is excavated as the wheel rotates and the design and tool equipment of the wheel are suited to the relevant ground properties. In hard rock, the term cutterhead is normally used.

Disc cutter (disc) Hard rock tool with a rotating hardened cutter ring, which can destroy the structure of the rock at the face.

Geotechnical report Description of the site investigation with characteristic values for the rock and rock mass parameters according to DIN 4020 Number 10.

Gripper Side-mounted, radially acting bracing apparatus in hard rock tunnel boring machines, intended to transfer the thrust forces into the surrounding rock mass, resist rolling and stabilise the TBM.

Ground behaviour Behaviour of the unsupported ground at the face and also at the sides of a tunnel without consideration of the construction process.

Ground profile Geometrical assumptions about the profile of natural formations or strata (DIN 4020 Appendix C2) or of homogeneous zones.

LCPC abrasiveness coefficient (ABR) Value from the test of the same name for the characterisation of the abrasiveness of soft ground or broken rock, named after the "Laboratoire Central des Ponts et Chaussées".

Liquefaction Loss of the shear strength of a soil due to positive pore water pressure.

Mechanical face support Support of the face with breasting plates.

Open mode In open mode, the excavation chamber is not under pressure.

Overcut Differential dimension between the bored radius and the shield radius measured at the shield blade. The overcut serves e.g. to improve driving round curves, to reduce the skin friction and relieve stress on the ground.

Primary wear Wear on the excavation tools solely due the excavation of the face; influenced by the strength, jointing and abrasiveness of the rock mass.

RMR Rock Mass Rating: Value for the classification of rock mass based on 6 rock mass parameters.

Rock mass behaviour Behaviour of the unsupported cavity in hard rock (referred to as ground behaviour in soft ground). The rock mass behaviour is determined by the properties of the rock and the jointing structure, the stress and formation water situation and the shape and size of the excavated cavity.

RQD index Rock Quality Designation index: value for the characterisation of rock quality based on the sum of the lengths of drill core pieces larger than 10 cm out of the total length of core taken according to ASTM D6032-02.

Surcharging Provision of an additional loading on the ground through e.g. filling above sections of tunnel with shallow overburden.

Secondary wear Secondary wear results from the rubbing and grinding action of the already excavated ground. Poor material flow and sticking increase the secondary wear.

Separation Description for the separation of fluid and solid in hydraulic material transport.

Stability Stability describes the stability of the ground including consideration of thee effect of the construction process. The stability is verified with calculations.

Standup time of the ground The length of time the ground can stand up without support. The verification through calculation of the stability of the face and sides of the excavation remain, however, decisive for the final evaluation and selection of the tunnelling machine.

Sticking Adhesion of excavated material to excavation tools and blocking of material removal passages and equipment in clay soil through adhesion, bridging, cohesion and insufficient dispersion capability.

System behaviour Behaviour of the overall system of ground/rock mass and tunnelling machine.

Temporary stability In the construction state, the temporary stability can be verified with reduced factors of safety on the actions.

19.3.2 Abbreviations

ABR	Abrasiveness coefficient according to LCPC
BR	Breakability according to LCPC
CAI	Cerchar Abrasivity Index
DSM	Double shield machine
EPB	Earth Pressure Balance (support)
ETBM	Enlargement tunnel boring machine, reamer
GV	Rock mass behaviour
LCPC	Laboratoire Central des Ponts et Chaussées abrasiveness test
RMR	Rock Mass Rating
RQD	Rock Quality Designation
SM	Shield machine
SM-V	Shield machine for full-face excavation
SM-T	Shield machine for partial excavation
SV	System behaviour
TVM	Tunnelling machine
TBM	Tunnel boring machine (for hard rock)

19.4 Application and structure of the recommendations

A procedure in seven steps is recommended for the selection of a tunnelling machine, as shown in Figure 19-1.

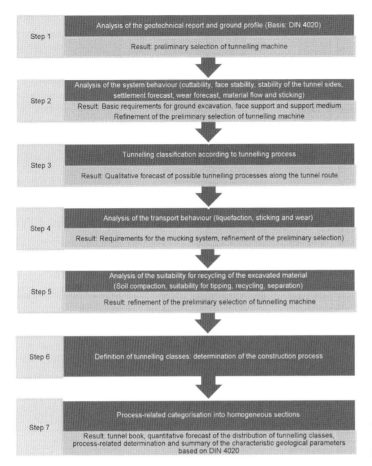

Figure 19-1 General diagram of the tunnelling machine selection process

The analysis in the geotechnical report, the ground profile (Step 1) and the system behaviour (Step 2) provide the basis for the selection of a suitable tunnelling machine. In Step 3, an exclusion procedure is used to find excavation processes, which are suitable in principle for individual tunnel sections. Taking into consideration the transport behaviour of the excavated material (Step 4) and the analysis of the recycling potential of the excavated material (Step 5), there then follows a further limitation to the optimal process for the specific technical and economic constraints of the project (Step 6). In the last step (Step 7), it is recommended to divide the tunnel into homogeneous sections for the selected process. A longitudinal section showing the tunnelling technologies is a particularly useful form of diagram, which can also be used as a contract document for construction planning and estimating and should thus also include all contract-relevant geotechnical parameters.

These recommendations define the key process-relevant geotechnical parameters for the analysis in the geotechnical report (Step 1) and are then restricted to the analysis of system behaviour (Step 2) and the preliminary selection of the tunnelling machine (Step 3). The subsequent steps should be carried out for the specific project, also considering the economic aspects.

The core of the recommendations can be found in Sections 19.5 and 19.6. In Section 19.5, the technical features of the tunnelling machines are explained and the various processes are categorised into types (Appendix 1). Section 19.6 explains the system behaviour of each machine type and includes details of the interaction between the machine and the surrounding ground. Then the significant ground/soil parameters used in the production and analysis of the geotechnical report (Step 1) are defined in relation to the processes (Appendix 2). Environmental aspects are dealt with in Section 19.7 and other significant constraints in Section 19.8. Finally, recommended applications are formulated in Section 19.9 for each type of machine based on the key parameters in the form of an application matrix (Appendix 3) for the selection of the tunnelling machine (Step 2).

19.5 Categorisation of tunnelling machines

19.5.1 Types of tunnelling machine (TVM)

Tunnelling machines either excavate the entire tunnel cross-section with a cutterhead or cutting wheel or excavate partial sections using appropriate excavation equipment.

Tunnelling machines can be differentiated into tunnel boring machines (TBM), double shield machines (DSM), shield machines (SM) and combination machines (KSM).

The machine is either continuously or intermittently driven forward as it excavates.

A systematic categorisation of tunnelling machines is shown in Figure 19-2 (see also Appendix 1 "Overview of tunnelling machines").

19.5.2 Tunnel boring machines (TBM)

Tunnel boring machines are used for the excavation of tunnels in stable hard rock. Active support of the face is not necessary and is not technically feasible. These machines can generally only bore circular cross-sections.

TBMs are differentiated into machines without shield skin (gripper TBM), enlargement tunnel boring machines or reamers (ETBM) and tunnel boring machines with shield skin (TBM-S).

19.5.2.1 Tunnel boring machines without shield (Gripper TBM)

Tunnel boring machines without shield are used in hard rock with medium to long standup time. They have no completely surrounding shield skin. The scope of economic application can be greatly influenced and limited by elaborate conventional support measures and the cost of wear to the excavation tools.

In order to be able to apply the contact force to the cutterhead, the machine is braced radially by plates (grippers) hydraulically driven against the sides of the tunnel. Rock is excavated, with little damage to the surrounding rock and to an exact profile, by disc cutters mounted on a cutterhead. The machine fills most of the cross-section. Systematic support of the inner surface of the tunnel is normally only carried out behind the machine (10 to 15 m and more behind the face). In less stable rock, particularly when there is a danger of rockfall, it must be possible to install steel arches, poling plates and rock bolts as close as possible behind the cutterhead.

19.5 Categorisation of tunnelling machines

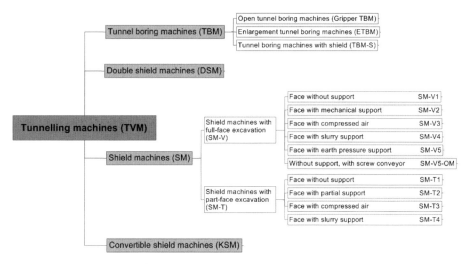

Figure 19-2 Categorisation of tunnelling machines

In case a shotcrete lining is required in the tunnel, this should first be applied in the back part of the backup in order to keep the contamination of the drive and steering units in the front part of the machine as small as possible. In exceptional cases, however, spraying of shotcrete must be possible as close behind the cutterhead as possible.

If poor rock or non-heterogeneous conditions (high degree of jointing, fault zones) are forecast, it is recommended to fit the machine with equipment for advance investigation and perhaps also rock improvement ahead of the machine.

The excavation of the face produces material in small pieces with associated dust development. These machines therefore require equipment for reducing the production of dust and for dedusting. This can be:

– spraying water at the cutterhead,
– dust shield behind the cutterhead,
– dust extraction with dedusting on the backup.

The material handling and supply of the machine require following backup facilities. These sometimes extend to a considerable length.

19.5.2.2 Enlargement tunnel boring machines (ETBM)

Enlargement tunnel boring machines (reamers) are used in hard rock to enlarge an already bored pilot tunnel to the planned final diameter. The enlargement to the full diameter is performed in one or two working stages by an appropriately designed cutterhead.

The main components of the machines are the cutterhead, the bracing and the thrust mechanism. The bracing in this special machine is ahead of the cutterhead and is supported by grippers in the pilot tunnel. The cutterhead of the machine is drawn towards the bracing as it bores. In disturbed rock formations, measures can be undertaken to stabilise fault zones from the previously bored pilot tunnel and the risks to the progress of the main tunnel can be minimised.

19.5.2.3 Tunnel boring machine with shield (TBM-S)

For rock with a short standup time or in rock liable to rockfall, tunnel boring machines are fitted with a shield skin. In this case, installation of the lining in the protection of the shield skin is appropriate (segments, pipes etc). As it advances, the machine can support itself off the lining, so the gripper equipment is usually omitted. Otherwise, the statements already made about tunnel boring machines apply correspondingly.

19.5.3 Double shield machines (DSM)

Double shield machines (DSM) consist of two parts one behind the other. The front part is equipped with the cutterhead and the main thrust cylinders, and in the back part are the auxiliary cylinders and the grippers. The front part of the machine can be extended forwards from the back part by a complete ring length with a telescoping mechanism.

In a stable rock mass, the grippers resist the torque and the thrust forces during advance. The back part of the machine is securely fixed by the grippers and the assembly of the segment ring can continue as the machine advances. In stable rock, the installation of segments may not be necessary.

In an instable rock mass, in which the grippers cannot grip adequately, the advance can be supported from the last ring of segments. The front and back parts of the machine in this case are telescoped together and the thrust forces are transferred to the segment ring by the auxiliary cylinders.

It is usually not possible to provide active support to the face and the excavated sides of the tunnel.

The rapid advance of the back part of the machine to reposition the grippers after the completion of a boring stroke means that the rock mass must be able stand up without support until the annular gap has been fully grouted or filled with pea gravel.

19.5.4 Shield machines (SM)

These are differentiated into shield machines for full-face excavation (with a cutting wheel; SM-V) and shield machines for partial excavation (with a cutting head, excavator; SM-T). Shield machines are used in soft ground above and below the groundwater table and the ground around the excavated cavity and the face normally have to be supported. Shield machines can be further categorised according to the type of face support (Fig. 2).

19.5.4.1 Shield machines for full-face excavation (SM-V)

Face without support (SM-V1)

If the face is stable, e.g. in clay with stiff consistency and sufficient cohesion or in solid rock, open shield machines can be used. A cutting wheel fitted with excavation tools excavates the ground and the excavated material is cleared on a conveyor belt.

In rock liable to rockfall, shields are mostly used, which are fitted with a largely closed cutterhead fitted with disc cutters and fully protected from instable ground by a shield skin. The thrust forces and the cutterhead torque are transferred to the last ring of segment by the thrust cylinders.

Face with mechanical support (SM-V2)

With tunnelling machines with mechanical support, the face is supported during excavation by elastically supported support plates arranged in the openings of the cutting wheel (between the spokes). In practice, however, experience shows that that no appreciable mechanical support of the face can be provided by the rotating cutting wheel. For this reason, this type of cutting wheel did not prove successful in unstable ground and is no longer in use today. Mechanical face support by the cutting wheel or the support plates should only be considered a supplementary safety measure and the supporting effect should not be taken into account in calculations to verify the stability of the face.

Face with compressed air support (SM-V3)

Shield machines of type SM-V3 can be used below the groundwater table even if it cannot be lowered or groundwater lowering is not allowed. In this case, the water at the face must be held back by compressed air. A precondition for the displacement of groundwater is the formation of an air flow to the surface. Impermeable strata above the tunnelling machine can retain the applied air and prevent effective displacement of the water (and thus the formation of an air flow). The permeability limit of the surrounding ground is therefore significant.

As no pressure difference can be built up at the face, compressed air cannot generally provide support against earth pressure, which applies particularly in permeable soil. The loss of apparent cohesion in non-saturated soil is also possible.

For the duration of tunnelling work, either the entire tunnel is pressurised or the machine is provided with a pressure bulkhead to maintain the excavation chamber under pressure. In both cases, air locks are required. Particular attention needs to be paid to compressed air bypassing the shield tail seal and lining. The recommendations and requirements for working under compressed air should complied with.

Any additional support of the face provided by the cutting wheel or support plates should be regarded solely as an additional security. It is not permissible to consider the supporting effect in calculations to verify the stability of the face.

Face with slurry support (SM-V4)

Tunnelling machines with slurry support provide support to the face through a pressurised fluid, which is specified depending on the permeability of the surrounding ground. It must be possible to vary the density and viscosity of the fluid, and bentonite suspensions have proved particularly successful for this purpose. In order to support the face, the working chamber is closed from the tunnel by a pressure bulkhead.

The required support pressure can be regulated very precisely with an air bubble behind a submerged wall and by adjusting the output of the feed and slurry pumps. The required and the maximum support pressures over the entire length to be bored should be calculated before the start of tunnelling (support pressure calculation).

The soil is excavated from the full face by a cutting wheel fitted with tools (open-mode or rimmed wheel) and removed hydraulically. Subsequent separation of the removed suspension is essential.

If it is necessary to enter the excavation chamber, for example to change tools, carry out repair work or to remove obstructions, the support slurry has to be replaced by compressed air. The support slurry then forms a low-permeability membrane on the face, which however is of limited durability (risk of drying out). The membrane permits the support of the face by compressed air and may need to be renewed regularly. The support slurry can be completely (empty) or only partially (lowering) replaced by compressed air. The maximum partial lowering is limited particularly by the requirement for sufficient working space. This should be chosen so large that safe working is possible at all times and an adequately large space is available for the workers to retreat.

If an open cutting wheel is used, it should be possible to mechanically close the face with shutter segments in the cutting wheel or with plates, which can be extended from behind, in order to protect the personnel working in the excavation chamber while the machine is stopped, which is also sensible due to the limited duration of the membrane effect.

Stones or rock benches can be reduced to a size that can be removed by discs in the cutting wheel and/or crushers in the working chamber.

In stable ground, the slurry shield can also be operated in open mode without pressurisation, with water being used for muck removal.

Any additional mechanical support of the face provided by the cutting wheel or support plates should be regarded solely as additional security and it is not permissible to consider the supporting effect in calculations to verify the stability of the face.

Face with earth pressure support (SM-V5)

Tunnelling machines with earth pressure balance support provide support to the face through remoulded excavated soil. The excavation chamber of the shield is closed from the tunnel by a pressure bulkhead. A cutting wheel, fitted with tools and more or less closed, excavates the soil. Mixing vanes on the back of the cutting wheel (rotors/back buckets) and on the pressure bulkhead (stators) assist the remoulding of the soil to a workable consistency. The pressure is checked with earth pressure cells, which are distributed on the front of the pressure bulkhead. A pressure tight screw conveyor removes the soil from the working chamber.

The support pressure is regulated by varying the revolution speed of the screw conveyor or through the injection of a suitable conditioning agent controlled according to pressure and volume. The pressure gradient between excavation chamber and tunnel is provided by friction in the screw conveyor. If the soil material in the screw conveyor cannot ensure the sealing of the discharge device, an additional mechanical device has to be installed. Complete support of the face, in particular the upper part of it, is only successful if the soil acting as a support medium can be remoulded to a soft or stiff-plastic mass. This is particularly influenced by the percentage content of fine grains smaller than 0.06 mm. The scope of application of earth pressure shields can be extended using soil conditioners such as bentonite, polymers or foam, but attention needs to be paid to the environmentally acceptable disposal of the material.

In stable ground, the earth pressure shield also can be operated in open mode without pressurisation with a partially filled excavation chamber (SM-V5-OM). In stable ground with water ingress, operation is also possible with partially filled excavation chamber and compressed air.

If the groundwater pressure is high and in ground liable to liquefaction, the critical location of the transfer of material from the screw conveyor to the conveyor belt can be replaced by a closed system (pumped material transport).

Any additional mechanical support of the face provided by the cutting wheel or support plates should be regarded solely as additional security and it is not permissible to consider the supporting effect in calculations to verify the stability of the face.

19.5.4.2 Shield machines with partial face excavation (SM-T)

Face without support (SM-T1)

This type of shield can be used with a vertically or steeply sloping and stable face. The machine consists of a shield skin and the excavation tools (excavator, milling head or ripper), the spoil removal equipment and the thrust cylinders. The excavated material is removed on a conveyor belt or scraper conveyor.

Face with partial mechanical support (SM-T2)

For partial support of the face, working platforms and/or breasting plates can be used. In platform shields, the tunnelling machine is divided into one or more platforms at the face. Natural slopes form on these, which support the face. The ground is excavated manually or mechanically. Platform shields have a low degree of mechanisation.

A disadvantage is the danger of large settlement resulting from uncontrolled face support. In shield machines with face support, the face is supported by breasting plates supported on hydraulic cylinders. In order to excavate the soil, the breasting plates are partially withdrawn. A combination of breasting plates and platforms is also possible. If support of the crown alone is sufficient, hinged breasting plates can be fixed at the crown.

Face with compressed air support (SM-T3)

The use of this type of machine (Fig. 1-12) is appropriate when types SM-T1 and SM-T2 are used in groundwater. Either the entire working area is pressurised, including the already excavated tunnel, or the machine is fitted with a pressure bulkhead (comparable to type SM-V3). The excavated material is transported hydraulically or through the lock in dry form.

Face with slurry support (SM-T4)

Many attempts have been made in the past to achieve face support using a support medium with partial face machines (e.g. thixshield). The excavation chamber in this case has to be completely filled. The soil can be excavated mechanically or by high-pressure jets.

As the excavation of the soil cannot be controlled sufficiently, this method of tunnelling did not prove successful and is no longer used.

19.5.5 Adaptable shield machines with convertible process technology (KSM)

Numerous tunnels run through very changeable ground conditions, which can range from rock to loosely consolidated soil. The process technology therefore has to be adapted to suit the geological conditions and appropriate adaptable shield machines have to be used. The various types are:

a) Shield machines, which can be operated with a different process without rebuilding:

– earth pressure balance shield SM-V5 ↔ compressed air shield SM-V3

b) Shield machines, which can be operated with a different process but have to be rebuilt. The following combinations have been tried:

– slurry shield SM-V4 ↔ shield without support SM-V1
– slurry shield SM-V4 ↔ earth pressure balance shield SM-V5
– earth pressure balance shield SM-V5 ↔ shield without support SM-V1

The rebuilding work normally lasts several shifts.

19.5.6 Special types

19.5.6.1 Blade shields

In blade shields, the shield skin is split into blades, which can be advanced independently. The ground is excavated by partial face machinery, cutting wheel or excavator. An advantage of blade shields is that they do not have to be circular and can, for example, drive a horseshoe-shaped section, in which case the invert is normally open. This is described as blade tunnelling. Because of various negative experiences in the past, however, blade shields are seldom used today.

19.5.6.2 Shields with multiple circular cross-sections

These shields are characterised by overlapping and non-concentric cutting wheels. The type of machine is currently only offered by Japanese manufacturers and mostly used to drive underground station cross-sections. The machines are difficult to steer and have not yet been used in Europe.

19.5.6.3 Articulated shields

Practically all types of shields can be equipped with an articulated joint to divide their length. This is provided particularly when the length of the shield skin is longer than the shield diameter in order to make the tunnelling machine easier to steer. The layout can also be necessary to drive very tight radius curves.

The description of the tunnelling machines is then according to one of the categories described above. A separate category of "articulated shields" is no longer usual.

19.5.7 Support and lining

With the tunnelling methods described here, the tunnelling machine and the support or lining are a combined process. For this reason, the most important support and lining methods are now described.

More detailed information about the various types of support and lining measures can be found in the appropriate standards, guidelines and recommendations (see Chapter 19.2).

19.5.7.1 Tunnel boring machines (TBM)

Due to the excavation process being relatively gentle to the surrounding rock mass and the favourable circular form, the extent of support measures is normally less than, for example, drilling and blasting. In less stable rock, the exposed surfaces have to be supported in good time in order to limit the loosening of the rock mass and thus mostly preserve the quality of the rock mass. If rupture occurs at the cutterhead, the extent of the support measures required can increase greatly.

Rock bolts

Rock bolts are normally installed radially in the cross-sectional plane of the cavity, and a layout oriented on the jointing can increase the effect of shear dowelling. If installed locally, they can hinder the spalling or breaking out of rock slabs, and if installed in a pattern they can reduce the loosening of the exposed sides of the tunnel. Rock bolts are particularly suitable for subsequent increasing of the support resistance, as they can also be installed later. The rock bolts are normally installed from the working platform behind the machine, or in special cases also directly behind the cutterhead shield.

Shotcrete

Shotcrete support is normally applied from a working platform at the back of the backup. The shotcrete serves to partially or completely seal the exposed surface of the rock mass (thickness 3 to 5 cm) or provide a load-bearing layer (thickness 10 to 30 cm). In order to increase the load-bearing action of the shotcrete layer, this is reinforced with one layer (rock side) or two layers (rock and air sides) of mesh reinforcement. Alternatively, steel fibre shotcrete can be used. The use of shotcrete robots enables high rates of spraying and is particularly beneficial from the point of view of health and safety.

Support arches

Support arches provide effective support to the rock mass and protection for the working space immediately after excavation and the installation of the arches. They are therefore mostly used in rock, which is liable to rockfall, instable or has squeezing characteristics. Either rolled steel profiles or lattice beams sections can be used for support arches. They are normally installed immediately behind the cutterhead in partial sections in the crown or as a closed ring.

19.5.7.2 Tunnel boring machines with shield (TBM-S), Shield machines (SM, DSM, KSM)

With tunnel boring machines with shield, or shield machines, the final support is installed in the protection of the shield skin or else the shield machine is operated at the front of a jacked pipe.

Precast elements installed in the shield tail (segments) serve to support the surrounding ground and as the abutment for the thrust force. The structural bond between the lining and the ground is created by grouting of the shield track as continuously as possible.

Segments and pipes are often used as a single-layer lining.

Concrete and reinforced concrete segments

The prefabricated elements used today are mostly precast concrete or reinforced concrete segments. The loading on the segment during transport and installation is often sufficient to require the installation of steel bars as reinforcement. Alternatively, segments with steel fibre reinforcement or a combination of rebars and fibres can be used. Steel fibres are particularly useful for the strengthening of the edges and corners, which are difficult to reinforce sufficiently with rebars.

SGI lining and steel segments

Spheroidal Graphite Iron (SGI) lining are now scarcely ever used because of the cost and fire protection problems. When the ground conditions are especially difficult, particularly if the bedding is poor, there is a danger of high deformation (convergence) or ring offsets. Welded steel segment rings, which are stiff in bending, are also often used to cope with the unusually high and asymmetrical loading at crosscuts, niches and other openings.

Hybrid segments

The hybrid segment is a combination of reinforced concrete and steel segments and offers an economic alternative to the use of full steel segments. These can be welded steel compartmented constructions filled with concrete or conventional reinforced concrete segments with integrated steel boxes bolted into the longitudinal or ring joint. This increases the stiffness of the system and the deformations are reduced.

Extruded concrete

Extruded concrete is a concrete tunnel lining, which is placed in a continuous process as unreinforced or fibre-reinforced concrete behind the tunnelling machine between the shield tail and a travelling inner formwork. The extruded concrete thus already supports the surrounding rock mass in the wet state. The use of extruded concrete is also possible below the groundwater table. Elastically supported face formwork, which is pushed forwards by the wet concrete pressure, ensures constant support pressure in the wet concrete.

Timber lagging

In ground without water, the primary support can consist of timber or steel lagging, which is installed between steel profiles in the protection of the shield tail (ribs and lagging). After the steel profiles have left the shield tail, they are braced against the ground by hydraulic cylinders, thus providing support. The tunnelling machine can be pushed forward against this braced support.

This method of support is not used in Europe due to the lack of fire protection during the construction period.

Pipes

Pipe jacking is a special process, in which reinforced concrete or steel pipes are jacked forward from a shaft and serve as support and final lining. These are normally circular but the use of rectangular sections is also possible.

Reinforced concrete

As with shotcrete, in-situ reinforced concrete can also be used with tunnelling machines to support the sides of the tunnel. As no thrust force can be transferred to the support, this type of support is only used with blade shields. The reinforced concrete is placed conventionally with a travelling formwork unit in 2.50 to 4.50 m wide sections in the protection of the following blades, which are still in contact with the last section to be concreted.

This process is no longer used in Central Europe on grounds of cost.

19.5.7.3 Advance support

The use of advance support measures, which are installed in the ground from a shield tunnelling machine, should only be used for short sections in emergency, since the implementation is technically laborious due to the poor accessibility and uneconomic due to the interruption of tunnelling advance. All possibilities of providing ground improvement from the surface should be exhausted first.

It is generally possible with current technology to implement rock bolts, pipe screens, drilled grouting and inclined or horizontal high-pressure grouting.

In order to make this possible, the tunnelling machine should be supplied with the necessary equipment, since later installation of drilling equipment is very expensive. Drilling booms are usually mounted on the segment erector and can drill through inclined pipes passing forwards through the shield skin (minimum angle to the shield centreline: about 8°).

Holes can also be drilled into the face through sealed openings in the pressure bulkhead. It should, however, be noted that broken drilling rods, which cannot be recovered, will lead to a severe obstruction of further advance.

The production of closed grouting bodies from the machine should not be provided, as this process is not practical for geometrical reasons. There is a basic risk in grouting that the grout can penetrate uncontrolled into the annular gap or the excavation chamber and thus lead to a failure of the face. When consolidation of the face is required, the excavation chamber should therefore be filled with a soil substitute first.

19.5.7.4 Support next to the tunnelling machine

If the ground is insufficiently stable, a tunnelling machine with active support and a shield should be selected. Next to the tunnelling machine, the ground is then supported by the shield skin, which however also completely obstructs access to the ground. Better passive support is provided by a short cylindrical shield than by a longer tapered shield.

19.6 Ground and system behaviour

19.6.1 Preliminary remarks

The system behaviour is of essential importance for the selection of a tunnelling machine, i.e. the behaviour of the overall system consisting of ground and selected tunnelling process [Austrian standard ÖNORM B 2203-2]. When a tunnelling machine is used, the ground behaviour criteria are fundamentally different from those in conventional tunnelling.

The geotechnical investigations are generally carried out based on DIN 4020. The determination of the characteristic values, the display and evaluation of the results of the geotechnical investigation and the conclusions, recommendations and notes should already be matched to the (probable) later tunnelling process early in the design phase.

More extensive and meaningful the preliminary investigations provide better preconditions for the selection of a process and a tunnelling machine. In this regard, it is recommended to include consideration of the entire process chain from excavation of the face, clearance of the muck and the final tipping or recycling of the excavated material in the planning of the geotechnical investigation.

The essential geotechnical parameters are summarised in Appendix 2 for each process and can serve as a rule of thumb for the selection of a tunnelling machine. They should be determined on each project for the relevant ground conditions. It should be noted that deviations of these ground parameters from the assumed values can result in complex and inconsistent consequences for the process chain. It is therefore recommended to provide appropriate provisions in the contract.

It is helpful and practical to display the expected ground conditions in a geotechnical longitudinal section and assign sections to relevant tunnelling classes.

There now follow basic remarks about the process-oriented analysis of system behaviour. A summary of the required characteristic values – split into soft ground and hard rock – is given in Appendices 2.1 and 2.2.

19.6.2 Ground stability and face support

The stability of the ground is the primary criterion for the selection of a type of tunnelling machine. The basis is the global and local stability of the face.

For an initial approximate evaluation of the stability (Austrian standard ÖNORM B2203, 1994, Tab 1), the following assignments can be defined according to the RMR classification:

A1 "stable": RMR 81–100

A2 "liable to rockfall": RMR 61–80

B1 "brittle": RMR 51–60

B2 "very brittle": RMR 41–50

B3 "non-cohesive": RMR 21–40

C squeezing rock": RMR < 20

For tunnels under built-up areas, a statement should also be made about the expected ground deformation or surface settlement with appropriate verification through calculations.

19.6.3 Excavation

The advance rate depends not only on the characteristic values of the ground but also on the selection of excavation tools, the geometry and design of the cutterhead/cutting wheel and the operating parameters of the machine. Changes in the geotechnical parameters can be unfavourable but also favourable for the advance. Because of the extremely complicated interactions between ground and tunnelling process, detailed analyses should be performed to clarify the causes.

Sticking in the excavation chamber and increased wear on the excavation tools in particular are the most frequent causes of disappointing progress and increased costs. These are now described in detail.

19.6.3.1 Sticking

The inclination of the soil to stick can have a decisive effect on the advance rate in mechanised tunnelling. Sticking reduces the advance rate because, for example, the excavation chamber of slurry-support machines has to be flushed or time-consuming manual cleaning leads to unplanned stoppages. In addition, sticking in combination with a high content of minerals liable to cause wear can lead to heavy wear on the cutting wheel and excavation chamber. Any propensity of the ground to stick should therefore always be described in geotechnical reports.

Soft ground with clay content, but also solid rock containing clay minerals, can result in considerable delays through sticking. Clays with pronounced plasticity and sedimentary rocks containing clay, like for example conglomerates/breccias with clay mineral content, siltstones and particularly claystones have proved particularly susceptible to sticking. Sticking often occurs in combination with water, which can come from natural formation water with open and earth pressure balance machines or process water (support suspension, soil conditioning, cutterhead jetting in hard rock).

The hindrance of progress through sticking can best be countered by recognising a potential sticking problem before the start of construction and appropriately adapting the equipment of the machine and the planned advance rate to take the problem into consideration. Geotechnical reports should provide the following information in this regard:

- Determination of the Atterberg limits and the consistency of the sol as an indication of the sticking potential according to DIN 18 122 for soft ground,
- Clay mineralogical analyses for the determination of the content of the most significant minerals (montmorillonite, kaolinite, illite, smectite, quartz etc.),
- Closer pattern of investigation in areas containing clay minerals for the more precise determination of the affected sections and the content of clay constituents at the face.

19.6.3.2 Wear

The wear on excavation and mucking components depends on the abrasiveness of the ground, the type of mechanical loading, the selection of tool materials and the operating parameters of the machine.

In soft ground, the mineralogical composition and the strength are relevant for tool wear but also the grading distribution, the grain shape and particularly the content of boulders and blocks. The test of the Laboratoire Central des Ponts et Caussées (LCPC test) offers one method of evaluating the abrasiveness of samples of soft ground with various mineralogical compositions and also takes the breakability of the grains into account. The verbal classification based on the ABR value used in the tables in Appendix 3 was not intended specifically for mechanised tunnelling and is currently being checked and revised in research programmes. Wear forecasts should therefore not be based on a verbal description of the abrasiveness shown by the values in the tables in Appendix 3, but use the index value (ABR value). In addition, the mineralogical composition, cutting wheel design, type of tool and process-related aspects of the excavation process should be taken into account.

In rock, wear can vary widely depending on rock strength, mineralogical composition, jointing and tunnel orientation to the texture of the rock mass. The Cerchar Abrasiveness Index (CAI) classifies the abrasiveness of rock. The most important parameters are the equivalent quartz content and the rock strength. High rock strength and correspondingly high CAI values lead to high primary wear in compact rock. In case abrasive, hard to break rocks are loosened out of the rock mass in an uncontrolled manner, the wear can increase over-proportionately due to impermissible shock loading. If the material flow is poor due to sticking or the design of the cutting wheel is unfavourable to material flow, a further increase of wear is likely (secondary wear). Further factors, which determine wear, are: brittleness, ductility, grain size, texture, porosity, mineral hardness, any foliation, the design of tools /cutting disc spacing, disc cutter diameter etc.), the materials used for tools, mode of operation and tool management (checking and replacement cycle).

In coarse- and mixed-grained soils, the primary wear is mainly determined by the breakability and strength of the coarse-grained fraction, boulders and blocks. The secondary wear increases with increasing equivalent quartz content and deterioration of the material flow and ease of excavation, particularly in wide-graded grain mixtures. Depending on the type of tunnelling machine used and the tools fitted, it is necessary to investigate whether breaking and grinding processes will occur in order to estimate wear rates.

19.6.3.3 Soil conditioning

The addition of additives in liquid or powder form, suspensions or water can be used to modify the properties of the excavated material. The concentration of the conditioning agent used can be estimated from experience and the characteristic values of the ground. The design of the cutting wheel, technical machine parameters and the required support pressure also have to be considered. Products should be chosen so that they do not flow uncontrolled into the surrounding ground but enable a homogeneous soil mixture.

For shield machines with slurry support to the face, conditioning can be in the form of liquid additives. For shield machines with earth pressure-supported face, conditioning in non-cohesive soft ground is normally provided by tenside foams with the possible addition of polymers, while in cohesive soft ground, polymer, bentonite or clay suspensions or even water can be used.

The purpose of any conditioning agent is the alteration of the properties of the excavated material to ensure the most trouble-free and economic tunnelling possible. For shields with slurry-supported face, this can mean that sticking and separation in the excavation chamber are reduced or prevented. For shield machines with earth pressure-supported face, non-cohesive soil can be processed into a plastic material by conditioning, sandy clays can be conditioned for less abrasive properties and in clay, conditioning is often used to reduce sticking and adhesion problems.

The additives added to soft ground should comply with the following minimum criteria:

– Simple and controllable dosage (ensured by the use of liquid additives),
– Avoidance of blockages in the additive feed and in the pipeline pumping the conditioned material out of the excavation chamber,
– Rapid development of effectiveness, in order to be able to react to geological alterations,
– Avoidance of environmental hazards.

19.6.3.4 Soil separation

On a tunnel project with slurry-supported face, the soil is separated from the transport medium (typically bentonite suspension) in a separation plant. Boulders and gravely and sandy soil contents are mechanically removed from the suspension on screens (coarse stage), cyclones and oscillating dewaterers (medium stage). Grain sizes below the sand fraction are separated from the suspension by chamber filter presses, centrifuges or high-performance cyclones (fine stage). Separation in centrifuges is improved by the previous addition of flocculants.

The configuration and dimensioning of the separation plant is mainly based on the grading distribution and the suspension circulation quantity. It should be borne in mind that ground improvement measures and breaking and grinding processes during excavation of the soil can increase the fines content and can worsen the properties of the suspension. High suspension densities and abrasive minerals increase wear to the excavation tools and the hydraulic mucking equipment.

19.6.3.5 Soil transport and tipping

In order to fully investigate cost-effectiveness for the selection of a tunnelling machine, muck transport and tipping also have to be considered. The characteristic parameters of the ground can be significantly altered by excavation, any soil conditioning and the individual control of the tunnelling machinery.

Uncontrolled ingress of formation water in shields without active face support can lead to liquefaction of the muck, which should be taken into account in the planning of transport and tipping.

Further information about tipping and conditioning is contained in Section 19.7.

19.7 Environmental aspects

Outside factors, which do not derive from the system behaviour (ground/tunnelling process, see Section 19.6), can sometimes also influence the selection of a tunnelling machine. Particularly when two different processes of equal technical value are possible, the factor "environmental impact" can be decisive. Particularly the suitability of the excavated material for recycling or landfill can be of great significance. The soil conditioning used with EPB shields, such as the addition of foams or polymers, can rule out the filling of the material in certain landfill sites.

Conditioning

The purpose of conditioning agents is described in Section 19.6.3.3. They only penetrate slightly into the subsoil, or not at all, but are transported out of the tunnel with the muck and thus have a significant effect on the suitability of the material for recycling or tipping.

Conditioning agents can be classified into various categories. These include water pollution classes (WGK 0, WGK 1, WGK 2, WGK 3), degradability (min. 60 % primary biological degradability and min. 80 % biological degradability) and the toxological threshold values for mammals (LD50) and water organisms (EC50).

Because of the wide range of conditioning agents, the composition of soft ground and its properties, no general classification of conditioned soil is possible. It is necessary to investigate on a case-by-case basis which threshold values are complied with and how the conditioned material should be processed. Information is given in the regulations in Section 19.2. According to the threshold, applicability of a conditioning agent can be so severely limited that the result can affect the tunnelling process (see also suitability for tipping).

Separation

In the separation plant, the excavated soil is separated from the transport medium, as described in Section 19.6.3.4.

When bentonite suspension is used as a transport medium, some residual bentonite content will always remain in the separated soil. This bentonite does not, however, alter the LAGA class (see below) of the soil. Recycling of separated soil is therefore possible, or not possible, according to the LAGA class of the excavated soil.

The separation of the fines in centrifuges is assisted by the previous addition of flocculants. Because of the number of flocculants available on the market, no general statements about the environmental acceptability of these products are possible. Information on this point can be found in the safety data sheet supplied by the manufacturer.

The material produced by centrifuges, filter presses and high-performance cyclones is very fine-grained and mostly of a pasty consistency. Recycling is therefore impractical, so the material has to be tipped. The same applies to used bentonite suspension, which should be passed to appropriate plants as liquid waste.

Working in groundwater

Bentonite suspension and additives for soil conditioning have both been used for many years in tunnels below the groundwater table. The same applies to the grout used for grouting the annular gap and the biologically degradable grease used for the sealing of the annular gap.

Tipping

The material removed from the tunnel should be processed and recycled if at all possible. If this is not possible, the soil will have to be tipped. When conditioning agents are used, attention should be paid to whether the excavated and conditioned soil complies with the chemical and physical requirements for tipping.

The tipping of conditioned soil is regulated in Germany by the guidelines of the Länderarbeitsgemeinschaft Abfall (LAGA) (States working collaboration on waste) and particularly by Guideline 20 "Requirements for the material recycling of mineral residues/wastes – Technical rules". This guideline governs the recycling of excavated soil and thus the tipping of excavated material from tunnelling. Only when the analytically determined value of chemical content rules out open tipping according to LAGA Guideline Nr. 20 (tipping classes Z0 to Z2), does the material have to be tipped in a regulated landfill site or even a tip for special waste (tipping classes Z3 to Z5). This is regulated in the Technical instructions for recycling, treatment and other disposal of municipal waste (TA Municipal waste). For ecological reasons, unrestricted or restricted open tipping is preferable.

Material from the coarse and medium stages of separating plants can normally be recycled. The fines content is mostly less than 5 %. However, these soils have special soil mechanics properties. The bentonite residue can swell again on contact with water and result in material with similar properties to cohesive soil. The material should therefore only be tipped in locations protected from water. For example, it can be used for backfilling a road tunnel beneath the carriageway. Alternatively, a further stage can be provided in the separating plant to wash the material. The simplest method is to spray the soil with water on the oscillating dewaterer. This can significantly reduce the residual bentonite content in the soil, which increases the quality of the soil and the possibilities for recycling.

Muck transported as sludge and the product of band filter presses and centrifuges cannot generally be tipped without further processing, as it tends to plastic flow. Conceivable ways of improving the suitability for tipping are consolidation through the addition of lime or storing on an intermediate stockpile until the material has dried and thus gained strength.

Concerning the suitability for tipping of soils, which have been treated with additives, the information about biological degradability supplied by the manufacturer is not sufficient on its own. The relevant regulations concerning pollutant content for each tipping class should be complied with to check the permissibility of tipping, particularly with regard to the residual content of hydrocarbons. Not least for economic reasons, the use of additives with slurry shields and earth pressure balance shields should be reduced to a minimum.

The soil to be tipped should already be classified into various categories during the design phase. This could be assignment to the classes according to LAGA (Z0, Z1, Z2, Z3, Z4, Z5) and the tipping classes (DK 0, DK I, DK II, DK III).

19.8 Other project conditions

In addition to the requirements resulting from the ground conditions and the location of the project in the surroundings, questions concerned with legal approvals and health and safety can also influence the selection of a tunnelling machine, and some of these are discussed below. This is not an exclusive list but examples are given, which could be of importance for the selection of a tunnelling machine in practice.

Planning decisions, requirements under water protection laws

The requirements and conditions of official approval, as can be attached to a planning decision, often restrict the selection of a tunnelling machine. For example, the temporary pumping of groundwater and the resulting lowering of the groundwater table may be limited or even forbidden, so that a tunnelling machine capable of operation under water pressure has to be used instead of an open machine. Another aspect could be conditions regarding the discharge of water into sewers or rivers.

Settlement and tunnelling beneath buildings

In urban areas, particularly when tunnelling beneath buildings and infrastructure, the permissible deformation of the ground at the surface is normally limited. In addition to the maximum absolute value of deformation, the extent and gradient of the settlement trough are to be included as criteria. Considering these "threshold values", which have to be calculated during the design phase, a suitable tunnelling machine should be selected to comply with the limits.

Material transport, restriction of construction traffic

The material can be transported in the tunnel by rail, truck, hydraulic slurry transport, sludge transport or on conveyor belts. In addition to the available space in the cross-section of the tunnel, the selection of a method is mostly based on the tunnel length, the possibilities of vertical transport in shafts and follow-up transport on the surface.

For transport above ground, the varying degrees of nuisance for local inhabitants are often significant (particularly in inner-city areas). The permissible limits for emissions, construction traffic restrictions (e.g. a night transport ban), duration of traffic disruption and vibration are the essential factors for transport, and these normally have to be considered in the approval process.

Occupational health and safety

The regulations concerning the protection of health and safety on construction sites (Baustellenverordnung – BaustellV) in addition to the workplace regulations serve to implement the Council Directive 92/57/EEC concerning the minimum health and safety regulations for construction sites, which are of limited duration or mobile. These regulations apply to all construction sites and thus apply to underground construction. Their application means that health and safety has to be considered in the design phase, which can well affect the selection of a tunnelling process.

The "Code of practice for the planning and implementation of a health and safety plan for underground construction sites" from DAUB and the national tunnelling associations of Austria and Switzerland (D-A-CH) is based on the regulations mentioned above among others and includes detailed requirements for the operation of tunnelling machines. In order to evaluate safety at work, a risk analysis is to be produced including consideration of the construction process and local conditions. The results of this risk analysis are then included in the decision process to select a tunnelling machine with a heavy weighting.

If, for example, the occurrence of gas like methane or argon in the ground is to be expected, the construction ventilation must be designed to cope with it or firedamp-safe machinery will have to be used. The presence of asbestos in the rock also demands special attention; appropriate continuous monitoring measuring devices should be installed permanently in the machine and in the tunnel and combined with an optical and acoustic alarm, which is activated automatically in case a critical value is measured. Closed machine types with active face support (SM-V 5, SM-V4) with closed material transport systems are advantageous, and the requirements for the sealing of segment gaskets should be defined. A two-layer lining system should be considered for the completed tunnel.

19.9 Scope of application and selection criteria

The recommendations for the scope of application and selection criteria are summarised for each type of machine in Tables 1 to 11 (Appendix 3).

19.9.1 General notes about the use of the tables

The feasibility of a system is first evaluated based on the key geotechnical parameters and processes, and economic evaluation criteria remain largely ignored. The tables are suitable for a preliminary selection on the exclusion principle. In case more than one type of machine would be possible, the final overall evaluation of suitability is then undertaken after an analysis of all project-specific parameters and processes, including consideration of economic and environmental aspects.

19.9.1.1 Core area of application

The fields marked black (Symbol "+") denote ranges, in which the type of machine has already been successfully used without many supplementary measures being required. The technical performance of the machine can vary among manufacturers, and the experience of the contracting company is also significant. The main areas of application shown for one parameter may be extended or restricted by the inclusion of other parameters.

19.9.1.2 Possible areas of application

The use of a tunnelling machine in the fields marked dark grey (Symbol "0") may require special technical measures, but the technical feasibility has been demonstrated. The achievable advance rates and cost-effectiveness may be reduced in comparison to the core area.

19.9.1.3 Critical areas of application

The use of a tunnelling machine in the fields marked light grey (Symbol "–") will probably require considerable additional measures or modification of the ground, otherwise difficulties should be expected. The achievable advance rates and cost-effectiveness will be considerably reduced in comparison to the core area. A founded analysis of the technical, economic and contractual risks and a comparison of variants with other tunnelling processes are strongly recommended.

19.9.1.4 Classification in soft ground

The grading distribution represents the direct and indirect evaluation criteria for the stability and permeability of the ground. Based on the shear strength parameters and the water pressure, and including consideration of the grading distribution, the stability of the ground is evaluated and the required support pressure is determined. The technical requirements placed on the machine increase with increasing ground and groundwater pressure.

19.9.1.5 Classification in rock

The table recommendations serve primarily to select the tunnelling machine and not to assess the cuttability. The rock mass classification and evaluation of stability are undertaken based on the RMR system. It is recommended to analyse the tunnelling machine system and all six project-specific parameters of the RMR system. Calculations to verify the stability and determine the support pressure are also recommended.

19.9.2 Notes about each type of tunnelling machine

19.9.2.1 TBM (Tunnel boring machine)

The main area of application is rock classed as stable to liable to rockfall, and water ingress from strata and joints can be overcome. The uniaxial compressive strength σ_D should be between 25 and 250 MN/m². Higher strengths, toughness of the rock and a higher content of mineral resistant to wear represent economic limits to application. A restricted ability of the machine to brace itself may also make the use of a machine impractical. For the assessment of the rock, the tensile strength and the RQD value are used. With a degree of fracturing of the rock mass with RQD from 50 to 100 % and a joint spacing of > 0.6 m, the use of a TBM seems assured. If the fracturing is worse, the stability should be checked. In soft ground or solid rock with similar properties to soil, the use of a TBM is impossible.

19.9.2.2 DSM (Double shield machines)

Double shield machines are mainly used for tunnel projects with long stretches through stable rock but also short stretches of rock classified as liable to rockfall to brittle. In a stable rock mass (see the requirements for the use of a TBM), the machine can work in continuous mode using the grippers for bracing. In fault zones or areas of lower rock strength, where the grippers cannot be used, the shield joint is retracted and the machine pushes itself using the auxiliary thrust cylinders against the last completed ring of segments.

19.9.2.3 SM-V1 (full-face excavation, face without support)

This type of machine can only be used in stable, predominantly water-impermeable, cohesive soft ground with high fines content. The stability of the face should be verified by calculation. It should also be verified that the sides of the excavation are temporarily stable until the final tunnel lining has been installed. Loosening of the ground, which could reduce the bedding, should be ruled out. If there is building on the surface susceptible to settlement, deformation of the subsoil and loosening should be verified using the usual damage classes (e.g. gradient of the settlement trough).

In rock, this type of machine can be used in rock classed as liable to rockfall to brittle, also with water in strata or joints. The strength of the rock mass can be greatly reduced even if the rock strength is good. This corresponds to a joint spacing of $\approx$ 0.6 to 0.06 m and a RQD value between approx. 10 and 50 %. In general, however, this type can be used in rock with compression strengths less than 5 MN/m^2, for example strongly weathered rock.

The stability of the face and the sides of the excavated cavity should be verified with calculations. In case of high water ingress, appropriate measures should be planned.

19.9.2.4 SM-V2 (full-face excavation, face with mechanical support)

Due to its failure on numerous projects, this type of machine is no longer recommended.

19.9.2.5 SM-V3 (Full-face excavation, face with compressed air application)

The application of compressed air enables machine type SM-V1 to be used in stable ground even under the groundwater table. The air permeability of the ground or the air consumption, the verification of the formation of an air flow and safety against blowouts are the essential criteria for the use of this type of machine. The groundwater table should be above the tunnel crown with an adequate margin of safety.

19.9.2.6 SM-V4 (full-face excavation, face with slurry support)

The main area of application of slurry shields is in coarse- and mixed-graded soil types. The groundwater table should also be above the tunnel crown with an adequate safety margin. As the ground is excavated, a fluid under pressure, e.g. bentonite suspension, supports the face. Highly permeable soils impede the formation of a membrane. At a permeability of over $5 \cdot 10^{-3}$ m/s, there is a danger that the bentonite flows uncontrolled into the ground. The scope of application can be extended by adding fine-grained material and filler or additives for the improvement of the rheological properties. Alternatively, additional measures to reduce the permeability of the soil (for example filling the pores) can be necessary. Boulders and blocks too large to be pumped can be broken by a crusher in front of the inlet. A high fines content can lead to difficulties with the separation. It should be borne in mind that the rheological properties of the support fluid are worsened by fine-grained material, as the separation of clay fractions and bentonite is not possible.

19.9.2.7 SM-V5 (full-face excavation, face with earth pressure balance support)

Machine types with earth pressure balance support are particularly suitable in soils with fines content (< 0.06 mm) of over 30 %. In coarse- and mixed-grained soils and rock, the contact force and the cutting wheel torque increase over-proportionately with increasing support pressure. The flow behaviour of the excavated muck can be improved with suitable conditioning agents like e.g. bentonite, polymers or foam. Soil conditioning with foam is recommended for active support pressure control and to ensure low settlement outside the predestined area of application.

Earth pressure balance shields have the advantage that operation is possible without modifying the process technology with partially filled and unpressurised excavation chamber in open mode (SM-V5-OM) without active face support. It should be noted that in this case the cutting wheel and screw conveyor combination will grind the excavated soil/rock considerably more than with a conveyor belt through the centre (SM-V1). If the soil tends to sticking, hindrance and increased wear have to be reckoned with. In order to improve the material flow and reduce the tendency to stick, conditioning agents should be used. Particularly unfavourable for earth pressure balance shields, both in soft ground and in rock, is a combination of high support pressure, high permeability, high abrasiveness and difficulty in breaking the grain structure.

19.9.2.8 SM-T1 (partial excavation, face without support)

This type of machine can be used above the groundwater table if the face is sufficiently stable, see here SM-V1.

Partial machines always offer good access to the face, so the process can be very advantageous if obstructions are to be expected.

19.9.2.9 SM-T2 (partial excavation, face with mechanical support)

This type of machine can be used when the support provided by material piling on the platforms at its natural angle of repose is sufficient for tunnelling with a limited degree of settlement control. Breasting plates can be installed for additional support in the crown and on the platforms. The main area of application is weakly to non-cohesive gravel-sand soils above the groundwater table with the corresponding angle of friction.

19.9.2.10 SM-T3 (partial excavation, face with compressed air application)

The use of this type of machine is appropriate when types SM-T1 and -T2 are to be used in the groundwater. The entire working area, including the already completed tunnel or just the working chamber, is pressurised.

19.9.2.11 SM-T4 (Partial excavation, face with slurry support)

Partial excavation machines with slurry-filled excavation chamber are no longer used.

19.9.2.12 KSM (Convertible shield machines)

Convertible machines combine the areas of application of each machine type in changeable ground conditions. Their area of application is therefore extended to both sets of criteria.

The number of conversions from one tunnelling process to another should be kept as low as possible, as rebuilding takes a long time and is expensive.

19.10 Appendices

Appendix 1 Overview of tunnelling machines

Information about the tunnelling machines and their areas of application can be found in Section 19.9.2 of these recommendations.

Brief description	Illustration (example)	
Tunnel boring machines (TBM)		
TBM Tunnelbohrmaschine ohne Schild Tunnel boring machine		
ETBM Erweiterungstunnelbohrmaschine Enlargement tunnel boring machine		
Double shield machines (DSM)		
DSM Doppelschildmaschine Double shield machine		
Shield machines (SM)		
SM-V1 ohne Stützung Without support		
SM-V2 mechanische Stützung Mechanical support		
SM-V3 Druckluftbeaufschlagung Full-face and compressed air application		
SM-V4 Flüssigkeitsstützung Full-face and slurry support		

SM-V5 Erddruckstützung Full-face and earth pressure balance support		
SM-T1 ohne Stützung Partial excavation and without support		
SM-T2 Teilstützung Partial excavation and partial support		
SM-T3 Druckluftbeaufschlagung Partial excavation and compressed air application		
SM-T4 Flüssigkeitsstützung Partial excavation and slurry support		

Convertible shield machines (KSM)

KSM Kombinationsschildmaschinen Convertible shield machines	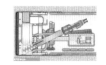	

Legend:

1 cutting wheel
2 shield skin
3 cylinders
4 erector
5 rear support
6 gripper
7 excavation chamber
8 pressure bulkhead

9 openings
10 excavation tool
11 muck clearance
12 carriage
13 air bubble
14 feed line
15 slurry line
16 pilot tunnel

Appendix 2.1   Process-related geotechnical parameters for soft ground

Process-related geotechnical parameters for soft ground	Brief description	Unit	TBM*	DSM*	SM-V1	SM-V2	SM-V3	SM-V4	SM-V5	SM-T1	SM-T2	SM-T3	SM-T4
Ortsbruststützung + Senkungsanalyse / **Face support + settlement analysis**													
Kornverteilung / Grain size distribution		%			x		x	x	x	x	x	x	
Dichte / Dichte unter Auftrieb / Soil density wet / submerged density	γ / γ'	kN/m³					x	x		x	x		
Lagerungsdichte / Compactness of the packing	D	–			x		x	x	x	x	x	x	
Reibungswinkel / Friction angle	φ'	°			x		x	x	x	x	x	x	
Kohäsion / Cohesion	c'	kN/m²			x	not recommended	x	x	x	x	x	x	not recommended
E-Modul / Elasticity modulus	E	MN/m²			x		x	x	x	x	x	x	
Dilatationswinkel / Dilatancy angle	Ψ	°			x		x	x	x	x	x	x	
Porenanteil / Pore content	n	–			x		x	x	x	x	x	x	
Porenzahl / Void ratio	e	–			x		x	x	x	x	x	x	
Durchlässigkeit / Permeability	k	m/s			x		x	x	x	x	x	x	
Erddruckbeiwert (horizontal) / Coefficient of lateral earth pressure	k_h	–			x		x	x	x	x	x	x	
Grundwasserdruck / Water pressure	p_{GW}	kN/m²			x		x	x	x	x	x	x	
Bodenabbau / **Soil removal**													
Verklebung / Sticking													
Plastizitätszahl ($I_P = w_L - w_P$) / Plasticity index	I_P	%			x		x	x	x	x	x	x	
Konsistenzzahl / Consistancy index	I_c	–			x	not recommended	x	x	x	x	x	x	not recommended
Fließgrenze / Liquid limit	w_L	%			x		x	x	x	x	x	x	
Ausrollgrenze / Rolling limit	w_P	%			x		x	x	x	x	x	x	
Wassergehalt / Water content	w	%			x		x	x	x	x	x	x	

19 DAUB recommendations for the selection of tunnelling machines

Process-related geotechnical parameters for soft ground	Brief description	Unit	TBM*	DSM*	SM-V1	SM-V2	SM-V3	SM-V4	SM-V5	SM-T1	SM-T2	SM-T3	SM-T4
Mineralogie Mineral composition					x		x	x	x	x	x	x	
Tonanteil (Siebrückstand < 0,002 mm) Percentage of clay		%			x		x	x	x	x	x	x	
Verschleiß **Wear**													
Abrasivität LCPC-Index Abrasiveness LCPC-Index	ABR	g/t			x		x	x	x	x	x	x	
Brechbarkeit LCPC-Index Breakability LCPC-Index	BR	%			x		x	x	x	x	x	x	
Quarzanteil Equivalent quarz index	äQu	%			x	not recommended	x	x	x	x	x	x	not recommended
Steinanteil Stone proportion		%			x		x	x	x	x	x	x	
Blockanteil Boulder proportion		%			x		x	x	x	x	x	x	
Druckfestigkeit Uniaxial compressive strength	σ_c	kN/m²			x		x	x	x	x	x	x	
Scherfestigkeit Shear strength					x		x	x	x	x	x	x	
Lagerungsdichte Compactness of the packing	D	–			x		x	x	x	x	x	x	
Bodenkonditionierung **Soil conditioning**													
Kornverteilung Grain size distribution		%						x	x				
Tonanteil (Siebrückstand < 0,002 mm) Percentage of clay		%						x	x				
Schluffanteil (Siebrückstand < 0,06 mm) Percentage of silt		%						x	x				
Plastizitätszahl ($I_P = w_L - w_P$) Plasticity index	I_P	%						x	x				
Konsistenzzahl Consistancy index	I_c	–						x	x				
Stützdruck Confinement pressure	p_s	bar				not recommended		x	x				not recommended
Porenanteil Pore content	n	–						x	x				
Durchlässigkeit Permeability	k	m/s						x	x				
chemische Grundwasseranalyse Chemical analysis of groundwater								x	x				
Anteil an organischen Substanzen (Kationen) Portion of organic substances (cations)		%						x	x				

Process-related geotechnical parameters for soft ground	Brief description	Unit	TBM*	DSM*	SM-V1	SM-V2	SM-V3	SM-V4	SM-V5	SM-T1	SM-T2	SM-T3	SM-T4
Bodenseparierung / Soil separation													
Restbentonitgehalt / Residual content of bentonite		%	n.r.	n.r.				x	x				n.r.
Restgehalt an chemischen Additiven / Residual content of chemical additives		%	n.r.	n.r.				x	x				n.r.
Tonanteil (Siebrückstand < 0,002 mm) / Percentage of clay		%	n.r.	n.r.				x	x				n.r.
Schluffanteil (Siebrückstand < 0,06 mm) / Percentage of silt		%	n.r.	n.r.				x	x				n.r.
Konsistenzzahl / Consistancy index	I_c	–	n.r.	n.r.				x	x				n.r.
undrainierte Kohäsion / Undrained cohesion	c_u	kN/m²	n.r.	n.r.				x	x				n.r.
Bodentransport und -deponierung / Soil transport and landfill													
Kornverteilung / Grain size distribution		%	n.r.	n.r.				x	x				n.r.
Schluffanteil (Siebrückstand < 0,06 mm) / Percentage of silt		%	n.r.	n.r.				x	x				n.r.
Tonanteil (Siebrückstand < 0,002 mm) / Percentage of clay		%	n.r.	n.r.				x	x				n.r.
Reibungswinkel / Friction angle	φ'	°	n.r.	n.r.	x	n.r.	x	x	x	x	x	x	n.r.
Kohäsion / Cohesion	c'	kN/m²	n.r.	n.r.	x	n.r.	x	x	x	x	x	x	n.r.
Plastizitätszahl ($I_P = w_L - w_P$) / Plasticity index	I_P	%	n.r.	n.r.				x	x				n.r.
Konsistenzzahl / Consistancy index	I_c	–	n.r.	n.r.				x	x				n.r.
E-Modul / Elasticity modulus	E	kN/m²	n.r.	n.r.	x	n.r.	x	x	x	x	x	x	n.r.
Restbentonitgehalt / Residual content of bentonite		%	n.r.	n.r.				x	x				n.r.
Restgehalt an chemischen Additiven / Residual content of chemical additives		%	n.r.	n.r.				x	x				n.r.
Wassergehalt / Water content	w	%	n.r.	n.r.				x	x				n.r.
Druckfestigkeit / Uniaxial compressive strength	σ_c	kN/m²	n.r.	n.r.	x	n.r.	x	x	x	x	x	x	n.r.
max. Kantenlänge / Max. block size		mm	n.r.	n.r.				x	x				n.r.

* TBM and DSM are only used in hard rock.

Appendix 2.2 Process-related geotechnical parameters for rock

Process-related geotechnical parameters for rock	Brief description	Unit	TBM	DSM	SM-V1	SM-V2	SM-V3	SM-V4	SM-V5	SM-T1	SM-T2	SM-T3	SM-T4
Ortsbruststützung + Senkungsanalyse / **Face support + settlement analysis**													
Gefüge / Texture			x	x	x		x	x	x	x	x	x	
Verwitterungsgrad / Weathering	W	–	x	x	x		x	x	x	x	x	x	
Zerlegung / Ratio matrix / fragmentation			x	x	x	not recommended	x	x	x	x	x	x	not recommended
Anisotropie / Anisotropy			x	x	x		x	x	x	x	x	x	
Porosität / Porosity	Φ	–	x	x	x		x	x	x	x	x	x	
Quellverhalten / Swelling capacity			x	x	x		x	x	x	x	x	x	
Diskontinuitäten / **Discontinuities**													
Einlagerungen / Infilling			x	x	x		x	x	x	x	x	x	
Orientierung / Discontinuity orientation			x	x	x		x	x	x	x	x	x	
Kluftabstand / Normal spacing of discontinuity sets			x	x	x		x	x	x	x	x	x	
Zerlegungsgrad / Degree of fracturing – Discontinuity frequency			x	x	x	not recommended	x	x	x	x	x	x	not recommended
Felsdruckbeiwert / Coefficient of lateral rock pressure	k_h	–	x	x	x		x	x	x	x	x	x	
Gebirgswasserzufluss / Water inflow	Q_W	l/s	x	x	x		x	x	x	x	x	x	
Gebirgswasserdruck / Water pressure	p_{GW}	kN/m²	x	x	x		x	x	x	x	x	x	
Bodenabbau / **Performance**													
Druckfestigkeit / Uniaxial compressive strength	σ_c	kN/m²	x	x	x	not recommended	x	x	x	x	x	x	not recommended
Spaltzugfestigkeit / Tensile strength (SPZ)	σ_z	MN/m²	x	x	x		x	x	x	x	x	x	
RQD / Rock Quality Designation	RQD	–	x	x	x		x	x	x	x	x	x	
Verwitterungsgrad / Weathering	W	–	x	x	x		x	x	x	x	x	x	

Process-related geotechnical parameters for rock	Brief description	Unit	TBM	DSM	SM-V1	SM-V2	SM-V3	SM-V4	SM-V5	SM-T1	SM-T2	SM-T3	SM-T4
RSR Rock Structure Rating	RSR	–	x	x	x		x	x	x	x	x	x	
RMR Rock Mass Rating	RMR	–	x	x	x		x	x	x	x	x	x	
GSI Geological Strength Index (Hoek Brown)	GSI	–	x	x	x		x	x	x	x	x	x	
RMI Rock Mass Index	RMI	–	x	x	x	not recommended	x	x	x	x	x	x	not recommended
Q-Index Q-value	Q-value	–	x	x	x		x	x	x	x	x	x	
Mineralogie Mineral composition			x	x	x		x	x	x	x	x	x	
Karbonat-Anteil Carbonate portion			x	x	x		x	x	x	x	x	x	
Verfestigung Cementation			x	x	x		x	x	x	x	x	x	
Diskontinuitäten Discontinuities													
Einlagerungen Infilling			x	x	x		x	x	x	x	x	x	
Blockgröße Block size			x	x	x		x	x	x	x	x	x	
Orientierung Discontinuity orientation			x	x	x	not recommended	x	x	x	x	x	x	not recommended
Kluftabstand Normal spacing of discontinuity sets			x	x	x		x	x	x	x	x	x	
Zerlegungsgrad Degree of fracturing – discontinuity frequency			x	x	x		x	x	x	x	x	x	
Verklebung Sticking													
Wassergehalt Water content	w	%	x	x	x		x	x	x	x	x	x	
Gebirgswasserzufluss Water inflow	Q_W	l/s	x	x	x	not recommended	x	x	x	x	x	x	not recommended
Mineralogie Mineral composition			x	x	x		x	x	x	x	x	x	

Process-related geotechnical parameters for rock	Brief description	Unit	TBM	DSM	SM-V1	SM-V2	SM-V3	SM-V4	SM-V5	SM-T1	SM-T2	SM-T3	SM-T4
Verschleiß / **Wear**													
Abrasivität (Cerchar Abrasivity Index) / Abrasiveness	CAI	–	x	x	x		x	x	x	x	x	x	
Quarzanteil / Equivalent quartz content	äQu	%	x	x	x		x	x	x	x	x	x	
Abrasivität RAI / Rock Abrasivity Index (RAI = Equ · UCS)	RAI	–	x	x	x		x	x	x	x	x	x	
Druckfestigkeit / Uniaxial rock compressive strength (UCS)	σ	MN/m²	x	x	x	not recommended	x	x	x	x	x	x	not recommended
Spaltzugfestigkeit / Tensile strength (SPZ)	σ_z	MN/m²	x	x	x		x	x	x	x	x	x	
Scherfestigkeit / Shear strength			x	x	x		x	x	x	x	x	x	
tonmineralische Zusammensetzung / Clay mineral composition			x	x	x		x	x	x	x	x	x	
Verwitterungsgrad / Weathering	W	–	x	x	x		x	x	x	x	x	x	
Verfestigung / Cementation			x	x	x		x	x	x	x	x	x	
Bodentransport und -deponierung / **Soil transport and landfill**													
max. Kantenlänge / Max. block size		mm	x	x	x	not recommended	x	x	x	x	x	x	not recommended
Druckfestigkeit / Uniaxial compression	σ_c	kN/m²	x	x	x		x	x	x	x	x	x	
Gebirgswasserzufluss / Water inflow	Q_W	l/s	x	x	x			x	x	x	x		
Wassergehalt / Water content	w	%	x	x	x		x	x	x	x	x	x	

Anlage 3.1 Areas of application and selection criteria TBM

Geotechnische Kennwerte Geotechnical values	TUNNELBOHRMASCHINE (TBM) Tunnel Boring Machine (TBM)				
	Lockergestein (Soft soil)				
Feinkornanteil (< 0,06 mm) DIN 18196 Fine grain fraction (< 0,06 mm)	< 5 %	5 – 15 %	15 – 40 %	> 40 %	
Durchlässigkeit k nach DIN 18130 [m/s] Permeability k [m/s]	sehr stark durchlässig very highly permeable > 10^{-2}	stark durchlässig strongly permeable $10^{-2} – 10^{-4}$	durchlässig permeable $10^{-4} – 10^{-6}$	schwach durchlässig slightly permeable 10^{-6}	
Konsistenz (Ic) nach DIN 18122 Consistency (Ic)	breiig pasty 0 – 0,5	weich soft 0,5 – 0,75	steif stiff 0,75 – 1,0	halbfest semi-hard 1,0 – 1,25	fest hard 1,25 – 1,5
Lagerungsdichte nach DIN 18126 Storage density	dicht dense	mitteldicht fairly dense	locker loose		
Stützdruck [bar] Supporting pressure [bar]	0	0		2 – 3	3 – 4
Quellverhalten Swelling behaviour	kein none	gering little	mittel fair	hoch high	
Abrasivität LCPC-Index ABR [g/t] Abrasiveness LCPC-index ABR [g/t]	sehr schwach abrasiv very low abrasive 0 – 500	schwach abrasiv low abrasive 500 – 1000	mittel abrasiv medium abrasive 1000 – 1500	stark abrasiv high abrasive 1500 – 2000	sehr stark abrasiv very high abrasive > 2000
Brechbarkeit LCPC-Index BR [%] Breakability LCPC-index BR [%]	sehr schwach very low 0 – 25	schwach low 25 – 50	mittel medium 50 – 75	stark high 75 – 100	sehr stark very high > 100
	Festgestein (Hard rock)				
Gesteinsfestigkeit [MPa] Rock compressive strength [MPa]	0 – 5 –	5 – 25 o	25 – 50 +	50 – 100 +	100 – 250 +	> 250 o
Bohrkern- Gebirgsqualität [RQD] Core sample - rock quality designation [RQD]	sehr gering very poor 0 – 25 –	gering poor 25 – 50 o	mittel fair 50 – 75 +	gut good 75 – 90 +	ausgezeichnet excellent 90 – 100 +
Rock Mass Ratio [RMR] Rock Mass Ratio [RMR]	sehr schlecht very poor < 20 –	schlecht poor 21 – 40 –	mäßig fair 41 – 60 o	gut good 61 – 80 +	sehr gut very good 81 – 100 +
Wasserzufluss je 10 m Tunnel [l/min] Waterinflow per 10 m tunnel [l/min]	0 +	0 – 10 +	10 – 25 +	25 – 125 o	> 125 –
Abrasivität (CAI) Abrasiveness (CAI)	kaum abrasiv not very abrasive 0,3 – 0,5 +	schwach abrasiv slightly abrasive 0,5 – 1 +	abrasiv abrasive 1 – 2 +	stark abrasiv very abrasive 2 – 4 o	extrem abrasiv extremely abrasive 4 – 6 –
Quellverhalten Swelling behaviour	kein none +	gering poor +	mittel fair o	hoch high –	
Stützdruck [bar] Supporting pressure [bar]	0 +	0 – 1 –	1 – 2 –	2 – 3 –	3 – 4 –

+ Haupteinsatzbereich / Main field of application
o Einsatz möglich / Application possible
– Einsatz kritisch / Application critical

Einsatz nicht empfohlen / Application not recommended

Anlage 3.2 Areas of application and selection criteria DSM

Geotechnische Kennwerte / Geotechnical values	DOPPELSCHILDMASCHINE (DSM) / Double Shield Machine (DSM)					
Lockergestein (Soft soil)						
Feinkornanteil (< 0,06 mm) DIN 18196 / Fine grain fraction (< 0,06 mm)	< 5 %	5 – 15 %	15 – 40 %	> 40 %		
Durchlässigkeit k nach DIN 18130 [m/s] / Permeability k [m/s]	sehr stark durchlässig / very highly permeable > 10^{-2}	stark durchlässig / strongly permeable $10^{-2} – 10^{-4}$	durchlässig / permeable $10^{-4} – 10^{-6}$	schwach durchlässig / low permeable < 10^{-6}		
Konsistenz (Ic) nach DIN 18122 / Consistency (Ic)	breiig / pasty 0 – 0,5	weich / soft 0,5 – 0,75	steif / stiff 0,75 – 1,0	halbfest / semi-hard 1,0 – 1,25	fest / hard 1,25 – 1,5	
Lagerungsdichte nach DIN 18126 / Storage density	dicht / dense	mitteldicht / fairly dense	locker / loose			
Stützdruck [bar] / Supporting pressure [bar]	0			2 – 3	3 – 4	
Quellverhalten / Swelling behaviour	kein / none	gering / little	mittel / fair	hoch / high		
Abrasivität LCPC-Index ABR [g/t] / Abrasiveness LCPC-index ABR [g/t]	sehr schwach abrasiv / very low abrasive 0 – 500	schwach abrasiv / low abrasive 500 – 1000	mittel abrasiv / medium abrasive 1000 – 1500	stark abrasiv / high abrasive 1500 – 2000	sehr stark abrasiv / very high abrasive > 2000	
Brechbarkeit LCPC-Index BR [%] / Breakability LCPC index BR [%]	sehr schwach / very low 0 – 25	schwach / low 25 – 50	mittel / medium 50 – 75	stark / high 75 – 100	sehr stark / very high > 100	
Festgestein (Hard rock)						
Gesteinsfestigkeit [MPa] / Rock compressive strength [MPa]	0 – 5	5 – 25	25 – 50	50 – 100	100 – 250	> 250
	○	○	+	+	○	○
Bohrkern- Gebirgsqualität (RQD) / Core sample - rock quality designation (RQD)		sehr gering / very poor 0 – 25	gering / poor 25 – 50	mittel / fair 50 – 75	gut / good 75 – 90	ausgezeichnet / excellent 90 – 100
		○	+	+	○	○
Rock Mass Ratio (RMR)		sehr schlecht / very poor < 20	schlecht / poor 21 – 40	mäßig / fair 41 – 60	gut / good 61 – 80	sehr gut / very good 81 – 100
		○	+	+	○	○
Wasserzufluss je 10 m Tunnel [l/min] / Waterinflow per 10 m tunnel (l/min)		0	0 – 10	10 – 25	25 – 125	> 125
		+	+	+	○	–
Abrasivität (CAI) / Abrasiveness (CAI)		kaum abrasiv / not very abrasive 0,3 – 0,5	schwach abrasiv / slightly abrasive 0,5 – 1	abrasiv / abrasive 1 – 2	stark abrasiv / very abrasive 2 – 4	extrem abrasiv / extremely abrasive 4 – 6
		+	+	+	○	–
Quellverhalten / Swelling behaviour		kein / none	gering / poor	mittel / fair	hoch / high	
		+	+	○	○	
Stützdruck [bar] / Supporting pressure [bar]		0	0 – 1	1 – 2	2 – 3	3 – 4
		+	–	–	–	–

+ Haupteinsatzbereich / Main field of application
○ Einsatz möglich / Application possible
– Einsatz kritisch / Application critical

Anlage 3.3 Areas of application and selection criteria SM-V1

Geotechnische Kennwerte / Geotechnical values	SCHILDMASCHINE mit Vollschnittabbau ohne Stützung (SM-V1) / Shield Machine with full-face and without support (SM-V1)				
	Lockergestein (Soft soil)				
Feinkornanteil (< 0,06 mm) DIN 18196 / Fine grain fraction (< 0,06 mm)	< 5 % –	5 – 15 % –	15 – 40 % o	> 40 % +	
Durchlässigkeit k nach DIN 18130 [m/s] / Permeability k [m/s]	sehr stark durchlässig very highly permeable > 10^{-2} –	stark durchlässig strongly permeable $10^{-2} – 10^{-4}$ –	durchlässig permeable $10^{-4} – 10^{-6}$ o	schwach durchlässig slightly permeable < 10^{-6} +	
Konsistenz (Ic) nach DIN 18122 / Consistency (Ic)	breiig pasty 0 – 0,5 –	weich soft 0,5 – 0,75 –	steif stiff 0,75 – 1,0 o	halbfest semi-solid 1,0 – 1,25 +	fest hard 1,25 – 1,5 +
Lagerungsdichte nach DIN 18126 / Storage density	dicht dense +	mitteldicht fairly dense o	locker loose –		
Stützdruck [bar] / Supporting pressure [bar]	0 +	0 – 1 –	1 – 2 –	2 – 3 –	3 – 4 –
Quellverhalten / Swelling behaviour	kein none +	gering little +	mittel fair o	hoch high –	
Abrasivität LCPC-Index ABR [g/t] / Abrasiveness LCPC-index ABR [g/t]	sehr schwach abrasiv very low abrasive 0 – 500 +	schwach abrasiv low abrasive 500 – 1000 +	mittel abrasiv medium abrasive 1000 – 1500 +	stark abrasiv high abrasive 1500 – 2000 +	sehr stark abrasiv very high abrasive > 2000 o
Brechbarkeit LCPC-Index BR [%] / Breakability LCPC-index BR [%]	sehr schwach very low 0 – 25 +	schwach low 25 – 50 +	mittel medium 50 – 75 +	stark high 75 – 100 +	sehr stark very high > 100 o
	Festgestein (Hard rock)				
Gesteinsfestigkeit [MPa] / Rock compressive strength [MPa]	0 – 5 o	5 – 25 o	25 – 50 +	50 – 100 +	100 – 250 o	> 250 o
Bohrkern- Gebirgsqualität (RQD) / Core sample - rock quality designation (RQD)	sehr gering very poor 0 – 25 o	gering poor 25 – 50 +	mittel fair 50 – 75 +	gut good 75 – 90 o	ausgezeichnet excellent 90 – 100 o
Rock Mass Ratio (RMR)	sehr schlecht very poor < 20 o	schlecht poor 21 – 40 +	mäßig fair 41 – 60 +	gut good 61 – 80 o	sehr gut very good 81 – 100 o
Wasserzufluss je 10 m Tunnel [l/min] / Waterinflow per 10 m tunnel [l/min]	0 +	0 – 10 +	10 – 25 +	25 – 125 o	> 125 –
Abrasivität (CAI) / Abrasiveness (CAI)	kaum abrasiv not very abrasive 0,3 – 0,5 +	schwach abrasiv slightly abrasive 0,5 – 1 +	abrasiv abrasive 1 – 2 +	stark abrasiv very abrasive 2 – 4 o	extrem abrasiv extremely abrasive 4 – 6 o
Quellverhalten / Swelling behaviour	kein none +	gering poor +	mittel fair o	hoch high –	
Stützdruck [bar] / Supporting pressure [bar]	0 +	0 – 1 –	1 – 2 –	2 – 3 –	3 – 4 –

+ Haupteinsatzbereich / Main field of application
o Einsatz möglich / Application possible
– Einsatz kritisch / Application critical

Anlage 3.4 Areas of application and selection criteria SM-V2

Geotechnische Kennwerte / Geotechnical values	SCHILDMASCHINE mit Vollschnittabbau und mechanischer Stützung (SM-V2) / Shield Machine with full-face and with mechanical support (SM-V2)					
	Lockergestein (Soft soil)					
Feinkornanteil (< 0,06 mm) DIN 18196 / Fine grain fraction (< 0,06 mm)	< 5 %	5 – 15 %	15 – 40 %	> 40 %		
Durchlässigkeit k nach DIN 18130 [m/s] / Permeability k [m/s]	sehr stark durchlässig / very highly permeable > 10^{-2}	stark durchlässig / strongly permeable $10^{-2} – 10^{-4}$	durchlässig / permeable $10^{-4} – 10^{-6}$	schwach durchlässig / slightly permeable < 10^{-6}		
Konsistenz (Ic) nach DIN 18122 / Consistency (Ic)	breiig / pasty 0 – 0,5	weich / soft 0,5 – 0,75	steif / stiff 0,75 – 1,0	halbfest / semi-solid 1,0 – 1,25	fest / hard 1,25 – 1,5	
Lagerungsdichte nach DIN 18126 / Storage density	dicht / dense	mitteldicht / fairly dense	locker / loose			
Stützdruck [bar] / Supporting pressure [bar]	0	0 – 1	1 – 2	2 – 3	3 – 4	
Quellverhalten / Swelling behaviour	kein / none	gering / little	mittel / fair	hoch / high		
Abrasivität LCPC-Index ABR [g/t] / Abrasiveness LCPC-index ABR [g/t]	sehr schwach abrasiv / very low abrasive 0 – 500	schwach abrasiv / low abrasive 500 – 1000	abrasiv / abrasive 1000 – 1500	stark abrasiv / high abrasive 1500 – 2000	sehr stark abrasiv / very high abrasive > 2000	
Brechbarkeit LCPC-Index BR [%] / Breakability LCPC-index BR [%]	sehr schwach / very low 0 – 25	schwach / low 25 – 50	mittel / fair 50 – 75	stark / high 75 – 100	sehr stark / very high > 100	
	Festgestein (Hard rock)					
Gesteinsfestigkeit [MPa] / Rock compressive strength [MPa]	0 – 5	5 – 25	25 – 50	50 – 100	100 – 250	> 250
Bohrkern- Gebirgsqualität [RQD] / Core sample - rock quality designation [RQD]	sehr gering / very poor 0 – 25	gering / poor 25 – 50	mittel / fair 50 – 75	gut / good 75 – 90	ausgezeichnet / excellent 90 – 100	
Rock Mass Ratio [RMR]	sehr schlecht / very poor < 20	schlecht / poor 20 – 40	mäßig / fair 41 – 60	gut / good 61 – 80	sehr gut / very good 81 – 100	
Wasserzufluss je 10 m Tunnel [l/min] / Waterinflow per 10 m tunnel [l/min]		0 – 10	10 – 25	25 – 125	> 125	
Abrasivität (CAI) / Abrasiveness (CAI)	kaum abrasiv / not very abrasive 0,3 – 0,5	schwach abrasiv / slightly abrasive 0,5 – 1	abrasiv / abrasive 1 – 2	stark abrasiv / very abrasive 2 – 4	extrem abrasiv / extremely abrasive 4 – 6	
Quellverhalten / Swelling behaviour	kein / none	gering / poor	mittel / fair	hoch / high		
Stützdruck [bar] / Supporting pressure [bar]	0	0 – 1	1 – 2	2 – 3	3 – 4	

Haupteinsatzbereich / Main field of application
Einsatz möglich / Application possible
Einsatz kritisch / Application critical

(Watermark: Ein Einsatz: nicht empfohlen / Application: not recommended)

19.10 Appendices

Empfehlung zur Auswahl von Tunnelvortriebsmaschinen (Stand 10/2010)

Anlage 3.5 Areas of application and selection criteria SM-V3

Geotechnische Kennwerte / Geotechnical values	SCHILDMASCHINE mit Vollschnittabbau und Druckluftstützung (SM-V3) / Shield Machine with full-face and compressed air application (SM-V3)					
	Lockergestein (Soft soil)					
Feinkornanteil (< 0,06 mm) DIN 18196 / Fine grain fraction (< 0,06 mm)	< 5 % –	5 – 15 % o	15 – 40 % +	> 40 % +		
Durchlässigkeit k nach DIN 18130 [m/s] / Permeability k [m/s]	sehr stark durchlässig / very highly permeable > 10^{-2} –	stark durchlässig / strongly permeable $10^{-2} - 10^{-4}$ –	durchlässig / permeable $10^{-4} - 10^{-6}$ o	schwach durchlässig / slightly permeable < 10^{-6} +		
Konsistenz (Ic) nach DIN 18122 / Consistency (Ic)	breiig / pasty 0 – 0,5 –	weich / soft 0,5 – 0,75 o	steif / stiff 0,75 – 1,0 +	halbfest / semi-solid 1,0 – 1,25 +	fest / hard 1,25 – 1,5	
Lagerungsdichte nach DIN 18126 / Storage density	dicht / dense +	mitteldicht / fairly dense o	locker / loose –			
Stützdruck [bar] / Supporting pressure [bar]	0 o	0 – 1 +	1 – 2 +	2 – 3 o	3 – 4 –	
Quellverhalten / Swelling behaviour	kein / none +	gering / little +	mittel / fair o	hoch / high –		
Abrasivität LCPC-Index ABR [g/t] / Abrasiveness LCPC-index ABR [g/t]	sehr schwach abrasiv / very low abrasive 0 – 500 +	schwach abrasiv / low abrasive 500 – 1000 +	mittel abrasiv / medium abrasive 1000 – 1500 o	stark abrasiv / high abrasive 1500 – 2000 –	sehr stark abrasiv / very high abrasive > 2000 –	
Brechbarkeit LCPC-Index BR [%] / Breakability LCPC-index BR [%]	sehr schwach / very low 0 – 25 +	schwach / low 25 – 50 +	mittel / medium 50 – 75 o	stark / high 75 – 100 –	sehr stark / very high > 100 –	
	Festgestein (Hard rock)					
Gesteinsfestigkeit [MPa] / Rock compressive strength [MPa]	0 – 5 o	5 – 25 o	25 – 50 o	50 – 100 o	100 – 250 o	> 250 o
Bohrkern- Gebirgsqualität [RQD] / Core sample - rock quality designation [RQD]	sehr gering / very poor 0 – 25 o	gering / poor 25 – 50 o	mittel / fair 50 – 75 o	gut / good 75 – 90 o	ausgezeichnet / excellent 90 – 100 o	
Rock Mass Ratio [RMR]	sehr schlecht / very poor < 20 o	schlecht / poor 21 – 40 o	mäßig / fair 41 – 60 o	gut / good 61 – 80 o	sehr gut / very good 81 – 100 o	
Wasserzufluss je 10 m Tunnel [l/min] / Waterinflow per 10 m tunnel [l/min]	0	0 – 10 +	10 – 25 +	25 – 125 o	> 125	
Abrasivität (CAI) / Abrasiveness (CAI)	kaum abrasiv / not very abrasive 0,3 – 0,5 o	schwach abrasiv / slightly abrasive 0,5 – 1 o	abrasiv / abrasive 1 – 2 o	stark abrasiv / very abrasive 2 – 4 –	extrem abrasiv / extremely abrasive 4 – 6 –	
Quellverhalten / Swelling behaviour	kein / none +	gering / poor +	mittel / fair o	hoch / high –		
Stützdruck [bar] / Supporting pressure [bar]	0 o	0 – 1 +	1 – 2 +	2 – 3 o	3 – 4 –	

+ Haupteinsatzbereich / Main field of application
o Einsatz möglich / Application possible
– Einsatz kritisch / Application critical

Empfehlung zur Auswahl von Tunnelvortriebsmaschinen (Stand 10/2010)

Anlage 3.6 Areas of application and selection criteria SM-V4

Geotechnische Kennwerte / Geotechnical values	SCHILDMASCHINE mit Vollschnittabbau und Flüssigkeitsstützung (SM-V4) / Shield Machine with full-face and fluid support (SM-V4)				
	Lockergestein (Soft soil)				
Feinkornanteil (< 0,06 mm) DIN 18196 / Fine grain fraction (< 0,06 mm)	< 5 % [+]	5 – 15 % [+]	15 – 40 % [+]	> 40 % [o]	
Durchlässigkeit k nach DIN 18130 [m/s] / Permeability k [m/s]	sehr stark durchlässig / very highly permeable $> 10^{-2}$ [+]	stark durchlässig / strongly permeable $10^{-2} – 10^{-4}$ [+]	durchlässig / permeable $10^{-4} – 10^{-6}$ [+]	schwach durchlässig / slightly permeable $< 10^{-6}$ [o]	
Konsistenz (Ic) nach DIN 18122 / Consistency (Ic)	breiig / pasty 0 – 0,5 [–]	weich / soft 0,5 – 0,75 [o]	steif / stiff 0,75 – 1,0 [+]	halbfest / semi-solid 1,0 – 1,25 [o]	fest / hard 1,25 – 1,5 [o]
Lagerungsdichte nach DIN 18126 / Storage density	dicht / dense [+]	mitteldicht / fairly dense [+]	locker / loose [o]		
Stützdruck [bar] / Supporting pressure [bar]	0 [o]	0 – 1 [+]	1 – 2 [+]	2 – 3 [+]	3 – 4 [+]
Quellverhalten / Swelling behaviour	kein / none [+]	gering / little [+]	mittel / fair [o]	hoch / high [–]	
Abrasivität LCPC-Index ABR [g/t] / Abrasiveness LCPC-index ABR [g/t]	sehr schwach abrasiv / very low abrasive 0 – 500 [+]	schwach abrasiv / low abrasive 500 – 1000 [+]	mittel abrasiv / medium abrasive 1000 – 1500 [+]	stark abrasiv / high abrasive 1500 – 2000 [o]	sehr stark abrasiv / very high abrasive > 2000 [o]
Brechbarkeit LCPC-Index BR [%] / Breakability LCPC-index BR [%]	sehr schwach / very low 0 – 25 [o]	schwach / low 25 – 50 [+]	mittel / medium 50 – 75 [+]	stark / high 75 – 100 [+]	sehr stark / very high > 100 [o]
	Festgestein (Hard rock)				
Gesteinsfestigkeit [MPa] / Rock compressive strength [MPa]	0 – 5 [o]	5 – 25 [o]	25 – 50 [o]	50 – 100 [o]	100 – 250 [o]	> 250 [o]
Bohrkern- Gebirgsqualität (RQD) / Core sample - rock quality designation (RQD)	sehr gering / very poor 0 – 25 [o]	gering / poor 25 – 50 [o]	mittel / fair 50 – 75 [o]	gut / good 75 – 90 [o]	ausgezeichnet / excellent 90 – 100 [o]
Rock Mass Ratio (RMR)	sehr schlecht / very poor < 20 [o]	schlecht / poor 21 – 40 [o]	mäßig / fair 41 – 60 [o]	gut / good 61 – 80 [o]	sehr gut / very good 81 – 100 [o]
Wasserzufluss je 10 m Tunnel [l/min] / Waterinflow per 10 m tunnel [l/min]	0 [o]	0 – 10 [+]	10 – 25 [+]	25 – 125 [+]	> 125 [+]
Abrasivität (CAI) / Abrasiveness (CAI)	kaum abrasiv / not very abrasive 0,3 – 0,5 [+]	schwach abrasiv / slightly abrasive 0,5 – 1 [+]	abrasiv / abrasive 1 – 2 [o]	stark abrasiv / very abrasive 2 – 4 [o]	extrem abrasiv / extremely abrasive 4 – 6 [o]
Quellverhalten / Swelling behaviour	kein / none [+]	gering / poor [+]	mittel / fair [o]	hoch / high [–]	
Stützdruck [bar] / Supporting pressure [bar]	0 [o]	0 – 1 [+]	1 – 2 [+]	2 – 3 [+]	3 – 4 [+]

[+] Haupteinsatzbereich / Main field of application
[o] Einsatz möglich / Application possible
[–] Einsatz kritisch / Application critical

Anlage 3.7 Areas of application and selection criteria SM-V5

Geotechnische Kennwerte / Geotechnical values	SCHILDMASCHINE mit Vollschnittabbau und Erddruckstützung (SM-V5) / Shield Machine with full-face and earth pressure balance support (SM-V5)				
	Lockergestein (Soft soil)				
Feinkornanteil (< 0,06 mm) DIN 18196 / Fine grain fraction (< 0,06 mm)	< 5 % −	5 – 15 % −	15 – 40 % ○	> 40 % +	
Durchlässigkeit k nach DIN 18130 [m/s] / Permeability k [m/s]	sehr stark durchlässig / very highly permeable > 10^{-2} −	stark durchlässig / strongly permeable $10^{-2} – 10^{-4}$ −	durchlässig / permeable $10^{-4} – 10^{-6}$ ○	schwach durchlässig / slightly permeable < 10^{-6} +	
Konsistenz (Ic) nach DIN 18122 / Consistency (Ic)	breiig / pasty 0 – 0,5 ○	weich / soft 0,5 – 0,75 +	steif / stiff 0,75 – 1,0 +	halbfest / semi-solid 1,0 – 1,25 ○	fest / hard 1,25 – 1,5 ○
Lagerungsdichte nach DIN 18126 / Storage density	dicht / dense +	mitteldicht / fairly dense +	locker / loose +		
Stützdruck [bar] / Supporting pressure [bar]	0 +	0 – 1 +	1 – 2 +	2 – 3 ○	3 – 4 −
Quellverhalten / Swelling behaviour	kein / none +	gering / little +	mittel / fair ○	hoch / high −	
Abrasivität LCPC-Index ABR [g/t] / Abrasiveness LCPC-index ABR [g/t]	sehr schwach abrasiv / very low abrasive 0 – 500 +	schwach abrasiv / low abrasive 500 – 1000 +	mittel abrasiv / medium abrasive 1000 – 1500 ○	stark abrasiv / high abrasive 1500 – 2000 ○	sehr stark abrasiv / very high abrasive > 2000 −
Brechbarkeit LCPC-Index BR [%] / Breakability LCPC-index BR [%]	sehr schwach / very low 0 – 25 +	schwach / low 25 – 50 +	mittel / medium 50 – 75 ○	stark / high 75 – 100 ○	sehr stark / very high > 100 −
	Festgestein (Hard rock)				
Gesteinsfestigkeit [MPa] / Rock compressive strength [MPa]	0 – 5 ○	5 – 25 ○	25 – 50 ○	50 – 100 −	100 – 250 −	> 250
Bohrkern- Gebirgsqualität [RQD] / Core sample - rock quality designation [RQD]	sehr gering / very poor 0 – 25 +	gering / poor 25 – 50 ○	mittel / fair 50 – 75 ○	gut / good 75 – 90 −	ausgezeichnet / excellent 90 – 100 −
Rock Mass Ratio [RMR]	sehr schlecht / very poor < 20 +	schlecht / poor 21 – 40 ○	mäßig / fair 41 – 60 ○	gut / good 61 – 80 −	sehr gut / very good 81 – 100 −
Wasserzufluss je 10 m Tunnel [l/min] / Waterinflow per 10 m tunnel [l/min]	0 ○	0 – 10 ○	10 – 25 ○	25 – 125 ○	> 125 ○
Abrasivität (CAI) / Abrasiveness (CAI)	kaum abrasiv / not very abrasive 0,3 – 0,5 +	schwach abrasiv / slightly abrasive 0,5 – 1 +	abrasiv / abrasive 1 – 2 ○	stark abrasiv / very abrasive 2 – 4 ○	extrem abrasiv / extremely abrasive 4 – 6
Quellverhalten / Swelling behaviour	kein / none +	gering / poor +	mittel / fair ○	hoch / high 	
Stützdruck [bar] / Supporting pressure [bar]	0 ○	0 – 1 +	1 – 2 	2 – 3 −	3 – 4 −

+ Haupteinsatzbereich / Main field of application
○ Einsatz möglich / Application possible
− Einsatz kritisch / Application critical

Anlage 3.8 Areas of application and selection criteria SM-T1

Geotechnische Kennwerte Geotechnical values	SCHILDMASCHINE mit Teilschnittabbau ohne Stützung (SM-T1) Shield Machine with part heading and without support (SM-T1)				
	Lockergestein (Soft soil)				
Feinkornanteil (< 0,06 mm) DIN 18196 Fine grain fraction (< 0,06 mm)	< 5 % −	5 – 15 % − ○	15 – 40 % +	> 40 % +	
Durchlässigkeit k nach DIN 18130 [m/s] Permeability k [m/s]	sehr stark durchlässig very highly permeable > 10^{-2} −	stark durchlässig strongly permeable $10^{-2} – 10^{-4}$ ○	durchlässig permeable $10^{-4} – 10^{-6}$ ○	schwach durchlässig slightly permeable < 10^{-6} +	
Konsistenz (Ic) nach DIN 18122 Consistency (Ic)	breiig pasty 0 – 0,5 −	weich soft 0,5 – 0,75 −	steif stiff 0,75 – 1,0 ○	halbfest semi-solid 1,0 – 1,25 +	fest hard 1,25 – 1,5 +
Lagerungsdichte nach DIN 18126 Storage density	dicht dense +	mitteldicht fairly dense ○	locker loose −		
Stützdruck [bar] Supporting pressure [bar]	0 +	0 – 1 −	1 – 2 −	2 – 3 −	3 – 4 −
Quellverhalten Swelling behaviour	kein none +	gering little +	mittel fair ○	hoch high −	
Abrasivität LCPC-Index ABR [g/t] Abrasiveness LCPC-index ABR [g/t]	sehr schwach abrasiv very low abrasive 0 – 500 +	schwach abrasiv low abrasive 500 – 1000 +	mittel abrasiv medium abrasive 1000 – 1500 +	stark abrasiv high abrasive 1500 – 2000 +	sehr stark abrasiv very high abrasive > 2000 ○
Brechbarkeit LCPC-Index BR [%] Breakability LCPC-index BR [%]	sehr schwach very low 0 – 25 +	schwach low 25 – 50 +	mittel medium 50 – 75 +	stark high 75 – 100 +	sehr stark very high > 100 ○
	Festgestein (Hard rock)				
Gesteinsfestigkeit [MPa] Rock compressive strength [MPa]	0 – 5 +	5 – 25 +	25 – 50 +	50 – 100 ○	100 – 250 −	> 250 −
Bohrkern- Gebirgsqualität (RQD) Core sample - rock quality designation (RQD)	sehr gering very poor 0 – 25 ○	gering poor 25 – 50 +	mittel fair 50 – 75 +	gut good 75 – 90 ○	ausgezeichnet excellent 90 – 100 ○
Rock Mass Ratio (RMR) Rock Mass Ratio (RMR)	sehr schlecht very poor < 20 ○	schlecht poor 21 – 40 +	mäßig fair 41 – 60 +	gut good 61 – 80 ○	sehr gut very good 81 – 100 ○
Wasserzufluss je 10 m Tunnel [l/min] Waterinflow per 10 m tunnel [l/min]	0 +	0 – 10 +	10 – 25 +	25 – 125 ○	> 125 −
Abrasivität (CAI) Abrasiveness (CAI)	kaum abrasiv not very abrasive 0,3 – 0,5 +	schwach abrasiv slightly abrasive 0,5 – 1 +	abrasiv abrasive 1 – 2 +	stark abrasiv very abrasive 2 – 4 ○	extrem abrasiv extremely abrasive 4 – 6 ○
Quellverhalten Swelling behaviour	kein none +	gering poor +	mittel fair ○	hoch high −	
Stützdruck [bar] Supporting pressure [bar]	0 +	0 – 1 −	1 – 2 −	2 – 3 −	3 – 4 −

+ Haupteinsatzbereich / Main field of application
○ Einsatz möglich / Application possible
− Einsatz kritisch / Application critical

19.10 Appendices

Empfehlung zur Auswahl von Tunnelvortriebsmaschinen (Stand 10/2010)
Seite 46 von 48

Anlage 3.9 Areas of application and selection criteria SM-T2

Geotechnische Kennwerte / Geotechnical values	SCHILDMASCHINE mit Teilschnittabbau und Teilstützung (SM-T2) / Shield Machine with part heading and partial support (SM-T2)				
	Lockergestein (Soft soil)				
Feinkornanteil (< 0,06 mm) DIN 18196 / Fine grain fraction (< 0,06 mm)	< 5 %	5 – 15 %	15 – 40 %	> 40 %	
	+	+	+	o	
Durchlässigkeit k nach DIN 18130 [m/s] / Permeability k [m/s]	sehr stark durchlässig / very highly permeable > 10^{-2}	stark durchlässig / strongly permeable $10^{-2} - 10^{-4}$	durchlässig / permeable $10^{-4} - 10^{-6}$	schwach durchlässig / slightly permeable < 10^{-6}	
	o	o	+	o	
Konsistenz (Ic) nach DIN 18122 / Consistency (Ic)	breiig / pasty 0 – 0,5	weich / soft 0,5 – 0,75	steif / stiff 0,75 – 1,0	halbfest / semi-solid 1,0 – 1,25	fest / hard 1,25 – 1,5
	–	o	o	o	o
Lagerungsdichte nach DIN 18126 / Storage density	dicht / dense	mitteldicht / fairly dense	locker / loose		
	+	+	o		
Stützdruck [bar] / Supporting pressure [bar]	0	0 – 1	1 – 2	2 – 3	3 – 4
	+	–	–	–	–
Quellverhalten / Swelling behaviour	kein / none	gering / little	mittel / fair	hoch / high	
	+	+	o	–	
Abrasivität LCPC-Index ABR [g/t] / Abrasiveness LCPC-index ABR [g/t]	sehr schwach abrasiv / very low abrasive 0 – 500	schwach abrasiv / low abrasive 500 – 1000	mittel abrasiv / medium abrasive 1000 – 1500	stark abrasiv / high abrasive 1500 – 2000	sehr stark abrasiv / very high abrasive > 2000
	+	+	+	+	o
Brechbarkeit LCPC-Index BR [%] / Breakability LCPC-index BR [%]	sehr schwach / very low 0 – 25	schwach / low 25 – 50	mittel / medium 50 – 75	stark / high 75 – 100	sehr stark / very high > 100
	+	+	+	+	o
	Festgestein (Hard rock)				
Gesteinsfestigkeit [MPa] / Rock compressive strength [MPa]	0 – 5 / 5 – 25	25 – 50	50 – 100	100 – 250	> 250
	+ +	+	o	–	–
Bohrkern- Gebirgsqualität [RQD] / Core sample - rock quality designation [RQD]	sehr gering / very poor 0 – 25	gering / poor 25 – 50	mittel / fair 50 – 75	gut / good 75 – 90	ausgezeichnet / excellent 90 – 100
	+	+	+	o	o
Rock Mass Ratio [RMR]	sehr schlecht / very poor < 20	schlecht / poor 21 – 40	mäßig / fair 41 – 60	gut / good 61 – 80	sehr gut / very good 81 – 100
	+	+	+	o	o
Wasserzufluss je 10 m Tunnel [l/min] / Waterinflow per 10 m tunnel [l/min]	0	0 – 10	10 – 25	25 – 125	> 125
	+	+	+	o	–
Abrasivität (CAI) / Abrasiveness (CAI)	kaum abrasiv / not very abrasive 0,3 – 0,5	schwach abrasiv / slightly abrasive 0,5 – 1	abrasiv / abrasive 1 – 2	stark abrasiv / very abrasive 2 – 4	extrem abrasiv / extremely abrasive 4 – 6
	+	+	+	o	o
Quellverhalten / Swelling behaviour	kein / none	gering / poor	mittel / fair	hoch / high	
	+	+	o	–	
Stützdruck [bar] / Supporting pressure [bar]	0	0 – 1	1 – 2	2 – 3	3 – 4
	+	–	–	–	–

+ Haupteinsatzbereich / Main field of application
o Einsatz möglich / Application possible
– Einsatz kritisch / Application critical

Empfehlung zur Auswahl von Tunnelvortriebsmaschinen (Stand 10/2010)

Anlage 3.10 Areas of application and selection criteria SM-T3

Geotechnische Kennwerte / Geotechnical values	SCHILDMASCHINE mit Teilschnittabbau und Druckluftstützung (SM-T3) / Shield Machine with part heading and compressed air application (SM-T3)				
	Lockergestein (Soft soil)				
Feinkornanteil (< 0,06 mm) DIN 18196 / Fine grain fraction (< 0,06 mm)	< 5 % −	5 – 15 % ○	15 – 40 % +	> 40 % +	
Durchlässigkeit k nach DIN 18130 [m/s] / Permeability k [m/s]	sehr stark durchlässig / very highly permeable > 10^{-2} −	stark durchlässig / strongly permeable $10^{-2} – 10^{-4}$ −	durchlässig / permeable $10^{-4} – 10^{-6}$ ○	schwach durchlässig / slightly permeable < 10^{-6} +	
Konsistenz (Ic) nach DIN 18122 / Consistency (Ic)	breiig / pasty 0 – 0,5 −	weich / soft 0,5 – 0,75 ○	steif / stiff 0,75 – 1,0 ○	halbfest / semi-solid 1,0 – 1,25 ○	fest / hard 1,25 – 1,5 ○
Lagerungsdichte nach DIN 18126 / Storage density	dicht / dense +	mitteldicht / fairly dense ○	locker / loose −		
Stützdruck [bar] / Supporting pressure [bar]	0 ○	0 – 1 +	1 – 2 +	2 – 3 ○	3 – 4 −
Quellverhalten / Swelling behaviour	kein / none +	gering / little +	mittel / fair ○	hoch / high −	
Abrasivität LCPC-Index ABR [g/t] / Abrasiveness LCPC-index ABR [g/t]	sehr schwach abrasiv / very low abrasive 0 – 500 +	schwach abrasiv / low abrasive 500 – 1000 +	mittel abrasiv / medium abrasive 1000 – 1500 +	stark abrasiv / high abrasive 1500 – 2000 +	sehr stark abrasiv / very high abrasive > 2000 ○
Brechbarkeit LCPC-Index BR [%] / Breakability LCPC-index BR [%]	sehr schwach / very low 0 – 25 +	schwach / low 25 – 50 +	mittel / medium 50 – 75 +	stark / high 75 – 100 +	sehr stark / very high > 100 ○
	Festgestein (Hard rock)				
Gesteinsfestigkeit [MPa] / Rock compressive strength [MPa]	0 – 5 +	5 – 25 +	25 – 50 +	50 – 100 ○	100 – 250 −	> 250 −
Bohrkern- Gebirgsqualität [RQD] / Core sample - rock quality designation [RQD]	sehr gering / very poor 0 – 25 +	gering / poor 25 – 50 +	mittel / fair 50 – 75 +	gut / good 75 – 90 ○	ausgezeichnet / excellent 90 – 100 ○
Rock Mass Ratio [RMR] / Rock Mass Ratio [RMR]	sehr schlecht / very poor < 20 +	schlecht / poor 21 – 40 +	mäßig / fair 41 – 60 +	gut / good 61 – 80 ○	sehr gut / very good 81 – 100 ○
Wasserzufluss je 10 m Tunnel [l/min] / Waterinflow per 10 m tunnel [l/min]	0	0 – 10	10 – 25	25 – 125	> 125
	kein Zufluss – Vortrieb im Grundwasser / no waterinflow – excavation below groundwater level				
Abrasivität (CAI) / Abrasiveness (CAI)	kaum abrasiv / not very abrasive 0,3 – 0,5 +	schwach abrasiv / slightly abrasive 0,5 – 1 +	abrasiv / abrasive 1 – 2 +	stark abrasiv / very abrasive 2 – 4 ○	extrem abrasiv / extremely abrasive 4 – 6 ○
Quellverhalten / Swelling behaviour	kein / none +	gering / poor +	mittel / fair ○	hoch / high −	
Stützdruck [bar] / Supporting pressure [bar]	0 ○	0 – 1 +	1 – 2 +	2 – 3 ○	3 – 4 −

+ Haupteinsatzbereich / Main field of application
○ Einsatz möglich / Application possible
− Einsatz kritisch / Application critical

Empfehlung zur Auswahl von Tunnelvortriebsmaschinen (Stand 10/2010)

Anlage 3.11 Areas of application and selection criteria SM-T4

Geotechnische Kennwerte *Geotechnical values*	SCHILDMASCHINE mit Teilschnittabbau und Flüssigkeitsstützung (SM-T4) *Shield Machine with part heading and fluid support (SM-T4)*				
	Lockergestein (Soft soil)				
Feinkornanteil (< 0,06 mm) DIN 18196 *Fine grain fraction (< 0,06 mm)*	< 5 %	5 – 15 %	15 – 40 %	> 40 %	
Durchlässigkeit k nach DIN 18130 [m/s] *Permeability k [m/s]*	sehr stark durchlässig *very highly permeable* > 10⁻²	stark durchlässig *strongly permeable* 10⁻² – 10⁻⁴	durchlässig *permeable* 10⁻⁴ – 10⁻⁶	schwach durchlässig *slightly permeable* < 10⁻⁶	
Konsistenz (Ic) nach DIN 18122 *Consistency (Ic)*	breiig *pasty* 0 – 0,5	weich *soft* 0,5 – 0,75	steif *stiff* 0,75 – 1,0	halbfest *semi-stiff* 1,0 – 1,25	fest *hard* 1,25 – 1,5
Lagerungsdichte nach DIN 18126 *Storage density*	dicht *dense*	mitteldicht *fairly dense*	locker *loose*		
Stützdruck [bar] *Supporting pressure [bar]*	0	0 – 1	1 – 2	2 – 3	3 – 4
Quellverhalten *Swelling behaviour*	kein *none*	gering *little*	mittel *fair*	hoch *high*	
Abrasivität LCPC-Index ABR [g/t] *Abrasiveness LCPC-index ABR [g/t]*	sehr schwach abrasiv *very low abrasive* 0 – 500	schwach abrasiv *low abrasive* 500 – 1000	mittel abrasiv *medium abrasive* 1000 – 1500	stark abrasiv *high abrasive* 1500 – 2000	sehr stark abrasiv *very high abrasive* > 2000
Brechbarkeit LCPC-Index BR [%] *Breakability LCPC-index BR [%]*	sehr schwach *very low* 0 – 25	schwach *low* 25 – 50	mittel *medium* 50 – 75	stark *high* 75 – 100	sehr stark *very high* > 100
	Festgestein (Hard rock)				
Gesteinsfestigkeit [MPa] *Rock compressive strength [MPa]*	0 – 5	5 – 25	25 – 50	50 – 100	100 – 250	> 250
Bohrkern- Gebirgsqualität (RQD) *Core sample - rock quality designation (RQD)*	sehr gering *very poor* 0 – 25	gering *poor* 25 – 50	mittel *fair* 50 – 75	gut *good* 75 – 90	ausgezeichnet *excellent* 90 – 100
Rock Mass Ratio (RMR)	sehr schlecht *very poor* < 20	schlecht *poor* 21 – 40	mäßig *fair* 41 – 60	gut *good* 61 – 80	sehr gut *very good* 81 – 100
Wasserzufluss je 10 m Tunnel [l/min] *Waterinflow per 10 m tunnel [l/min]*	0	0 – 10	10 – 25	25 – 125	> 125
Abrasivität (CAI) *Abrasiveness (CAI)*	kaum abrasiv *not very abrasive* 0,3 – 0,5	schwach abrasiv *slightly abrasive* 0,5 – 1	abrasiv *abrasive* 1 – 2	stark abrasiv *very abrasive* 2 – 4	extrem abrasiv *extremely abrasive* 4 – 6
Quellverhalten *Swelling behaviour*	kein *none*	gering *poor*	mittel *fair*	hoch *high*	
Stützdruck [bar] *Supporting pressure [bar]*	0	0 – 1	1 – 2	2 – 3	3 – 4

+ Haupteinsatzbereich / Main field of application
O Einsatz möglich / Application possible
− Einsatz kritisch / Application critical

Bibliography

[1] Aich, H.; Meseck, H.: Schildvortrieb im Bergsenkungsgebiet. Tiefbau BG 103 (1991), 738-748.

[2] Anagnostou, G.; Kovári, K.: Die Stabilität der Ortsbrust bei Erddruckschilden. MITTEILUNGEN der Schweizerischen Gesellschaft für Boden- und Felsmechanik. Frühjahrstagung, 29. April 1994, Zürich.

[3] Anheuser, L.: Der Vortriebsschild im Tunnelbau. Stahlbau 46 (1977), 133-137.

[4] Anheuser, L.: Der WF-Hydrojetschild im Rohrleitungsbau. Beiträge zur 5. Fachtagung Talsperrenbau. Wiss. Zeitschrift der Hochschule für Architektur und Bauwesen Weimar 30 (1984), 119-123.

[5] Anheuser, L.: Feststofftrennung aus Fördersuspensionen. Tagungsbericht Internat. Kongress Leitungsbau, Hamburg (1987), 617-622.

[6] Anheuser, L.: Gemessene Setzungen über mit dem Hydroschild aufgefahrenen Tunneln. Forschung + Praxis 27 (1982), 120-128.

[7] Anheuser, L.: Neuzeitlicher Tunnelausbau mit Stahlbetonfertigteilen. Beton- und Stahlbetonbau 76 (1981), 145-150.

[8] Apel, F.: Tunnel mit Schildvortrieb. 1. Aufl., Werner Verlag, Düsseldorf 1968.

[9] Arbeitsgemeinschaft 4. Röhre Elbtunnel: Die 4. Röhre des Elbtunnels in Hamburg. Hamburg 2004.

[9a] Azer, H.: Der Bau des Nord-Süd-Tunnels der Fernbahn in Berlin, ETR Eisenbahntechnische Rundschau 51 (2003) Nr. 6, S. 326-333.

[9b] Babendererde, L.: TBM mit Slurry- bzw. Erddruckstützung – Einsatzbereiche und Zuverlässigkeitsanalyse. Felsbau 21 (2003) Nr. 5, 155-160. Essen, Glück auf Verlag 2003.

[10] Babendererde, S.: Stand der Technik und Entwicklungstendenzen beim maschinellen Tunnelvortrieb im Lockerboden. Forschung + Praxis 33 (1990), 78-84.

[11] Babendererde, S.: Tunneling machines in soft ground. In: Civil engineering for underground rail transport. London, Butterworth 1990.

[12] Babendererde, S.: Tunnelvortrieb mit Messerschild und Spritzbetonausbau. Forschung + Praxis 19 (1976), 50-55.

[13] Babendererde, S.: Vorteile der Erddruckstützung maschineller Vortriebe bei Verwendung von Polymeren. Vortrag STUVA-Tagung, Hamburg 1993.

[14] Baldauf, H.; Timm, U.: Betonkonstruktionen im Tiefbau. Verlag Ernst & Sohn, Berlin 1988.

[15] Balmer, P.: Das speziell konzipierte Mixschildsystem für den Bau des zweispurigen Eisenbahntunnels Grauholz. Felsbau 10 (1992), 32-38.

[16] Baumann, Th.: Tunnelauskleidungen mit Stahlbetontübbingen. Bautechnik 69 (1992), Nr. 1, 33-35.

[17] Baumann, L.; Zischinski, U.: Neue Lösungen und Ausbautechniken zur maschinellen „Fertigung" von Tunneln in druckhaftem Fels. Felsbau 12 (1994), Nr. 1, 25-29.

[18] Baustellenprospekt: Arge Kajima, Tobishima, Toda, Maeda, Aoki: Underground regulation pond works, Loop 7. Tokyo 1992.

[19] Beckmann, U.: Tunnelvortriebsmaschinen und ihr Einsatz im Lockergestein. In: Taschenbuch für den Tunnelbau 1987. Essen, Glückauf Verlag 1986.

[20] BEG: Hauptbaumaßnahme H3-4 Münster – Wiesing, www.beg.co.at/.../unterinntalbahn.../hauptbaumassnahme-h3-4/.

[21] BEG: Hauptbaumaßnahme H8 Jenbach, www.beg.co.at/.../unterinntalbahn.../hauptbaumassnahme-h8/.

[22] Berger, K.: Straßentunnel unter dem Westbahnhof in Aachen. Züblin-Rundschau, 1983.

[23] Berufsgenossenschaft der Bauwirtschaft (BG BAU): BG-Information Handlungsanleitung für sicheres Arbeiten im Tunnelbau (BGI 5014). 2007.

[24] Berufsgenossenschaft der Bauwirtschaft (BG BAU): BG-Regel Bauarbeiten unter Tage (BGR 160). Oktober 1994.

[25] Berufsgenossenschaft der Bauwirtschaft (BG BAU): Druckluftverordnung. Juni 2005.

[26] Berufsgenossenschaft der Bauwirtschaft (BG BAU): Gefährdungsbeurteilung Maschineller Tunnelvortrieb. April 2009.

[27] Berufsgenossenschaft der Bauwirtschaft (BG BAU): Unfallverhütungsvorschrift Bauarbeiten (BGV C 22). Oktober 2002.

[28] Berufsgenossenschaft der Bauwirtschaft (BG BAU): Unfallverhütungsvorschrift Grundsätze der Prävention (BGV A 1). Januar 2004.

[29] Bielecki, A.; Schreyer, J.: Eignungsprüfungen für den Tübbingausbau der 4. Röhre des Elbtunnels, Hamburg. Tunnel 15 (1996), Nr. 3, 32-39.

[30] Biggart, A.; Rivier, J. P.; Sternath, R.: Storebaelt Railway Tunnel Construction. Int. Symposium on technology of bored tunnels under deep waterways. Kopenhagen (1993) 63-93.

[31] Bochon, A.; Rescamps, Y.; Chantron, L.: La Détection des Anomalies d'excavation au Tunnelier de Pression de boue: méthode mise au point sur le chantier EOLE, 1997.

[32] Boenert, L.: Baukostensenkung durch Anwendung innovativer Wettbewerbsmodelle, Artikel aus der Zeitschrift BFT Betonwerk + Fertigteil-Technik, ISSN 0373-4331, Jg. 69, Nr.5, 2003, Seite 28-35.

[33] Bolliger, J.: Erfahrungen TBE Vortrieb mit Hinterschneidtechnik. ETH-Symposium Tunnelbau-Hinterschneidtechnik Zürich (2005).

[34] Boscardin, M.D.; Cording, E. J.: Building Response to Excavation-Induced Settlement, Journal of Geotechnical Engineering 114 (1989), Nr. 1, 1-21.

[35] Braach, O.: Extrudierbetonbauweise für Tunnelauskleidungen. In: Taschenbuch für den Tunnelbau 1994, 211-237, Essen, Glückauf Verlag 1993.

[36] Braach, O.: Extrudierbetonbauweise im Tunnelbau; Verfahren und Ausführungsbeispiele. Beton 38 (1988) 95-98.

[37] Braach, O.: Risiken und Chancen beim mechanisierten Tunnelbau. Forschung + Praxis 34 (1992) 41-45.

[38] Braach, O.: Tunnel Freudenstein. In: Taschenbuch für den Tunnelbau 1994, 228-230. Essen, Glückauf Verlag 1993.

[38a] Braach, O.; Otten, B.: Der Westerscheldetunnel – Außerplanmäßige Ereignisse beim Vortrieb und deren Bewältigung, Forschung + Praxis 40 (2003), 188-195.

[39] Brinkgreve, R. B. J.; Broere, W. (EDT.): Handbuch: Plaxis 3D Tunnel – Version 2. Delft. Plaxis bv 2004.

[40] Brunar, G.; Powondra, F.: Nachgiebiger Tübbingausbau mit Meypo-Stauchelementen. Felsbau 3 (1985), Nr. 4, 225-229.

[41] Bundesanstalt für Straßenwesen: Zusätzliche Technische Vertragsbedingungen und Richtlinien für Ingenieurbauten ZTV-ING, Teil 5 Tunnelbau, Abschnitt 3 Maschinelle Schildvortriebsverfahren.

[42] Bundesministerium für Verkehr: Zusätzliche technische Vertragsbedingungen und Richtlinien für den Bau von Straßentunneln, Teil 1, Geschlossene Bauweise. 1995.

[43] Bundesverband der deutschen Beton- und Fertigteilindustrie: Handbuch für Rohre aus Beton, Stahlbeton, Spannbeton. Wiesbaden, Bauverlag 1978.

[44] CEN TC 114 N196 Temperature of Touchable Surfaces; Ergonomic Data to Establish Tempurature Limit Values for Hot Surfaces.

[45] CEN TC 151/WG4 N22 Tunnelling Machines – Air Locks; Safety Requirements (Vorentwurf).

[46] CEN TC 151/WG4 N8 Tunnelling Machines; Shield Machines, Horizontal Thrustboring Machines, Lining Erection Equipment; Safety Requirements (Vorentwurf vom 28.2.1992).

[47] Chew, A.: Alliancing in delivery of major infrastructure projects and outsourcing services – An overview of legal issues, Malleson Stephen Jaques, April 2005.

[48] Comulada, M., Wingmann, J.: Technisches Prozesscontrolling im Schildvortrieb. Vortrag STUVA 2010 – 50 Jahre STUVA, Düsseldorf, 17./18. Juni 2010.

[49] Conrad, E. U.: Funktionsmechanismen und Gestaltungskriterien von Erddruckschilden. Straßen- und Tiefbau (1987) 14-21.

[50] Crow, S.C.: Jet tunnelling machines; a guide for design. Tunnels & Tunnelling 7 (1975) 23-37.

[51] Dahl, J.: Anwendung des Stahlfaserbetons im Tunnelbau. In: Fachseminar Stahlfaserbeton, 04.03.1993. Institut für Baustoffe, Massivbau und Brandschutz, TU Braunschweig, Heft 100, 1993.

[52] Dahl, J.; Nußbaum, G.: Neue Erkenntnisse zur Ermittlung der Grenztragfähigkeit im Bereich der Koppelfugen. In: Taschenbuch für den Tunnelbau 1997, 291-319. Essen, Glückauf Verlag 1996.

[53] Darling, P.: How the TBMs have made it possible. Special issue: The Channel Tunnel; the French side. Tunnels & Tunnelling 23 (1991) 24-27.

[54] DAUB: Empfehlungen zur Auswahl von Tunnelvortriebsmaschinen. Herausgegeben vom Deutschen Ausschuss für unterirdisches Bauen e. V. (DAUB). Köln, 2010.

[55] DAUB: Empfehlungen zur Planung, Ausschreibung und Vergabe von schildgestützten Tunnelvortrieben. Herausgegeben vom Deutschen Ausschuss für unterirdisches Bauen e. V. (DAUB). Köln, Tunnel 2/93.

[56] DAUB, Empfehlungen zur Risikoverteilung in Tunnelverträgen. Herausgegeben vom Deutschen Ausschuss für unterirdisches Bauen e. V. (DAUB). Köln, Tunnel 3/98.

[57] DAUB: Empfehlung Statische Berechnung Schildvortriebsmaschinen. 2005.

[58] DAUB: Funktionale Leistungsbeschreibung für Verkehrstunnelbauwerke – Möglichkeiten und Grenzen für die Vergabe und Abrechnung. Herausgegeben vom Deutschen Ausschuss für unterirdisches Bauen e. V. (DAUB). Köln, Tunnel 4/97.

[59] DAUB: Leitfaden für die Planung und Umsetzung eines Sicherheits- und Gesundheitskonzeptes auf Untertagebaustellen. STUVA Extraausgabe 2007.

[60] DE – PS 3216919: Rohrabschnitt zum Herstellen eines geschlossenen Rohrbauwerks im Rohrvorpressverfahren.

[61] DE – PS 3222880: Verfahren und Einrichtung zum Herstellen einer Rohrleitung im unterirdischen Vortrieb.

[62] Deutsche Bahn AG: Richtlinie 853, Eisenbahntunnel planen, bauen und instandhalten. 1998.

[63] Deutsche Bundesbahn: Vorschrift für Eisenbahntunnel (VTU). DS 853, Vorausgabe 1984 und Ergänzungsbestimmung EzVTU 16 (Schildvortrieb), neuer Entwurf 1993.

[64] Deutscher Beton-Verein e. V. (Hrsg.): Merkblatt Bemessungsgrundlagen von Stahlfaserbeton im Tunnelbau. Wiesbaden: Eigenverlag 1996.

[65] Deutscher Beton-Verein e. V. (Hrsg.): Merkblatt Stahlfaserbeton. In Bearbeitung.

[66] Dewner, W.: Moderner Tunnelbau in Frankreich (Teil 2). Tiefbau BG 100 (1988) 4-21.

[67] Diete, P.; Kalthoff, D.; Stewering, T.: Stadtbahntunnel Duisburg TA 6 Duissern. Tiefbau BG 103 (1991) 68-79.

[68] Diete, P.; van de Linde, W.: Erfahrungen beim Auffahren des Westerschelde-Tunnels = Findings while driving the Westerschelde Tunnel. Tunnel 20 (2001) Nr. 5, S. 38-49a. Gütersloh, 2001.

[69] Dietrich, J.: Zur Qualitätsprüfung von Stahlfaserbeton für Tunnelschalen mit Biegezugbeanspruchung. Diss. Ruhr-Universität Bochum. Technisch-wissenschaftliche Mitteilungen des Instituts für Konstruktiven Ingenieurbau der Ruhr-Universität Bochum, Nr. 92-4, 1992.

[70] Dietz, W.; Kohle, R.: Metro Taipei – Vergleich der Erfahrungen beim Einsatz verschiedener Erddruckschildsysteme. Vortrag STUVA-Tagung, Hamburg, 1993.

[71] Dietz, W.; Raisch, D.: Untersuchungen zur Verwendung von zusammengesetzten Stahlbetonfertigteilen beim Bau großer Tunnel im Durchpressverfahren. Tiefbau Ingenieurbau Straßenbau 21 (1979) 991-995.

[72] DIN EN 12 110 Tunnelbaumaschinen – Druckluftschleusen – Sicherheitstechnische Anforderungen, Oktober 2002.

[73] DIN EN 12 336 Tunnelbaumaschinen – Schildmaschinen, Pressbohrmaschinen, Schneckenbohrmaschinen, Geräte für die Errichtung der Tunnelauskleidung – Sicherheitstechnische Anforderungen. März 2010.

[74] DIN 1054: Baugrund, Sicherheitsnachweise im Erd- und Grundbau.

[75] DIN 4085: Baugrund – Berechnung des Erddrucks. Februar 1987.

[76] DIN 15 262: Stetigförderer; Schneckenförderer für Schüttgut; Berechnungsansätze. Januar 1983

[77] Distelmeier, H.: Entwicklungstendenzen beim Bau von Fernbahntunnel in Deutschland – im Spiegel der Unternehmerinteressen, Vortrag STUVA-Tagung 1995, Tiefbau 10/1996.

[78] Donovan, H. J.: Expaded tunnel linings. Tunnels & Tunnelling 6 (Februar 1974) 46-53.

[79] Double-O-Tube Gesellschaft: Double-O-Tube Method. Tokyo.

[80] Duddeck, H.: Analysis of linings for shielddriven tunnels. In: Proceedings of the international symposium AFTES, Lyon (1984) 235-243.

[81] Duddeck, H.: Empfehlungen zur Berechnung von schildvorgetriebenen Tunneln (1973). Herausgegeben vom Arbeitskreis Tunnelbau der Deutschen Gesellschaft für Erd- und Grundbau e.V., Essen. In: Taschenbuch für den Tunnelbau 1977. Essen, Glückauf 1976.

[82] Duddeck, H.: Empfehlungen zur Berechnung von Tunneln in Lockergestein (1980). Herausgegeben vom Arbeitskreis Tunnelbau der Deutschen Gesellschaft für Erd- und Grundbau e.V., Essen, Bautechnik 57 (1980) 349-356.

[83] Duddeck, H.: Tunnelauskleidungen aus Stahl. Taschenbuch für den Tunnelbau 1978, 159-194, Essen, Glückauf 1977.

[84] Dumont, Ph.: Channel Tunnel Project. French Works. In: Proceedings of the RETC (1991) 763-773.

[85] Dyckerhoff und Widmann AG: Vorschubtechnik der Dyckerhoff und Widmann AG; Dammdurchpressung Westtangente Bochum. Firmeninformation.

[86] East Japan Railway Company; Kumagai Gumi: Keiyosen Kyobashi Tunnel – Multi Face Shield Method.

[87] Elmer, H.: Stadtbahn Mülheim a. d. Ruhr, BA 8; Ruhrunterquerung: Anpassung der Vortriebs- und Sicherungsarten an die Baugrund- und Grundwasserverhältnisse. In: Tunnelbauten im Rhein-Ruhr-Gebiet, Uni-GH Essen (1991) 49-71.

[88] EN 23411: Earth-moving machinery – Human physical dimensions of operators & minimum operator space envelope.

[89] EN 23449: Earth-moving machinery – Falling object protective structures (FOPS) – Laboratory test and performance requirements.

[90] EN 292: Safety of machinery – Basic concepts; General principles of design.

[91] EN 3: Portable Fire Extinguishers.

[92] EN 418: Safety of machinery – Emergency stop equipment.

[93] EN 50014: Electrical apparatus for potentially explosive atmospheres – General requirements.

[94] Englert, K.: Rechtlich richtige Beweisführung im Zusammenhang mit Bauleistungen, Mängeln und Schäden bei Untertagebauarbeiten, Forschung + Praxis 42 (2002), 216-220.

[95] EP 0538718: Improvements in a tunnel lining method and apparatus suitable for the purpose.

[96] Eves, R. C. W.; Curtis, D. J.: Tunnel lining design and procurement. In: The Channel Tunnel. Part 1: Tunnels. Civil Engineering Special Issue 1992, 127-143.

[97] Fang, Y.-S.; Lin, S.-J.; Lin, J.-S.: Time and settlement in EPB shield tunnelling. Tunnels & Tunnelling 25 (November 1993) 27-28.

[98] Farmer, I.; Garritti, Ph.; Glossop, N.: Operational characteristics of full face tunnel boring machines. In: Proceedings of the RETC (1987) 188-201.

[99] Feyerabend, B.: Zum Einfluss unterschiedlicher Stahlfasern auf das Verformungs- und Rissverhalten von Stahlfaserbeton unter den Belastungsbedingungen einer Tunnelschale. Diss. Ruhr-Universität Bochum, Technisch-wissenschaftliche Mitteilungen des Instituts für Konstruktiven Ingenieurbau der Ruhr-Universität Bochum, Nr. 95-8. 1995.

[100] Fives-Cail-Babcock (FCB) Firmeninformationen.

[101] Fleck, H.; Sonntag, G.: Statische Berechnung gebetteter Hohlraumauskleidungen mit einem ortsveränderlichen, last- und verformungsabhängigen Bettungsmodul aus der Methode der Finiten Elemente. Bautechnik 54 (1977) 149-156.

[102] Frankfurter Maschinenbau AG: Taschenbuch für Druckluftbetrieb. 9. neu bearbeitete Auflage, Berlin: Springer 1970.

[103] Fulcher, B.; Bednarski, J.; Bell, M.; Burger, W.: Piercing the mountain and overcoming difficult ground and water conditions with two hybrid hard rock TBMs. Proceedings of the RETC (2007), 229-254.

[104] Gehbauer. F.: Lean-Organisation: Möglichkeiten eines erweiterten Potenzials des Last Planner Systems, Positionspapier VDI 2008.

[105] Gehle, B., Wronna, A.: Der Allianzvertrag – Neue Wege kooperativer Vertragsgestaltung, Baurecht 38 (2007), Nr.1, 2-11.

[106] Gehring, K.: Anpassung von Teilschnittvortriebsmaschinen an durch von Gebirge und Bauwerk vorgegebene Bedingungen. In: 4. Int. Congress on Rock Mechanics, Montreux 1979, 401-410.

[107] Gehring, K.: Besonderheiten geologisch-geotechnischer Voruntersuchungen beim Einsatz von Teilschnittmaschinen. In: 4. Nat. Tagung über Felsmechanik der DGEG, Aachen (1980), 115 – 134.

[107a] Girmscheid, G.: Baubetrieb und Bauverfahren im Tunnelbau. 2. Auflage 2008. Ernst & Sohn Verlag, Berlin 2008.

[108] Göbl, A.: Hydroshield drive H3-4 in the lower Inn valley – construction experience/Hydroschildvortrieb H3-4 im Unterinntal – Erfahrungen der Bauausführung. Geomechanics and Tunnelling 2 (2009), No. 1, 61-72.

[109] Götz, W.: Der Tunnel unter dem Ärmelkanal; Erkenntnisse aus den maschinellen Vortrieben. Forschung + Praxis 34 (1991) 12-17.

[110] Grandori, C.: Development and current experience with double shield TBM. Proceedings of the 8th RETC (1987), 509-514.

[111] Grandori, C.: Fully mechanized tunnelling machine and method to cope with the widest range of ground conditions – Experiences with a hard rock prototype machine. Proceedings of the 3rd RETC (1976), 355-376.

[112] Grandori, C.; Dolcini, G.; Antonini, F.: The Los Rosales water tunnel in Bogota. Proceedings of the 10th RETC (1991), 561-581.

[113] Grandori, C.; Jäger, M.; Antonini, F.; Vigl, A.: Evinos-Mornos-Tunnel – Greece. In: Proceedings of the 12th RETC (1995), 747-767.

[114] Haack, A.: Neue Entwicklungen auf dem Gebiet des Schildvortriebs. Straßen- und Tiefbau 34 (1980), 11-26.

[115] Haack, A.: Vergleich zwischen der einschaligen und zweischaligen Bauweise mit Tübbingen bei Bahntunneln für den Hochgeschwindigkeitsverkehr. Forschung + Praxis 36 (1996). 251-256.

[116] Haid, H. G.; Maidl, U.: Prozess Controlling zur Klärung der Risikosphäre bei den Vorträgen des Katzenbergtunnels, Felsbau 25 (2007) Nr. 5, 153-158.

[117] Hamburger, H.; Weber, W.: Tunnel boring of large cross sections with fullface and enlarging machines in hard rock. In: Proceedings of the 11. RETC (1993), 811-832.

[118] Häussermann, I.: Die Tellerfeder; ein interessantes Konstruktionselement im Ingenieurbau. Bauingenieur 57 (1982), 133-138.

[119] Hazama Gumi: Firmeninformation.

[120] Heijboer, J.; van den Hoonaard, J.; van de Linde, F. W. J.: The Westerschelde Tunnel – Approaching Limits. A. A. Balkema 2004, Lisse.

[121] Hennecke, J.; Lange, R.; Setzepfandt, W.: Hydraulische Bergeförderung beim maschinellen Vortrieb des Radau-Stollens. Glückauf 116 (1980), 426-431.

[121a] Herdina, J.: Die Unterinntaltrasse in der Bauausführung. Beton- und Stahlbetonbau 103 (2008) Nr. 10, S. 672-681.

[122] Herrenknecht AG: Firmeninformation.

[123] Herrenknecht GmbH: Firmeninformationen.

[124] Herrenknecht, M.: Laseranwendungen beim unterirdischen Bauen kleiner Querschnitte. Tiefbau Ingenieurbau Straßenbau 32 (1990), 310-317.

[125] Herzog, M.: Die Pressenkräfte bei Schildvortrieb und Rohrvorpressung im Lockergestein. Baumaschine + Bautechnik 32 (1985), 236-238.

[126] Heß, F.; Domzig, H.; Knüppel, H.: Der Düsseldorfer Rheintunnel. Baumaschine + Bautechnik 8 (1961) 87-94, 153-161.

[127] Hettwer, J.; Elmer, H.: Ruhrunterquerung mit Suspensionsschild im Hartgestein. Felsbau 11 (1993), 119-127.

[128] Hewett, B. H.; Johannesson, S.: Shield and Compressed Air Tunnelling. New York: McGraw 1922. Deutsche Übersetzung: Schild- und Drucklufttunnelbau. Düsseldorf, Werner Verlag 1960.

[129] Hirokawa, H.; Nishitake, S.; Sugiyama, M.; Watanabe, T.; Miyazawa, K.: Development of shield machine for mechanical shield docking method (Part 1). Proceedings of the International Congress Towards New Worlds in Tunnelling, Acapulco, 16-20 May 1992. Rotterdam, Balkema 1992.

[130] Hitachi Zosen Corporation: Shield Tunnelling Machines. Firmeninformationen.

[131] Hofmann, R.; Suda, J.; Poisel, R.: Interaction of EC7 with EC2 in Tunnelling/Interaktion des EC7 mit dem EC2 beim Tunnelbau. Geomechanik und Tunnelbau 3 (2010), No. 1, S. 11-23.

[132] Holzhäuser, J.; Hunt, S. W.; Mayer, C.: Global Experience with Soft Ground and Weak Rock Tunneling under Very High Groundwater Heads. RETC 2006.

[133] Holzmann, Ph. AG: Fördetunnel Kiel. Tiefbau BG 105 (1993), 216-221.

[134] Holzmann, Ph. AG: Tunnelbau mit dem Thixschild im Bergsenkungsgebiet. Firmeninformation.

[135] Home, L.: Improving advance rates of TBMs in adverse ground conditions/Verbesserungen von TBM-Vortriebsleistungen bei ungünstigen Gebirgsverhältnissen. Geomechanics and Tunnelling 2 (2009), No. 2, 157-167.

[136] Horn: Landeskonferenz der ungarischen Tiefbauindustrie, Budapest, 1961.

[137] Howaldswerke Deutsche Werft AG: Tunnelvortriebsmaschine Energieversorgungstunnel Kiel. In: Zukunftorientiertes Konzept umweltgerecht realisiert. Firmeninformationen.

[138] Hurtz, G.; Weber, W.: Ermittlung der Rohrreibung und Entwicklung eines Bohrgerätes. Tiefbau Ingenieurbau Straßenbau 23 (1981), 550-555.

[139] Huwar, H.; Maidl, R.; Melzer, K.-H.: Schildertüchtigung und bauverfahrenstechnische Zusatzmaßnahmen beim Einsatz eines Bentonitschildes für die Stadtbahn Duisburg. Tunnels & Tunnelling, Bauma Special Issue, (April 1992) 36-39.

[140] IBECO Bentonit-Technologie GmbH: Bentonit für Tunnelbau und unterirdische Bauverfahren – Ein Handbuch für die Baupraxis. Kapitel II Bentonitsuspensionen als Stütz- und Fördermedium bei Hydroschilden. Mannheim, November 2003.

[141] Ide, K.; Watanabe, T.; Sugiyama, M.: Development of shield machine for mechanical shield docking method (Part 2). Proceedings of the International Congress Towards New Worlds in Tunnelling, Acapulco, 16-20 May 1992. Rotterdam, Balkema 1992.

[142] Inokuma, A. u.a.: The design for a shield with an elliptical excavation face. Proceedings of the International Congress Towards New Worlds in Tunnelling, Acapulco, 16-20 May 1992. Rotterdam, Balkema 1992.

[143] Isendahl, H.; Wild, M.: Gusseisen im Tunnelbau. Taschenbuch für den Tunnelbau 1979, 213-239, Essen, Glückauf 1978.

[144] ISO 2860: Earth-moving machines – minimum access dimensions. 1983.

[145] ISO 2867: Earth-moving machinery – Access systems. 1980

[146] ISO 3457: Earth-moving machinery – Guards & shields; Definitions & specifications. 1986.

[147] ISO 3795: Road vehicles – Determination of burning behaviour of interior material for motor vehicles. 1989.

[148] ISO 3864: Safety colors and safety signs. 1984.

[149] ISO 4413: Hydraulic fluid power – General rules for the application of equipment to transmission and control systems. 1978.

[150] ISO 4414: Pneumatic fluid power – Recommendations for the application of equipment to transmission and control systems. 1982.

[151] Jacob, E.: Der Bentonitschild; Technologie und erste Anwendung in Deutschland. Forschung + Praxis 19 (1976), 30-38.

[152] Jacob, E.: Weiterentwicklung des Bentonitschildes, insbesondere Schildanlage, Fördersystem, Separiereinrichtung, Schwanzblechabdichtung. Forschung + Praxis 21 (1978), 50-54.

[153] Janßen, P.: Tragverhalten von Tunnelausbauten mit Gelenktübbings. Diss. TU Braunschweig. Berichte aus dem Institut für Statik der Technischen Universität Braunschweig, Nr. 83/41. 1983.

[154] Japan Tunnelling Association: Challenges and changes: Tunnelling activities in Japan 1992. Tokyo.

[155] Japan Tunnelling Association: Tunnelling activities in Japan 1990. Tokyo.

[156] John, M.; Crighton, S.: Der Kanaltunnel: Gesamtprojekt und Arbeitsfortschritt. Felsbau 9 (1991), 6-12.

[157] Jonker, J. H.; Maidl, U.: Betuweroute: Erfahrungen mit dem Einsatz innovativer Schildvortriebstechnik. Forschung + Praxis 39 (1996), 52-55.

[158] Kainuma, N.; Akiyama, Y.: Development of Multiple-Duct-Pipe Jacking Method for small-bore power conduits. Proceedings of the International Congress Towards New Worlds in Tunnelling, Acapulco, 16-20 May 1992. Rotterdam, Balkema 1992.

[159] Kanbe, Y. u. a.: Development and verification tests of flexible section shield tunneling method. Proceedings of the International Congress Towards New Worlds in Tunnelling, Acapulco, 16-20 May 1992. Rotterdam, Balkema 1992.

[160] Kasper, T.; Meschke, G.: A 3D finite element simulation model for TBM tunnelling in soft ground. International Journal for Numerical and Analytical Methods in Geomechanics, 28 (2004), 1441-1460.

[161] Kazanskij, I.; Hinsch, J.: Einfluss der Pumpenkennlinienänderung auf die Wirtschaftlichkeit eines Saugbaggers. In: Proceedings of the 6. International Hafenkongress K.V.I.V. (1974) 2.06/1-2.06/4.

[162] Kirkland, C. J.; Craig, R. N.: Precast Linings for High Speed Mechanised Tunnelling. Forschung und Praxis 36 (1996), 119-123.

[162a] Kirschke, D.: Approaches to technical solutions for tunnelling in swellable ground/Lösungsansätze für den Tunnelbau in quell- und schwellfähigem Gebirge. Geomechanics and Tunnelling 3 (2010), Nr. 5, 547-556.

[163] Klawa, N.; Haack, A.: Tiefbaufugen. Ernst & Sohn, Berlin 1989.

[164] Kordina, K.; Dobbernack, R.: Zum Bericht über den Brandversuch in einem Stadtbahntunnel aus gewellten Stahlblechen in Gelsenkirchen. Tunnel 7 (1988), 123-132.

[165] Korittke, N.: Vortrieb und Vermessung beim Bau des Eurotunnels. In: Taschenbuch für den Tunnelbau 1991, 85-108, Essen: Glückauf 1990.

[166] Krabbe, W.: Die Wasserhaltung beim Schildvortrieb durch Grundwasserabsenkung und Druckluft. In: Vorträge der Baugrundtagung 1968 in Hamburg, DGEG, Essen, 1969.

[167] Krabbe, W.: Tunnelbau mit Schildvortrieb. In: Grundbau Taschenbuch, Bd. I, Ergänzungsband. Ernst & Sohn, Berlin 1971, 218-292.

[168] Kramer, J.: Der Einfluss von Tunnelbauten auf die Geländeoberfläche. Essen, Haus der Technik (1973) Nr. 314, 51-60.

[169] Krause, Th.: Geeignete Stützflüssigkeiten für Schilde mit flüssigkeitsgestützter Ortsbrust (Teil 1 und 2). Baumaschine + Bautechnik 35 (1988) 63-68, 129-134.

[170] Krause, Th.: Japanische Schildsysteme (Teil 1). Baumaschine + Bautechnik 33 (1986) 329-333

[171] Krause, Th.: Schildvortrieb mit flüssigkeits- und erdgestützter Ortsbrust. Mitteilung des Instituts für Grundbau und Bodenmechanik, TU Braunschweig, 1987, Heft 24.

[172] Kretschmer, M.; Fliegner, E.: Unterwassertunnel in offener und geschlossener Bauweise, Ernst & Sohn, Berlin 1987.

[173] Küffner, G.: Über den Sieg entscheidet nicht nur die Geschwindigkeit. Frankfurter Allgemeine Zeitung 11.01.2005.

[174] Kuhnhenn, K.: Neue Ansätze zur Ermittlung der Pressenkräfte durch Schildvortriebe. Forschung + Praxis 21 (1978), 55-61.

[175] Küttner, K. H.; Beitz, W.: Taschenbuch für den Maschinenbau. Springer Verlag, Berlin 1987.

[176] Kuwahara, H.; Fukushima, K.: The first application for the hexagonal tunnel segment in Japan. In: Tunnels for People. Rotterdam, Balkema 1997, 405-410.

[177] Lächler, W.: Einfluss von Bentonitschmierung beim Vorpressen großer Querschnittselemente. Tiefbau Ingenieurbau Straßenbau 23 (1981), 612-622.

[178] Lenz, D.; Möller, H. J.: Beispiele für im Durchpressverfahren eingebaute große Leitungen aus Stahlbeton- und Spannbetonrohren. Beton- und Stahlbetonbau 65 (1970), 183-193.

[179] Lovat Inc.: Firmeninformation.

[180] Lucke, W. N.: Versuchs- und Entwicklungsarbeiten mit Tunnelauskleidungen in den USA. Forschung + Praxis 15 (1974), 41-48.

[181] Magnus, W.: Neue Bauverfahren mit Stahlfaserpumpbeton beim Sammlerbau Hamburg. Konstruktiver Ingenieurbau, Ruhr-Universität Bochum, 1980, Heft 34

[182] Magnus, W.; Gebhard, K.: Ein außergewöhnliches Tunnelbauverfahren zur Unterquerung der Südelbe in Hamburg. Bauingenieur 54 (1979), 153-156.

[183] Maidl, B.: Handbuch des Tunnel- und Stollenbaus, Band I: Konstruktion und Verfahren. Essen, Glückauf 1984.

[184] Maidl, B.: Handbuch des Tunnel- und Stollenbaus, Band II: Grundlagen und Zusatzleistungen für Planung und Ausführung. Essen, Glückauf 1988.

[185] Maidl, B.: Handbuch für Spritzbeton. Ernst & Sohn, Berlin 1992.

[186] Maidl, B.: Der Tunnelbau braucht Innovationen für die Aufgaben der Zukunft. Felsbau 12 (1994), 148-159.

[187] Maidl, B.: Die neue Generation der Vortriebsmaschinen. 1. Internationales Rohrleitungsbausymposium Moderne Bauverfahren. Berlin (1993), 39-57.

[188] Maidl, B.: Konstruktive und wirtschaftliche Möglichkeiten zur Herstellung von Tunneln in einschaliger Bauweise. Schlussbericht des Forschungsvorhabens im Auftrag des Bundesministeriums für Verkehr der Bundesrepublik Deutschland. 1993.

[189] Maidl, B.: Stahlfaserbeton. Ernst & Sohn, Berlin 1991.

[190] Maidl, B.; Berger, Th.: Empfehlungen für den Spritzbetoneinsatz im Tunnelbau. Bauingenieur 70 (1995), Nr. 1, 11-19.

[191] Maidl, B.; Dietrich, J.: Erster Multi-Brust-Schild in Japan erfolgreich eingesetzt. Bautechnik 66 (1989) 145-148.

[192] Maidl, B.; Gipperich, Ch.: Entwicklung von Vortriebsmaschinen für Mikrotunnel. In: Taschenbuch für den Tunnelbau 1994, 285-312. Essen, Glückauf 1993.

[193] Maidl, B.; Handke, D.: Beispiele zum Stand der Schildvortriebstechnik in Deutschland (Teil 1 und 2). Tiefbau Ingenieurbau Straßenbau 32/33 (1991) 856-860/14-21.

[194] Maidl, B.; Handke, D.: Deponierfähigkeit von Böden bei Schildvortrieben. Tiefbau Ingenieurbau Straßenbau 30 (1988), 479-483.

[195] Maidl, B.; Herrenknecht, M.; Möhring, K.: Berichte zum 2. Internationalen Symposium Mikrotunnelbau München am 08.04.1992. Rotterdam, Brookfield, Balkema 1992.

[196] Maidl, B.; Herrenknecht, M; Anheuser, L.: Maschineller Tunnelbau im Schildvortrieb. 1. Auflage, Ernst & Sohn, Berlin 1994.

[197] Maidl, B.; Jodl, H.G.; Schmid, L.; Petri, P.: Tunnelbau im Sprengvortrieb. Berlin, Springer 1997.

[198] Maidl, B.; Kirschke, D.; Heimbecher, F.; Schockemöhle, B.: Abdichtungs- und Entwässerungssysteme bei Verkehrstunnelbauwerken. Forschung Straßenbau und Straßenverkehrstechnik, Heft 773. Bonn, Bundesdruckerei 1999.

[199] Maidl, B.; Maidl, U.; Ruse, N.: Erfahrungen mit der FEM-Simulation im Rahmen des Prozesscontrollings beim Schildvortrieb. Bauingenieur 80 (2005), Nr. 7, 337-342.

[200] Maidl, B.; Maidl, U.: Maschineller Tunnelbau mit Tunnelvortriebsmaschinen. In: Zilch, K.; Diederichs, C.J.; Katzenbach, R. (Hrsg.): Handbuch für Bauingenieure. Berlin, Springer 2001.

[201] Maidl, B.; Ortu, M.: Einschalige Bauweise. In: Festschrift Prof. Falkner, Institut für Baustoffe, Massivbau und Brandschutz der TU Braunschweig, Heft 142. 1999.

[202] Maidl, B.; Overmeyer, M.; Maidl, U.: Planung und Bau des Tunnels unter dem Ärmelkanal. Tunnel (1991) 173-186.

[203] Maidl, B.; Schmid, L.; Ritz, W.; Herrenknecht, M.: Tunnelbohrmaschinen im Hartgestein. Ernst & Sohn, Berlin 2001.

[204] Maidl, U.: Aktive Stützdrucksteuerung bei Erddruckschilden. Bautechnik 74 (1997), Nr. 6, 376-380.

[205] Maidl, U.: Erweiterung des Einsatzbereiches von Erddruckschilden durch Konditionierung mit Schaum. Diss. Ruhr-Universität Bochum 1994. Technisch-Wissenschaftliche-Mitteilungen des Instituts für konstruktiven Ingenieurbau. 1995.

[206] Maidl, U: FEM-Simulation und wissensbasierte Entscheidungsfindung im Rahmen des Process-Controllings beim hoch mechanisierten Schildvortrieb. Vortrag Geotechnik Kolloquium Salzburg 2004, Salzburg, Österreich.

[207] Maidl, U.: Geotechnical and mechanical interactions using the earth-pressure balanced shield technology in difficult mixed face and hard rock conditions. Vortrag RETC 2003, New Orleans, Louisiana, 16.-18. Juni 2003.

[208] Maidl, U.: Injektionsmaßnahmen zur Abdichtung von Fels auf der französischen Seite des Ärmelkanals. Diplomarbeit. TU München.

[208a] Maidl, U.: Partnering im hochmechanisierten Tunnelbau innerhalb bestehender Vertragsgerüste. In: VDI-Initiative „Partnerschaft am Bau" – Positionspapier – Dezember 2009.

[209] Maidl, U.: Process-Controlling bei hoch mechanisierten Bauverfahren. Vortrag Ruhr-Universität Bochum 2003.

[210] Maidl, U.; Nellessen, Ph.: Zukünftige Anforderungen an die Datenaufnahme und -auswertung bei Schildvortrieben, Bauingenieur Ausg. 3, 2003.

[211] Mair, R. J.; Taylor, R. N.: Bored tunnelling in the urban environment. Proc. 14th Int. Conference on Soil Mechanics and Foundation Engineering, Hamburg. Vol. 4, 2353-2385. Rotterdam, Balkema-Verlag 1997.

[212] Mair, R.J.; Taylor, R.N.; Burland, J. B.: Prediction of ground movements and assessment of risk of building damage due to bored tunnelling. In: Geotechnical Aspects of Underground Construction in Soft Ground, Mair & Taylor (Edt.). Rotterdam, Balkema-Verlag 1996.

[213] Mathewson, A.; Laval, D.: Brunel's tunnel... and where it led. Published by Brunel Exhibition Rotherhithe, 1992.

[214] Matsumoto, Y.; Uchida, S.; Koyama, Y.; Arai, T.: Multi-circular face shield driven tunnel. Tunnels and Water, Serrano (ed.). Rotterdam: Balkema 1988, 511-518.

[215] Meidinger, Ch.: Eureka 360: ALS-Tübbingauskleidungssystem CARRY. Magazin für Bauwirtschaft und Handel (5/1990) 12-15.

[216] Meldner, V.: Zur Statik der Tunnelauskleidung mit Stahlbetontübbings. Festschrift zum 100-jährigen Bestehen der Wayss & Freytag AG 1875/1975, 231-237.

[217] Merkblatt GW 312: Statische Berechnung von Vortriebsrohren. DVGW, Jan. 1990.

[218] Meseck, H.; Hollstegge, W.: Separierung von Stützflüssigkeiten im kritischen Kornbereich. Forschung + Praxis 33 (1990) 150-156.

[219] Mitsubishi Heavy Industries Ltd.: Firmeninformationen.

[220] Moh, Z. C.; Hwang, R. N.: Underground construction of Taipei transit system. Proceedings of the 11. South-East Asian Geotechnical Conference, Singapore (1993) 15-24.

[221] Möhring, K.: Gesteuerter Rohrvortrieb, eine wirtschaftliche Alternative beim Bau kleiner Abwasserkanäle. Tiefbau BG 98 (1986), 724-739.

[222] Müller, F., Bauernfeind, P.: U-Bahnbau in Nürnberg. Rock Mechanics Supplementum 6 (1978), 160-191.

[223] Myers, A.; John, M.; Fugemann, J. C. D.; Laffard, G. M.; Purrer, W.: Planung und Ausführung der britischen Überleitstelle im Kanaltunnel. Felsbau 9 (1991), 37-47.

[224] Nitschke, A.: Tragverhalten für Stahlfaserbeton für den Tunnelbau. Diss. Ruhr-Universität Bochum. Technisch-wissenschaftliche Mitteilungen des Instituts für Konstruktiven Ingenieurbau der Ruhr-Universität Bochum, Nr. 98-5, 1998.

[225] Nußbaumer, M.: Vorpressen von Großrohren und dessen Grenzen. Forschung + Praxis 32 (1988) 55-61.

[226] ÖNORM ENV 1997-1 (Eurocode 7: Entwurf, Berechnung und Bemessung in der Geotechnik).

[227] ÖNORM B 2203-1, Untertagebauarbeiten Werkvertragsnorm, Teil 1: Zyklischer Vortrieb. Ausgabe 2001.

[228] ÖNORM B 2203-2, Untertagebauarbeiten Werkvertragsnorm, Teil 2: Kontinuierlicher Vortrieb. Ausgabe 2005.

[229] O'Reilly, M. P.; New, B. M.: Settlements above tunnels in the United Kingdom – their magnitude and prediction. Transport and Road Research Laboratory, Crowthorne. England.

[230] Ortu, M.: Rissverhalten und Rotationsvermögen von Stahlfaserbeton für Standsicherheitsuntersuchungen im Tunnelbau. Diss. Ruhr-Universität Bochum. Fortschritt-Bericht VDI, Reihe 4, Nr. 164. Düsseldorf, VDI 2000.

[231] Pantet, A.: Creusement de galeries à faible profondeur à l'aide d'un tunnelier à pression de boue. Mésure „in situ" et étude théorique du champ de déplacement. Dissertation, INSA Lyon, 1991.

[232] Partnering bei Bauprojekten, Arbeitskreis „Partnerschaftsmodelle in der Bauwirtschaft", Hauptverband der Deutschen Bauindustrie e. V.

[233] Peck, R. B.: Deep excavation and tunnelling in soft ground. State of the Art Report. Proceedings of the 7th ICSMFE, Mexico (1969) 255-284.

[234] Philipp, G.: Schildvortrieb im Tunnel- und Stollenbau, Teil B; Tunnelauskleidung hinter Vortriebsschilden. In: Taschenbuch für den Tunnelbau 1987, 211-274. Essen, Glückauf 1986.

[235] Philipp, G.: Schildvortrieb im Tunnel- und Stollenbau, Teil A. In: Taschenbuch für den Tunnelbau 1986, 309-370. Essen: Glückauf 1985.

[236] Pont-à-Mousson, S. A.: Revêtement pour tunnels et galeries avec voussoirs en fonte ductile. Firmeninformation.

[237] Pröbstl, A.; Rolle, J.; Paul, F.: Entwicklung eines Erddruckschildes – zugeschnitten auf die bodenmechanischen und hydrologischen Gegebenheiten im Ruhrgebiet. Forschung + Praxis 32 (1987), 38-42.

[238] RAB 25: Arbeiten in Druckluft (Konkretisierungen zur Druckluftverordnung). Stand 12.11.2003.

[239] Redaktionsnotiz: Boomer drill rigs. Tunnels & Tunnelling 25 (August 1993) 11.

[240] Redaktionsnotiz: Japan boasts first rotating shield. Tunnels & Tunnelling 25 (Juni 1993) 13.

[241] Redaktionsnotiz: Österreich/Frankreich: Verbindungstunnel für TGV in Paris. Tunnel 12 (1993), 240.

[242] Redaktionsnotiz: Robots for automatic assembly of bolted segments. Tunnels & Tunnelling 17 (Juli 1985) 49.

[243] Redaktionsnotiz: Undersea breakthrough in Kiel. Tunnels & Tunnelling 22 (Juni 1990) 9.

[244] Rehm, U.; Wehrmeyer, G.: Maschinenkonzept für den maschinellen Tunnelvortrieb unter schwierigsten geotechnischen Bedingungen im Unterinntal. Tunnel 1/2008.

[245] Reuter, G.: Die Auffahrung einer zweigleisigen Tunnelstrecke für die U-Stadtbahn in Essen mit dem Westfalia-Messerschild-Vortriebssystem. Westfalia Berichte.

[246] Richtlinien für Tunnelbauten nach der Verordnung über den Bau und Betrieb der Straßenbahnen (BOStrab, Tunnelbaurichtlinien). Bundesminister für Verkehr, Dortmund, Borgmann 1971.

[247] Riecken, W.; Regger, F. H.; Unterleutner, K.; Steiner, A.: Tunnelvortrieb unter Druckluft beim U-Bahnbau in München. Tiefbau BG 105 (1993), 590-596.

[248] Robbins, R. J.: Hard rock tunneling machines for squeezing rock conditions. Three machine concepts. In: Tunnels for People. Rotterdam: Balkema 1997.

[249] Rostami, J.; Ozdemir, L.: A new model for performance prediction of hard rock TBMs. Proceedings of the 11. RETC (1993) 793-809.

[250] Rowe, R.: Pros and cons of continuous haulage. Tunnels & Tunnelling 25 (März 1993) 31-34.

[251] Rziha, F.: Lehrbuch der gesamten Tunnelbaukunst. Berlin, Ernst & Korn 1867 und 1872.

[252] Salomo, K. P.: Experimentelle und theoretische Bestimmung der Pressenkräfte und der Bodenverformung beim Vortrieb eines Vorpressrohres im rolligen Boden. Dissertation, TU Berlin, 1979.

[253] Scherle, M.: Rohrvortrieb. Bd. I: Technik, Maschinen, Geräte; Bd. II: Statik, Planung, Ausführung. Wiesbaden, Bauverlag 1977.

[254] Schimazek, J.: Verschleiß der Abbauwerkzeuge beim Einsatz von Teil- und Vollschnittmaschinen im Tunnel- und Bergbau. Forschung + Praxis 27 (1981), 41-45.

[255] Schmid, L.: Versuche Blasversatz Gubristtunnel, 1979, unveröffentlicht.

[256] Schmitter, J. J.: Compressed air and slurry tunnelling at Mexico-City. Proceedings of the International Congress Towards New Worlds in Tunnelling, Acapulco, 16-20 May 1992. Rotterdam, Balkema 1992.

[257] Schneider, E.: Gestaltung von Tunnelbauverträgen – das österreichische Modell. Vortrag beim 5. Deggendorfer Bausymposium am 12.3.2004.

[258] Schockemöhle, B; Heimbecher, F.: Stand der Erfahrungen mit druckwasserhaltenden Tunnelabdichtungen in Deutschland. Bauingenieur 74 (1999), Nr. 2, 67-72.

[259] Scholkämper, P.: Dammdurchpressung Westtangente Bochum. Tiefbau BG 101 (1989) 4-9.

[260] Schretter, K.; Maidl, U.; Wingmann, J.; Labda, T.: Process controlling for the hydroshield drives in the lower Inn valley (H3-4 and H8)/Prozesscontrolling bei den Hydroschildvortrieben im Unterinntal (H3-4 und H8). Geomechanics and Tunnelling Vol. 2 (2009), No. 6, S. 709-720.

[261] Schreyer, J.; Winselmann, D.: Eignungsprüfungen für die Tübbingauskleidung der 4. Röhre Elbtunnel – Ergebnisse der Scher-, Abplatz-, Verdrehsteifigkeits- und Lastübertragungsversuche. Taschenbuch für den Tunnelbau 1999, 337-352. Essen, Glückauf 1998.

[262] Schreyer, J.; Winselmann, D.: Eignungsprüfungen für die Tübbingauskleidung der 4. Röhre Elbtunnel – Ergebnisse der Großversuche. Forschung + Praxis 38 (1985) 102-107.

[263] S.E.L.I. S. p. A.: Firmeninformation.

[264] SIA 198, Untertagebau, Ausgabe 1998.

[265] Sicherheitsmanagementplan Wienerwald Tunnel; HL-AG.

[266] Sievers, W.: Entwicklungen im Tunnelbau. Beton 34 (1984), 347-354.

[267] Sozio, L. E.: Settlements in a Sao Paulo shield tunnel. Tunnels & Tunnelling 10 (September 1978), 53-55.

[268] Stack, B.: Handbook of Mining and Tunnelling Machinery. Chichester: Wiley 1982.

[269] Stadt Mühlheim an der Ruhr: Stadtbahn Mühlheim an der Ruhr, Bauabschnitt 8/Ruhrunterquerung.

[270] Stadtwerke München: Informationen über das Projekt Hofoldinger Stollen.

[271] Stahl, W.; Stadler, R.: Systeme zum Ein- oder Austrag von Schüttgütern in oder aus Druckräumen. Chem. Ing.-Techn. 56 (1984), 755-768.

[272] Stein, D.; Möllers, K.; Bielecki, R.: Leitungstunnelbau: Neuverlegung und Erneuerung von Ver- und Entsorgungsleitungen in geschlossener Bauweise. Ernst & Sohn, Berlin 1988.

[273] Stein, D.; Niederehe, W.: Instandhaltung von Kanalisationen; 2. Auflage, Ernst & Sohn, Berlin 1992.

[274] Steiner, W.: Balzari & Schudel AG (CH); Firmeninformationen.

[275] Stockhausen, O.: Der Elbtunnel in Hamburg und sein Bau. Zeitschrift des Vereines deutscher Bauingenieure 1912 (56), 177-185.

[276] Strohhäusl, S.: Eureka Contun: TBM tunnelling with high overburden. Tunnels & Tunnelling 28 (1996), Nr. 5, 41-43.

[277] Szechy, K.: Tunnelbau. Springer, Wien, 1969.

[278] Tauber, H.: S-Bahn-Tunnel aus vorgefertigten Teilstücken im Vorpressverfahren. Beton- und Stahlbetonbau 70 (1975), 68-70.

[279] Tauschinger, A.; Hulka, G.: Die besondere Eignung des GMP-Modells für Fast-Track Projekte im Hochbau, Bauingenieur Bd. 77, Oktober 2002.

[280] The Robbins Company: Firmeninformation.

[281] Thewes, M.; Burger, W. (2003): Verklebungen beim Schildvortrieb in Tonformationen – Erkennen und Begrenzen technischer und vertraglicher Risiken. Forschung + Praxis 40, Vorträge der STUVA-Tagung 2003 in Dortmund, Bertelsmann, Gütersloh.

[282] Thompson, J. F. K.: Flexible All-Purpose Segmental Tunnelling by Tunnel Boring Machine. Forschungsantrag: FAST by TBM, Ruhr-Universität Bochum, 1996, unveröffentlicht.

[283] Timm, T.: Erfahrungen mit der Anwendbarkeit der WLF-Gleichung zur Abschätzung der Relaxation von Elastomeren im langzeitigen Einsatz. Kautschuk und Gummi-Kunststoffe (1973), 134-140.

[284] Trawinski, H.: Der Hydrozyklon – die Zentrifuge einfachster Bauart; Anwendungsgebiete für den Hydrozyklon. Chemie-Anlagen und Verfahren (1970).

[285] Vigl, A.: Planung Evinos-Tunnel. Felsbau 12 (1994), Nr. 6, 495-499.

[286] Vigl, L.; Gütter, W.; Jäger, M.: Doppelschild-TBM – Stand der Technik und Perspektiven. Felsbau 17 (1999), Nr. 5, 475-485.

[287] Vigl, L.; Jäger, M.: Tunnel Plave-Doblar, Slowenien, 7 m Doppelschild-TBM mit einschaligem, hexagonalem Volltübbing-Auskleidungssystem. Vortrag auf dem Österreichischen Betontag 2000, 27.-28. April 2000, Wien.

[288] Vigl, L.; Pürer, E.: Mono-shell segmental lining for pressure tunnels. In: Tunnels for People. Rotterdam: Balkema 1997, 361-366.

[289] v. Schenck, W. R.: Beitrag zur Beschreibung des Baugrundverhaltens beim Druckluftschildvortrieb. Mitteilung des Franzius-Instituts für Grund- und Wasserbau, TH Hannover, Heft 19, (1961), 153-219.

[290] v. Schenck, W. R.; Wagner, H.: Luftverbrauch und Überdeckung beim Tunnelvortrieb mit Druckluft. Bautechnik 40 (1963) 41-47.

[291] Voest Alpine: Firmeninformationen.

[292] Wagner, H.: Die Luftdurchlässigkeit wasserhaltiger Böden. Mitteilung der Hannoverschen Versuchsanstalt für Grundbau und Wasserbau, Franzius-Institut, TH Hannover, Heft 11, 178-233.

[293] Wagner, H.: Verkehrstunnelbau, Bd. I: Planung, Entwurf und Bauausführung. Ernst & Sohn, Berlin 1968.

[294] Wagner, H.; Schulter, A.; Strohhäusl, S.: Gleitsegmente für flache und tiefe Tunnel. Forschung + Praxis 36 (1996), 257-266.

[295] Wallis, S.: Bau und Finanzierung des Kanaltunnels im Rückblick. Tunnel 12 (1993), 86-90.

[296] Wayss & Freytag AG: Firmeninformation.

[297] Wayss & Freytag AG: Sielbaustelle Blohmstraße, Hamburg. Erster Einsatz des Hydrojet-Schildes. Firmeninformation.

[298] Wehrmeyer, G.: Förderkreisläufe beim maschinellen Tunnelvortrieb. In: Taschenbuch für den Tunnelbau 2007, 208–236. Essen, VGE 2006.

[299] Wehrmeyer, G.: Massenkontrolle bei Schildvortrieben – Stand und Erfahrungen. In: Taschenbuch für den Tunnelbau 2002, 184–227. Essen: Glückauf 2001.

[300] Wenzel, P.: Probleme bei der Umsetzung von Partneringmodellen im öffentlichen Bausektor, VDI-Arbeitskreis „Partnering-Modelle für große Bauprojekte", Fraport AG 2008.

[301] Westfalia Lünen: Ausrüstungen für den Tunnel- und Stollenbau. Firmeninformationen.

[302] Westfalia Lünen: Messerschilde. Firmeninformationen.

[303] Wikipedia: http://de.wikipedia.org/wiki/Key_Performance_Indicator.

[303a] Wittke, W.: Felsmechanik. Grundlagen für wirtschaftliches Bauen im Fels. Berlin, Springer, 1984.

[304] Wooge: Auffahrung des Erkundungsstollens Freudensteintunnel mit zwei Tunnelvortriebsmaschinen. Referat auf dem 21. Tiefbau Kolloquium, Essen, 12.-13. März 1987.

[305] Yi, X.: Ground Movement and Pore Pressures induced by Shield Tunneling in soft ground. Dissertation Ruhr-Universität Bochum. Fortschrittberichte VD, Reihe 4, Nr. 105. Düsseldorf 1991.

[306] Yoshida, S.; Hiraide, T.: Construction of a ring subway line in Tokyo. Proceedings of the International Congress Towards New Worlds in Tunnelling, Acapulco, 16-20 May 1992. Rotterdam: Balkema 1992.

[307] Yoshikawa, T.; Uto, Y.; Nonomura, S.: A study of shield tunnelling machine; soil conditions for pressurized slurry shield to be adapted (japanisch).

[308] Zell, S.: Tunnelbau mit dem Hydrojet-Schild. Entwicklung und Ersteinsatz eines neuartigen Vortriebsverfahrens. Beton 31 (1981) 180-184.

[309] Züblin AG: Hydraulischer Rohrvortrieb mit dem Membranschildverfahren. Firmeninformationen.

[310] Zwicky, P.: Erfahrungen mit Tunnelabdichtungen in der Schweiz von 1960 bis 1995. Forschung + Praxis 37 (1998), 130-132.

Index

A

Abrasiveness   266
ABR value   see LCPC abrasiveness coefficient
Access openings   362
Active balancing pipes   237, 245
Advance grouting   see Grouting scheme
Advance rate   242, 245, 250
– at the Arrowhead Tunnel   193
– at the Channel Tunnel   188
– at the Grauholz Tunnel   299
– daily   245
– early   189
Advance rock consolidation   17
Advance speed   see Advance rate
Agitators   236, 238, 251, 262
Air bubble   37, 216, 226, 231
Air escape   see Air loss
Air loss   213, 232, see also Blow-outs
Air monitoring equipment   367
Air requirement   209
Airside Road Tunnel   284
Akima Tunnel   155
Annular gap   37, 47, 122, 139, 163, 404
Annular gap, causes for   38
Annular gap grouting   see Grouting
Area influenced by settlement   42
Arrowhead Tunnel   191
Articulated shield   404, 414
Assembly inaccuracies   137
Auger   339
Auger conveyor   see Screw conveyor
Auger microtunnelling   339
Auxiliary thrust cylinders   195
Award   387

B

Backflow valve for segments   140
Backhoe   221, see also Excavator arm
Back-loading disc cutters   80
Baffle   see Submerged wall
Band filter press   114
Bearings   88
– of the cutting wheel   269
Bedding of the segment rings   162
Belt scales   105, 260
Bends   see Compressed air sickness
Bentonite   28, 108, 256
– cement suspensions   172
– filter cake   see Filter cake
– jets   98
– loss of   245

– lubrication   196, 323
– shield   224, see also Hydroshield
– suspension   28, 98, 216, 225, 262
Bill of quantities   383, 385
Blade   see Scrapers
Blade resistance   56
Blade shields   12, 154, 218, 309, 414
Blade shield combined with the extruded steel fibre concrete   314
Blade shield, expanding   311
Blade shields with in-situ concrete lining   312
Blade shield with shotcrete support   314
Blasting   92
Blind shield   72, 100, 255
Block segments   123
Blow-out safety   209
Blow-outs   27, 209, 219, 249, 404
BOOT   388
BOP seals   169
BOT   388
Botlek Tunnel   293
Boulders   84, 177
Box jacking   325
Braking system   361
Breasting plates   53, 74, 80, 180, 302, 404, see also Support plates
Breathing apparatus   369
Brush seal   240
Buckets   70, 95
Building block system   293, 295, 337
Bulk density of the earth slurry   273
Bulking factor   105, 260
Bullflex hose   241

C

Calibre area   67
Canopies   300
CAI, Cerchar Abrasivity Index   404, 420
Casing rotator   273
Cement suspension   171
Central flushing   245
Centre cone type   229, 270
Centre cutter   235, 238, 247
Centre shaft type   229
Centrifugal pumps   103
Centrifuges   114
Centring cones   136
Chainage   351
Chamber filter press   113, 303
Channel Tunnel   186, 321, 332
Chisel   64, 66, 229, see also Round-shafted Chisel

Chongming project 250
Claim management 357
Clay 236, 238
Clay suspension 264
Cleaning pins 344
Closed cutting wheel 256, 268, 285
Closed mode 266, 404
Coarse screen 110
Collapse of face 300, 362
Collector pipe 346
Combinations, possible shield type combinations 291
Compressed air 9, 18, 201, 224, 312, 360, 368, 427
Compressed air equipment 201, 217
Compressed air locks 19, 204, 359
Compressed air regulations 204, 207, 359
Compressed air shield 11, 201, 215, 291
Compressed air shield with part face excavation 215
Compressed air shield with unpressurised working space 214
Compressed air sickness 209
Compressed air supply 206
Compressed air support 9, 26, 58, 324, 411, 413
Compressor installations 206
Compressor station 206
Compressor types 207
Concrete segment see Reinforced concrete segment
Conditioning 256, 259, 420
Conditioning agent 32, 256
Conditioning tools 271
Conex lining system 127
Confined soil shield 255
Connection of segments 142
Consistency of the soil 259, 263
Consolidation see Filtrate water loss
Construction method, Berlin 346
Construction method, double-layer 118
Construction method, single-layer 118
Constructional elements 47
Contact forces 57, 68
Contact pressure, disc cutters 60
Contact pressure, excavation tools 57
Contact surfaces 132, 360
Continuous advance 200
Contract law 384
Contract models 383
Control 363
Control cabin 351
Control circuit of an earth pressure shield 257
Control devices 364
Control of support pressure 231, 260
Control position 339, 355, 363
Control stations
Control systems 364
Convergences 44, 128
Convertible shield 184, 232, 291, 414, 428
Copy cutter 251
Copy tools see Gauge tools
Cost drivers 386
Crossbeam 49

Crossover 189, 321
Crossover caverns 182
Crusher 97, 227, 232, 238, 293, 344
Curve radius 280, 354
Curved tunnels 351
Cutter arm 67
Cutterheads 79, 360, 404, see also Cutting wheels
Cutting wheels in multi-face shields 316
Cutting bars 65
Cutting edges 64
Cutting wheels 79, 229, 232, 234, 267, 281, 293, 316, 404
Cutting wheel drive see Drive, cutting wheel
Cutting wheel openings 235
Cutting wheel, accessible 85, 247
Cutting wheel, tilting 59
Cutting wheel, rim 80
Cutting wheel, spokes 79
Cyclone plant 110

D

Daily advance see Advance rate
Damage to segments 144
Data management 393
DAUB 5, 401
DAUB recommendations 401
Dedusting 7
Deflector bar 47
Density measurement 106
Dewaterer 110, 112
Dewatering, open 177
Digging, manual 73
Displacement cutterhead 346
Disc cutters 68, 227, 264, 269, 404
Disc cutters, contact pressure on 60
DOT shield 316
Double shield machines 7, 13, 195, 410, 426
Dowelled segments 127
Drag pick 66, 269
Drifting 352
Drilling channels 302
Drive, central 269
Drive, cutting wheel 89, 241, 269
Drum type 229, 269
Dry transport 95
Duisburg light railway 296
Dumper 102
Dust 367

E

Earth pressure 258
– active 53
– passive 53
Earth pressure balance shield see Earth pressure shield
Earth pressure cells 256
Earth pressure shield 11, 256, 277, 281, 285, 291
Earth pressure shield with foam injection 277
Earth pressure support 10, 256, 262, 324, 345, 411, 428

Earth plug   32
Earth slurry   259, 264, 271
Earth support   32, 58
Eccentricity of the lining   38
Elbe Tunnel fourth bore   247
Elbe Tunnel, Old   217
Electrical cut-out and safety devices   364
Electric power failure   365
Emergency seal   251
Emergency stop   365
Emergency switch   363
Energy supply tunnel under the Kiel Fjord   219
Enlargement machine   7, 409
Enlargement shield   331
Enlargements   319
Environmental impact   172, 422
EPB mode   266
EPB shield   see Earth pressure shield
Epoxy resins   172
Equipment for grouting   169
Erector   122
Erecting device   361
Eurotunnel   see Channel Tunnel
Excavation   71
Excavation chamber   47, 223, 229, 230, 271, 306
Excavation jets   233
Excavation process   71
Excavation tools   57, 64, 229
Excavation tools, contact pressure   57
Excavation tools, hand-held   64
Excavation, full-face, 3, 18   78, 181
Excavation, hydraulic, 3   91
Excavation, part-face   see Partial-face Excavation
Excavation, partial-face   see Partial-face Excavation
Excavation, shuttle   92
Excavator   75, 180, 310
Excavator arm   221
Excluders   161
Exhaust gases   367
Expanding segments   127
Extruded concrete   118, 149, 315, 416
Extrusion excavation   72
Extrusion process   4, 149, 1691
Extrusion shield   see Blind shield

F

Face, collapse of   249, 362
Face formwork   153, 161
Face support   3, 311, 418
Face support, active   21, 223
Face support, compressed air   26
Face support, earth   32
Face support, mechanical   26
Face support, natural   25
Face support, slurry   28
Face support, suspension   223
Factors influencing settlement   39
Failure body model   32
FEM simulation   397
Field production plants   323

Filters   113
Filter cake   29, 63, 108, 116, 226, 232, 233
Filtrate water loss   152, 162
Filtration   108
Fine separation   113
Finger shield   190
Fire extinguisher systems   368
Fire protection   206, 368
Flocculant   115
Flow measurement   107
Flow speed   105, 236
Fluid jets   91
Flushing concept   238
Flushing jet   91, 98
Foam   100, 256, 262
Foam conditioning   80, 273
Foam generator   281
Foam plant   274
Folding formwork   312
Follow-up shield, telescoping   153
Formwork   125
Formwork segment   314
Formwork tolerances   143
Freeing of the shield tail seal   169
Friction coefficients, shield skin   55
Friction forces at the shield skin   55
Front shield   47
Full-face excavation   3, 78, 181, 427
Fuses   see Electrical cut-out
Fuzzy logic   396

G

Galleria Aurelia   292
Gants Hill station   320
Gap width   see Annular gap
Gardner   21
Gaskets   119, 141
Gauge area   67
Gauge tools   90
Geological risk   385
Grading curves of soil suspensions   36, 113
Grauholz Tunnel   291, 297
Greathead   18, 177
Greathead shield   18
Grease chamber pressure   50
Grid buckets   225
Grippers   196, 361, 405
Gripper TBM   6, 408
Ground behaviour   384, 405, 418
Ground deformation   37
Ground displacement process   338
Ground freezing   214, 251, 289, 300
Ground heave   39
Ground improvement   169, 249
Groundwater lowering   39, 201, 338
Grout   168
Grout canopy   170
Grout lines   140, 166, 285
Grout mix   161
Grout pump   165

Grout recipes 168
Grouting 143, 162
Grouting equipment 157
Grouting pressure 44, 50, 164, 285
Grouting process 162
Grouting scheme 174
Grouting through the segments 164
Grouting through the shield tail 164
Gyroscope system 351

H

Hamburg-Wilhelmsburg main sewer 224
Hand shields 3, 177
Hard rock cutterhead; see Cutterhead
Hard rock tools see Disc cutters
Health and safety 424
Heave 46
Heavy loads 361
High density slurry 100, 115, 259
High-pressure compressor 206
High-pressure water 91, 344
Hose water level 350
Horseshoe cross-section 12, 178
Hydraulic fluids 368
Hydraulic hammers 93
Hydraulic transport 97, 102, 235, 266, 294, 344
Hydrocyclones 110
Hydrojet shield 91, 233
Hydroshield 87, 224, 230

I

Inclinometer 351
Injected concrete 149
Injection nozzles 93, 100, 267
Injection points 166
In-situ concrete 147, 312
Installation of segmental lining 124
Intake screen see Screen
Interjacks see Intermediate jacking stations
Intermediate jacking stations 221, 325
Intermediate platforms see Platforms
Intermediate spacers 136
Interventions 288

J

Jacked pipe 337
Jaw crusher see Crusher
Jenbach/Wiesing Tunnel 243
Joints
– with tongue and groove 135
– with pin and socket 138

K

Katzenberg Tunnel 276
Key segment see Keystone
Keystone 123
Kneading tools see Agitators
KPI, Key Performance Indicators 390, 394, 397
Kyobashi Tunnel 315

L

Laser 340, 351, 367
Laser methods 261
Laser tachymeters 350
Launching shaft 322, 346
LCPC abrasiveness coefficient 405, 420
LCPC test 420
Lining 117, 415
– double layer 118
– extruded 150
– final 155
– inner 118, 148
– in-situ concrete 147
– permanent 118
– segmental 121, 163
– single-layer 118
– yielding 128
Lining systems 4
Lining, eccentricity of the 38
Liquefaction 405
Load transfer plates 146
Loading assumptions 62
Loading bucket 75
Loading from the thrust cylinders 54
Loading on the pressure bulkhead 53
Loading on the shield 50
Loading on the shield skin 51
Loads, shield 50
Lobster claw loader 95
Lock layouts in compressed air tunnelling 202
Lock, combined 208
Locks see Compressed air locks
Longitudinal joints 123, 132
Low-pressure compressors 206
Lower Inn Valley railway 243
Lynacell seal 159

M

M-30 280
Maintenance 362, 369
Man locks see Compressed air locks
Mass balance 106
Mass-volume control 259
Material locks 205
Material transport 293, 424
Measurement equipment see Surveying
Measuring wheel 351
Membrane shield 201, 216
Membrane see Filter cake
Membrane model 29
Metal thickness of the shield 49
Metro Taipei 263
Meypo yielding element 131
Microshield 344
Microtunnelling 337
Milling cutting head 66
Minitunnel system, English 342
Mixed face conditions 288
Mixes for grout 172
Mixing and kneading arms 93

Mixing buckets 256
Mixshield 225, 232, 291, 296
Mobile miner 78
Monthly advance see Advance rate
Muck cars 105
Muck removal 5, 93
Muck ring 79, 95, 267
Muck transport 235, 264
Mucking chutes 95
Mucking trains 103
Mud pressurized shield 255
Muddy soil shield 255
Münster/Wiesing Tunnel 243
Multiple circular cross- 12, 414
Multi-cyclones 112
Multi-face shield 22, 315, 328
Multi mode machines 291
Multi-phase grouting 171

N

Nagatacho Station 320
Natural stress state 162
Navigation systems 349
Neuronal networks 396
Non-circular cross-sections 328

O

Obstructions 233, 292
Occupational safety see Workplace safety
Open mode 264, 266, 405
Open shields 92, 177, 291
Open transport 101
Operating instructions 369
Operating modes 264
Operational loading 50
Operator cabin 356
Outer brushes 161
Overcut 37, 63, 87, 90, 405
Overcutter see Gauge tools
Oversize blocks 225
Oxygen breathing apparatus 205

P

Part face machines see Partial-face machines
Partial-face excavation 3, 73, 413
Partial-face machines 27, 78, 179, 233
Partial safety factors 36
Partnering 390
Partnering, risk of 388
Pea gravel 119, 342
Peak resistance 56
Pellets 110
Penetration 68, 70, 84
Penetration model 29
Performance see Advance rate
Performance killers 387
Performance specifications 385
Peripheral drives 47, 292
Personnel lock 205
Pilot tube process 338

Pilot tunnel 319
Pipe eating 346
Pipe jacking 221, 243, 322, 340, 350
Pipe jacks 324
Pipe laying 337
Pipe run 346
Piston compressor 206
Piston pumps 105, 165, 285
Planet cutters 328
Plastic seals 158
Platforms 26, 57, 74, 178
Poling plates 227, 234
Polymer 256
Polymer foams 263
Polymer suspension 31, 259
Polyshield 291, 297
PPP 388
Prefabricated utility duct 347
Preliminary screen 110
Press thrust force see Thrust forces
Pressure bulkhead 47, 53, 202, 226, 230, 294
Pressure cells 273
Pressure chamber 95
Pressure holding shield 255
Price shield 19
Primary grouting 163
Prism 353
Process analysis 399
Process controlling 280, 357, 392, 393
Process optimisation 390
Process sequence scheme 394
Processing 93
ProCon 280, 283, 357
Production of segments 143
Profile saw 92
Project controlling 383
Protective roof 362

Q

Quantity determination 105

R

Reamer machine 331
Reaction forces 50
Reception shaft 346
Recirculation pumps 239
Rectangular chisel see Chisel
Rectangular cross section 325
Reduction of stress at the face 39, 178
Regenerators 108
Regulation of the support pressure see Control of support pressure
Regulations for shield design 62
Reinforced concrete segment 142, 155
Relay pumps 97
Removal of obstructions 216, 226, 229
Repair caverns see Enlargements
Repair of waterproofing 147
Residual water content 115
Resistance to advance 55

Revolution speed, working 89
Rim 268
Ring building time 119, 279
Ring box 49
Ring building 119
Ring convergence 349
Ring convergence measurement 352, 354
Ring former 247
Ring installation sequence 354
Ring joint 122, 135
Ring taper 354
Ripper 67
Ripping tooth 75
Risk evaluation 370
Risk management 349
Risk transfer 383s
RMR, Rock Mass Rating 405
Road header 66, 75, 310, 321
Rock bolts 415
Rock cutterhead see Cutterhead
Rock machines 77
Rock mass behaviour 405
Rolling 340, 353, 361
Roof shield see Segment shield
Rotary feeder 266, 273, 296
Rotation shields 331
Rotor 238, 256, 271
Round-shafted chisel 67, 269
RQD, Rock Quality Designation 405
Rubber seals 159

S
S1 seal 159
Safety chamber 285, 306
Safety equipment 369
Safety factors 62
Sand, foam-treated 274
Saw cut 92
Scope of application of earth pressure shields 263
Scope of application of Hydroshields 228
Scope of application of slurry shields 229
Scouring 219
Scrapers 65, 229, 264, 269
Scraper conveyor 95
Scraper conveyor, encapsulated 96
Screen 232
Screw casing rotator 273
Screw conveyors 100, 256, 271
Sealing block 240, 243
Sealing plug 273
Sealing systems 89, 241, 269
Sealing systems, screw conveyor 272
Sealing wall 241, 248
Seepage water 177
Secondary grouting 163
Sedimentation 108
Segment formwork 143
Segmental lining see Lining
Segment collars 194
Segment production 125, 143

Segment shield 320
Segment types for multi-face sections 319
Segment wheek lock 96
Segments 4, 121, 416
– hexagonal 126, 195
– invert, isolated 236
– rhomboidal 126
– spiral 125
– trapezoidal 126
Semi open mode 265
Separation 106, 405, 421
Separation capacity 115
Separation plant 31, 107, 245, 307, 340
Separation process 108
Setting head see Erector
Settlement 25, 37, 162, 245, 286, 304, 324, 424
Settlement basins 108, 262
Settlement curve 42
Settlement trough 37, 39
Settlement, area affected by 43
Settlement, minimisation of 162
Settlements, factors influencing 39
SGI lining 426
Shield blade 47, 63
Shield construction process 2
Shield design 62
Shield docking method 332
Shield machines 8, 410
Shield microtunnelling 340
Shield presses see Thrust cylinders
Shield skin 47
Shield skin, friction coefficients 55
Shield skin, friction forces 55, 194
Shield skin, loading on 51
Shield skin, lubrication of 61
Shield, thickness 47
Shield tail 49
Shield tail hood 314
Shield tail seals 37, 49, 157, 225
Shield track see Annular gap
Shield, basic principle 3
Shield, constructional elements 47
Shield, design basis 52
Shield, empirical values for dimensioning 60
Shield, external loading 50
Shield, for flexible cross-sections 328
Shield, historical development 16
Shield, universal 292
Shield, convertible 232
Shotcrete 118, 155, 314, 415
Shuttle transport 240
Shutters 181
Sickroom air locks 205
Silo theory 33
Single layer construction 118
Single shield 7
Single shield mode 196
Slime shield 255
Slipform 149
Slurry circuit 31, 97, 110, 340

Slurry shield   12, 223, 228, 291
Slurry support   9, 28, 58, 324, 411, 413, 427
Socatop   292, 305
Soil conditioning process   see Conditioning
Soil pressure shield   255
Soil slurry   see Earth slurry
Sophiaspoor Tunnel   200
Spacing   68
Spade   64
Spalling   132, 145
Special processes   309
Special shields   309
Spiral segment   125
Splitter   see Crusher
Spokes   79
Sprayed concrete   see Shotcrete
Squeezing rock   128
Stabilisation of the face   see Face support
Starting shaft   see Launching shaft
Stator   256, 271
Steel brush seals   160
Steel fibre concrete   see Steel fibre reinforced concrete
Steel fibre reinforced concrete   119, 139, 148, 314
Steering   349, 355
Steering cylinders   340, 346
Steering head   338
Steering jacks   325
Steering movements   51
Sticking   235, 250, 270, 300, 405, 419
Sticking of the cutting wheel   259
Sticking potential   236
Stirrer   see Agitator
Stone crusher   see Crusher
Stone traps   98
Stress redistribution   25, 42
Stresses, permissible for shield   62
Structural design   156
Structural design basis for shields   52
Structural verification   60
Styropor beads   129
Submerged wall   49, 226, 230, 231, 294
Suction effect   239
Suction intake   97
Support   see Lining
Support arches   415
Support fluid   224, see also Bentonite suspension
Support medium   see Face support
Support of the cavity   25, 37
Support of the face;   see Face support
Support plates   26, 64
Support pressure   25, 35, 44, 229
Support pressure calculation   32
Support pressure control   see Control of support pressure
Surveying   349
Suspension   see Bentonite suspension
Swelling rubber   141, 147
System launching shaft   344

T

Tail skin   see Shield tail
Taper   47, 125
Target units   350
Teeth   68
Telescopic joint   153
Telescopic pipe   103
Telescopic shield   196
Tender   384
Tenside   273
Thixshield   11, 73, 77, 233
Thixotropic fluids   31
Thrust cylinders   49, 60, 136, 322
Thrust cylinder forces   60, 122
Thrust forces   4, 54, 61, 361
Thrust pit   324
Thrust presses   see Thrust Cylinders
Thrust reaction ring   123
Tilting of the cutting wheel   59
Timber lagging   416
Tipping   115, 421, 423
Tolerances   349
– formwork   143
– segments   144
Tool changing   84, 247
Torque, cutting wheel   90, 229, 259
Total station   352
Towing connections   366
Trailing plates   314
Transport equipment   93, 101
Transport methods   3
Transported solids   106
Treatment chambers   205, 208
Tribrach   353
Tunnel and shaft transport   101
Tunnel jacking   325
Tunnel lining   see Lining
Tunnel support   118, see also Lining
Two-part grout   164

U

Ultra flexible shield   330
Ultrasound method   260
Undercutting   78, 86
Underground station   319, 331
Underground urban development   1
Universal ring   125

V

Ventilation   367
Vibration   37, 325
VOB   384
Volume loss   40, 286
Volume determination   107
Volume measurement   105

W

Walking mechanism 248, see also Grippers
Walkways 362
Water draining construction 119
Water ingress 192
Water lenses 177
Water permeability 221, 264
Waterproof concrete 120
Waterproofing canopy 248
Watertight construction 119
Wear 65, 235, 420
Westerschelde Tunnel 237
Westtangente Bochum 325
Wohlmeyer principle 86, 329
Working chamber 231
Workplace safety 359

Y

Yielding joint construction 130

Z

Zürich Thalwil 301